Essentials for Algebra
Concepts and Skills

AUTHORS

Robert E. Willcutt
Patricia R. Fraze
Francis J. Gardella

EDITORIAL ADVISER

Charles E. Allen

TEACHER CONSULTANTS

Dotty J. Garrett   Adrienne A. Kapisak   Patricia C. McNeil
Tommy Tomlinson   Marvin Weingarden

Houghton Mifflin Company · Boston

Atlanta   Dallas   Geneva, Ill.   Lawrenceville, N.J.   Palo Alto   Toronto

Authors

Robert E. Willcutt
Associate Professor, Science-Mathematics
Education Department, School of Education, Boston University

Patricia R. Fraze
Chairman, Mathematics Department
Ann Arbor Huron (Michigan) High School

Francis J. Gardella
Mathematics Supervisor
East Brunswick Public Schools, New Jersey

Editorial Adviser

Charles E. Allen
Instructional Specialist, Secondary Mathematics
Los Angeles City Schools, California

Teacher Consultants

Dotty J. Garrett
Mabry Middle School
Marietta, Georgia

Adrienne A. Kapisak
Gateway Senior High School
Monroeville, Pennsylvania

Patricia C. McNeil
Parkland Junior High School
Rockville, Maryland

Tommy Tomlinson
T. K. Gorman High School
Tyler, Texas

Marvin Weingarden
Detroit Public Schools
Michigan

Printed in U.S.A.

ISBN: 0-395-37880-X

DEFGHIJ-D-943210-89876

Table of Contents

Contents

4 Fractions

5 Solving Equations

6 Formulas

7 Ratio, Proportion, Percent

8 Integers and Rational Numbers

9 Graphing in the Coordinate Plane

10 Statistics

11 Special Triangles

12 Probability

13 Geometry

14 Programming in BASIC

1
Whole Numbers and Variables

How old is algebra? Older than the hills! When the ancient Mayan city of Chichén Itzá was settled around A.D. 530 the study of algebra was already about 2300 years old.

1-1 Variables

It costs $4 per hour to rent a canoe
at the Bantam Boatyard. The total
cost of any rental varies with the
amount of time spent canoeing. If
you let n equal the number of hours
spent canoeing, then $4 \times n$ stands
for the cost of the rental. We call n
a **variable** and $4 \times n$ an **expression.**

Here are some other ways you
can show multiplication.

$$4 \cdot n \qquad (4)(n) \qquad 4n$$

To evaluate the expression $4n$,
replace n with different numbers.

Example 1   Evaluate $4n$ when $n = 6$, when $n = 10$

Solution $\begin{aligned} 4n &= (4)(6) \\ &= 24 \end{aligned}$ $\begin{aligned} 4n &= (4)(10) \\ &= 40 \end{aligned}$

Your Turn   Evaluate the expression.
- $3x$, when $x = 8$
- $12 + (2a)$ when $a = 12$ ⟵ *Work inside the paren-*
  theses first.

You can often write expressions for word phrases.

Example 2   Write an expression for the phrase.
a number n added to 45

Solution $n + 45$ or $45 + n$ ⟵ *Either order*

Example 3 Write an expression to show that Alison's new salary is $10 more per week than her old salary.

Solution Let x = Alison's old salary.
Then $x + 10$ = her new salary.

Example 4 Write an expression to show that the distance from Boston to Springfield is 15 miles less than the distance from Boston to Northampton.

Solution Let d = the distance from Boston to Northampton. So, you should subtract 15 from d to find the distance from Boston to Springfield. The expression is $d - 15$.

Your Turn Complete.
- Ron is 7 years older than Nan.
  Let a = Nan's age.
  Then __?__ = Ron's age.
- The attendance at the evening performance was 3 times as great as at the matinee.
  Let n = the number of people at the evening performance.
  Then __?__ = the number of people at the matinee.

Class Exercises

What is the variable in the expression?

1. $4a + 7$ **2.** $32 + 3x$ **3.** $57b \div 2$ **4.** $37 + w$

5. $(20)(b) - 3$ **6.** $a + 72$ **7.** $68 \div p$ **8.** $175 - 3y$

Let $n = 9$. Evaluate the expression.

9. $6 + n$ **10.** $22 - n$ **11.** $n - 5$ **12.** $n + n$

13. $12 - n$ **14.** $54 \div n$ **15.** $n + 37$ **16.** $81 \div n$

Exercises

Evaluate the expression using the given value for the variable.

A   **1.** $(4n) + 6$; $n = 3$ **2.** $n + 7$; $n = 12$ **3.** $x \div 3$; $x = 33$

    4. $9 + (3a)$; $a = 8$ **5.** $75 - (5b)$; $b = 5$ **6.** $38 - (2n)$; $n = 15$

    7. $48 \div y$; $y = 6$ **8.** $16 - (2z)$; $z = 7$ **9.** $x - 15$; $x = 21$

Let $x = 5$, $y = 6$, and $z = 15$. Evaluate the expression.

B   **10.** $8x$ **11.** $x + y$ **12.** $y + 2$

   13. $(2x) + y$ **14.** $z \div x$ **15.** $(xy) - z$

   16. $x + y + z$ **17.** $(2y) + (3z)$ **18.** $(5y) \div z$

Write an expression for the phrase.

19. 15 added to x **20.** 47 times n

21. y minus 65   **22.** a plus 155

23. r divided by 26 .   **24.** z times 79

25. the number of hours in n days   **26.** the numbers of eggs in n dozen

27. the quotient of s divided by 8   **28.** the product of y and 37

29. n increased by 52   **30.** 84 less a number x

31. 8 more than y **32.** z decreased by 12

33. 125 decreased by t **34.** 85 greater than b

35. the cost of one gallon of gas if n gallons cost \$19.35   **36.** The cost of b plants if one plant costs \$4.98

Problems

Complete.

  1. A grasshopper can jump a distance that is 20 times its body length.
Let b = body length.
Then __?__ = the jumping distance.

  2. Charlie slept 4 fewer hours on Sunday than on Saturday.
Let h = the hours slept on Saturday.
Then __?__ = the hours slept on Sunday.

1-2 Addition

Numerical sentences containing the equality symbol, $=$, are called **equations.**

Example 1   Write and solve an equation to show that n is the sum of 897 and 675.

Solution   Let $n =$ the sum
Then, $n = 897 + 675$
$n = 1572$

> *Remember how to add whole numbers.*
> $$\begin{array}{r} \overset{1\,1}{897} \\ +675 \\ \hline 1572 \end{array}$$

Your Turn   Write an equation for the sentence. Solve the equation.
- The sum of 482 and 59 is x.
- 1535 plus 496 equals b.

Though some problems may be easy to solve without writing equations, they offer practice that will help you write and solve more difficult problems later.

Example 2   Sales are booming at Computers, Etc. Last month 659 personal computers were sold. The month before 398 were sold. How many computers were sold in the two months?

Solution   Let $n =$ the total number sold
Then, $n = 659 + 398$
$n = 1057$
A total of 1057 computers were sold.

Your Turn Write an equation using n for the variable. Solve the equation.
- Harry delivers 35 packages on Friday, and 37 on Saturday. How many packages does he deliver in all?
- Monica sells 40 lb of fish on Wednesday and 39 on Thursday. How many pounds does she sell in all?
- Gavin pumps 256 gal of gasoline on Tuesday and 460 on Wednesday. How many gallons are pumped in all?

The following properties can be used to make addition easier.

Commutative Property of Addition

Changing the order of the addends doesn't change the sum.

In Arithmetic   In Algebra
$5 + 6 = 6 + 5$ $a + b = b + a$

Associative Property of Addition

Changing the grouping of the addends doesn't change the sum.

In Arithmetic   In Algebra
$(5 + 6) + 7 = 5 + (6 + 7)$ $(a + b) + c = a + (b + c)$

You can combine the commutative and associative properties to make addition easier.

Example 3 Use the commutative and associative properties to add $(26 + 8) + 4$.

Solution Think: $(26 + 8) + 4 = 26 + 4 + 8$ ← ┐ *To add in your head*
$$= 30 + 8$$ *look for sums of 10,*
$$= 38$$ *20, 30, etc. to add first.*

Your Turn Use the commutative and associative properties to add.
- $(99 + 47) + 1$ $(88 + 66) + 12$ $(187 + 136) + 73$

5

You can also use the properties to make column addition easier.

Example 4 Add: $6 + 8 + 4 + 8$.

Solution

$$\begin{array}{r} 6 \\ 8 \\ 4 \\ +8 \\ \hline 26 \end{array}$$

$10 = 6 + 4$
$16 = 8 + 8$

First look for sums of 10 to make the addition easier.

Class Exercises

Find the sum.

1. 56 +44	**2.** 855 + 76	**3.** 1391 + 900	**4.** 99 +199	**5.** 67,481 + 3,892
6. 175 82 376 +198	**7.** 27 336 424 + 82	**8.** 89 15 +111	**9.** 6734 266 842 +1746	**10.** 991 1158 769 +4442

Exercises

Use the commutative and associative properties to do the addition in your head.

A **1.** $(33 + 15) + 7$ **2.** $(45 + 39) + 5$ **3.** $(96 + 89) + 4$

  4. $186 + 57 + 14$ **5.** $381 + 56 + 19$ **6.** $597 + 398 + 103$

Solve the equation.

B **7.** $n = 326 + 263$ **8.** $a = 465 + 132$ **9.** $465 + 932 = b$

  10. $375 + 975 = a$ **11.** $563 + 999 = m$ **12.** $r = 9642 + 8659$

  13. $t = 1235 + 427$ **14.** $9003 + 75 = z$ **15.** $3992 + 793 = b$

Write an equation for the sentence. Solve the equation.

16. The sum of 4862 and 57 is n.

17. 63 added to 598 is x.

18. The total of 60,845 and 72 is b.

19. The numbers 886 and 749 add to a sum of y.

20. 64 increased by 93 is y.

21. 158 plus 6839 is z.

Use the commutative and associative properties to solve the equation.

22. $a = (47 + 59) + 3$

23. $46 + (39 + 4) = x$

24. $n = (14 + 49) + 36$

25. $y = (56 + 192) + 8$

26. $b = 137 + (15 + 13)$

27. $n = 999 + (881 + 1)$

Use the map below. Write and solve an equation for the exercise.

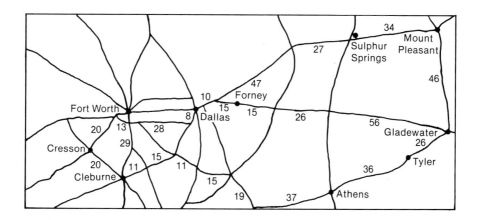

Find the total distance.

28. from Ft. Worth to Cleburne to Athens.

29. from Forney to Gladewater to Mount Pleasant.

30. from Mount Pleasant to Sulphur Springs to Dallas.

31. from Gladewater to Forney to Dallas.

32. from Cleburne to Cresson to Dallas.

Solve.

C 33. How many miles are covered in a round trip from Athens to Cleburne to Ft. Worth?

34. You can travel to Mount Pleasant from Dallas by way of Sulphur Springs or Gladewater. Which is the shorter route?

1-3 Subtraction

There are 1547 seats in the opera hall of the Sydney Opera House in Australia. At a recent performance 1459 seats were filled. How many seats were empty?

You can subtract to solve the problem.

$$\begin{array}{r} \overset{4\ 13\ 17}{1\cancel{5}\cancel{4}7} \\ -\ 1\ 4\ 5\ 9 \\ \hline 8\ 8 \end{array}$$

There were 88 empty seats.

Addition and subtraction are opposite operations.

In Arithmetic

If $8 + 11 = 19$, then $19 - 11 = 8$.

In Algebra

If $a + b = c$, then

$c - b = a$.

You can use addition to check subtraction. Since $1459 + 88 = 1547$ the answer above is correct.

Example   Write an equation to show that n is the difference between 687 and 29. Solve and check.

Solution $n = 687 - 29$ ←────┐ $\begin{array}{r} \overset{7\ 17}{6\cancel{8}\cancel{7}} \\ -\ \ \ 2\ 9 \\ \hline 6\ 5\ 8 \end{array}$

$n = 658$

Check: $658 + 29 = 687$

8

Your Turn   Write and solve an equation for the sentence.
- The difference 1500 minus 685 is y.
- The number of days in a year minus 87 equals n.

Class Exercises

Subtract.

1. $\begin{array}{r} 76 \\ -35 \end{array}$	**2.** $\begin{array}{r} 186 \\ -\ 56 \end{array}$	**3.** $\begin{array}{r} 59 \\ -49 \end{array}$	**4.** $\begin{array}{r} 168 \\ -\ 29 \end{array}$
5. $\begin{array}{r} 3000 \\ -\ 241 \end{array}$	**6.** $\begin{array}{r} 4001 \\ -1392 \end{array}$	**7.** $\begin{array}{r} 5050 \\ -\ 62 \end{array}$	**8.** $\begin{array}{r} 7777 \\ -\ 499 \end{array}$
9. $\begin{array}{r} 30{,}048 \\ -\ 6{,}829 \end{array}$	**10.** $\begin{array}{r} 100{,}064 \\ -\ 29{,}833 \end{array}$	**11.** $\begin{array}{r} \$684.00 \\ -\ 27.59 \end{array}$	**12.** $\begin{array}{r} \$847.00 \\ -199.99 \end{array}$

Exercises

Subtract. Use addition to check your answer.

A   **1.** $6384 - 4695$ **2.** $7635 - 3746$ **3.** $4000 - 2998$ **4.** $\$29.00 - 16.98$

  5. $\$30.00 - 26.59$ **6.** $\$85.00 - 25.55$ **7.** $30{,}060 - 7912$ **8.** $104{,}600 - 75{,}633$

Solve.

B   **9.** $n = 417 - 325$ **10.** $g = 387 - 211$ **11.** $m = 1529 - 299$

  12. $7006 - 378 = r$ **13.** $682 - 293 = m$ **14.** $769 - 590 = x$

  15. $k = 3652 - 937$ **16.** $p = 4300 - 411$ **17.** $r = 6340 - 798$

Write an addition or subtraction equation for the sentence.

18. 16 greater than 30 is x.

19. The difference between 4080 and 67 is x.

20. The sum of 278 and 439 is y.

21. x is equal to 6000 minus 87.

22. n is 239 more than 18.

23. a is 75 plus 46.

Evaluate the expression using the given value of the variable.

24. $x - 358; x = 1500$ **25.** $n - 27; n = 45$ **26.** $a - 986; a = 2000$

27. $472 - y; y = 98$ **28.** $6000 - b; b = 571$ **29.** $1003 - z; z = 744$

30. $8006 - d; d = 322$ **31.** $b - 403; b = 500$ **32.** $n - 59; n = 208$

The table shows the attendance at various performances during a Summer Arts Festival that toured several cities.

	City						
Activity	**Orleans**	**Newark**	**Joplin**	**Adelphi**	**Chester**	**Nichols**	**York**
Rock Concert	1351	625	2600	1758	4000	901	2177
Play	649	1400	399	637	768	539	426
Symphony	327	598	406	482	888	603	749
Circus	703	255	–	897	1006	591	986

Write and solve an equation to show the difference in attendance for the performance.

33. circus attendance in Adelphi and Newark

34. rock concert attendance in Chester and Orleans

35. play attendance in Joplin and York

Solve.

C **36.** Is subtraction commutative? Give examples to support your answer.

37. Is subtraction associative? Give examples to support your answer.

Self Test 1

Evaluate the expression using the given value of the variable.

1. $6a; a = 7$ **2.** $(15x) + 3; x = 2$ **3.** $18 - (2y); y = 8$ **1-1**

Solve the equation.

4. $n = 18 + 35$ **5.** $75 + 103 = b$ **6.** $y = 96 + (12 + 4)$ **1-2**

7. $4001 - 376 = x$ **8.** $b = 339 - 276$ **9.** $160 - 98 = a$ **1-3**

1-4 Problem Solving

The problem solving process can be broken down into several steps. Though the solutions to some problems may be obvious, others will be more easily seen if you take the problem apart.

Example 1   Harold had $68 in a savings account. When he decided to close the account and take out his money he received $97. How much interest had the money earned?

Solution   Break the problem down into smaller segments.
1. What do you know and what do you want to know? Harold had $68. He received $97. How much interest?
2. Make a plan. Write a subtraction equation to show that the interest is the difference between $97 and $68.
3. Use the plan.
      Let i = the interest earned
   Then $i = 97 - 68$
         $i = 29$
4. Show your answer and check to see that it is reasonable and answers the question.
   The money earned $29 interest.

Your Turn   Use the problem solving process to solve the problem.
- In June, Seth weighed 75 kg. By the time soccer practice began in September he was down to 69 kg. How much weight did he lose?

Class Exercises

A ▒ covers the numbers in the problems below. Match each problem to an equation and solve the equation.

1. Mary has 2 stamp albums. One contains ▒ stamps, the other ▒. How many stamps does she have in all?

2. A sports arena has seating for ▒ people. At a recent concert, ▒ seats were occupied. How many seats were empty?

3. Olivia earns ▒ each week as a computer programmer. Weekly taxes amount to ▒. How much does Olivia receive in her pay-check?

a. $y = 10{,}937 - 7465$
b. $p = 527 - 149$
c. $400 + 0 = p$
d. $376 + 978 = n$
e. $a = 47 - 15$

Exercises

Write an equation. Solve.

A 1. There are 480 freshmen, 521 sophomores, 511 juniors, and 609 seniors at Watson High School. What is the school's total enrollment?

2. During a recent football game, the Benson High team scored 16, 27, 13, and 37 points. What was the team's total score for the game?

3. Wally won the 400 m race with a time of 72 seconds. Harley, the second place finisher, completed the race in 94 seconds. How much better was Wally's time than Harley's?

4. "I drove 579 mi last week," said Estelle, "and 623 mi the week before." How many more miles did Estelle drive the first week?

B 5. Regal Repair received payments of $350 for repair of a computer terminal and $158 for repair of a video disk player. From the total, an assistant's salary of $198 was deducted. How much money was left?

6. Miriam took two rolls of film at a class picnic. One roll contained 36 shots and the other contained 24. Of the photos taken, 17 did not come out. Should Miriam buy an album that holds 50 photos or one that holds 70?

7. On Tuesday, John had a balance of $475 in his checking account. He paid a bill for $298 and then deposited a check for $965. What is John's new balance?

8.

> Memorandum   ABC Industries
>
> To: Cashier   From: Evelyn Clanden
>
> Subject: Travel Expenses
>
> Enclosed is an itemized report of expenses for my recent business trips. The total for the Washington trip was $789. The total for the Chicago trip was $367. I received a cash advance of $1500.

How much money is owed? To whom is the money owed?

Use the information in the phrases to write and solve a problem.

C **9.** Air fare—$479.82
Hotel—$688.50
Meals—$285.79

10. Food Bills
Jan.—$365.47
Feb.—$308.51
Mar.—$386.32

11. Earned—$1508.17
Spent— $68.37
Saved— $150.00

12. Trip distance—1407 mi
Drove—528 mi

13. Stamp Collection
Have—478
Sold— 99
Bought— 35

14. Car Sales
On hand—72
Sold—13
New shipment— 9

1-5 Multiplication

The Waukegan Widget Company produces 3542 widgets per day. At this rate how many widgets will they produce in 21 days? You can use a multiplication equation to solve the problem.

Let $n = 3542 \times 21$ ◄

Then, $n = 74{,}382$

In 21 days 74,382 widgets will be produced.

> *Remember how to multiply whole numbers.*
>
> $$\begin{array}{r} 3542 \\ \times 21 \\ \hline 3542 \\ 70840 \\ \hline 74{,}382 \end{array}$$

Like addition, multiplication has a commutative and associative property.

Commutative Property of Multiplication

Changing the order of the factors does not change the product.

In Arithmetic

$3 \cdot 8 = 8 \cdot 3$

In Algebra

$ab = ba$

Associative Property of Multiplication

Changing the grouping of the factors does not change the product.

In Arithmetic

$(6 \cdot 4)5 = 6(4 \cdot 5)$

In Algebra

$(ab)c = a(bc)$

Multiplication also has another special property.

The Distributive Property of Multiplication over Addition and Subtraction

In Arithmetic

$6(5 + 3) = 6 \cdot 5 + 6 \cdot 3$

$3(4 - 2) = 3 \cdot 4 - 3 \cdot 2$

In Algebra

$a(b + c) = ab + ac$

$a(b - c) = ab - ac$

You can check to see if the distributive property is true.

$$6(5 + 3) \stackrel{?}{=} 6 \cdot 5 + 6 \cdot 3 \quad \longleftarrow$$

$6 \cdot 8$	$30 + 18$
48	48

We read $a \stackrel{?}{=} b$ as "is a equal to b?"

$48 = 48$, so $6(5 + 3)$ is equal to $6 \cdot 5 + 6 \cdot 3$.

Using the properties of multiplication can make it easier to do some multiplication.

Example 1 Solve.
$$n = (25 \cdot 6)2$$

Solution
$$n = (25 \cdot 6)2$$
$$= (25 \cdot 2)6 \quad \longleftarrow \text{ \textit{Associative and commutative properties}}$$
$$= 50 \cdot 6$$
$$= 300$$

Your Turn Use the properties of multiplication to help make the multiplication easier.
- $x = (2 \cdot 97)5$
- $y = (25 \cdot 86)4$
- $n = 50(2 \cdot 77)$
- $a = 6(50 + 7)$

When a factor is multiplied by itself it is raised to a **power.** A raised number called an **exponent** is used to indicate the number of times the factor is used.

Example 2 Write $3 \cdot 3 \cdot 3 \cdot 3$ using an exponent.

Solution $3^4 \quad \longleftarrow$ *The exponent tells you that the factor 3 will be multiplied by itself 4 times.*

Example 3 Write $a \cdot a \cdot a \cdot a \cdot a$ using an exponent.

Solution $a^5 \longleftarrow$ *The factor a is multiplied by itself 5 times.*

Example 4 What is the value of $2x^3$ if $x = 3$?

Solution $2x^3 = 2(3^3) \longleftarrow$ *Do all exponentiations first.*
$= 2(27)$
$= 54$

Your Turn Write the product.
- $y \cdot y \cdot y \cdot y \cdot y$
- 4^2
- $3y^2$ when $y = 4$

Class Exercises

Multiply.

1. 14	**2.** 75	**3.** 37	**4.** 432	**5.** 1006
$\times 3$	$\times 9$	$\times 6$	$\times 25$	$\times 328$

Exercises

Multiply.

A
1. 38	**2.** 47	**3.** 671	**4.** 587	**5.** 3316
$\times 26$	$\times 19$	$\times 39$	$\times 75$	$\times 919$

Solve.

B
6. $n = 36 \cdot 9$	**7.** $z = 12^3$	**8.** $p = 59 \cdot 9$
9. $x = 3^2$	**10.** $w = 20^4$	**11.** $r = (77 \cdot 31)3$
12. $p = 4^5$	**13.** $x = 128(95 \cdot 2)$	**14.** $d = 6^6$
15. $d = 6(7 + 2)$	**16.** $y = 25(7 + 10)$	**17.** $m = 14(8 + 9)$

Evaluate the expression using the given value of the variable.

18. $m \cdot 27$; $m = 10$ **19.** $58a$; $a = 7$ **20.** $17y$; $y = 58$

21. x^3; $x = 7$ **22.** $w \cdot 63$; $w = 40$ **23.** $10m^2$; $m = 6$

24. $35p^2$; $p = 2$ **25.** $5 + y^2$; $y = 9$ **26.** $a^5 - 10$; $a = 2$

Write an addition, subtraction, or multiplication expression.

27. 5086 multiplied by b

28. the sum of x and 4075

29. r increased by 3000

30. the product of t and 375

31. 7215 more than a

32. y used as a factor 6 times

33. A parking garage has 9 levels. Let n = the number of spaces on each level. Then _?_ = the number of spaces in the garage.

34. A contractor ordered 88 spools of wire. Let f = the number of feet of wire on each spool. Then _?_ = the total amount of wire ordered.

Problems

Use the table to answer the questions.

Ticket Prices for Strings'n'Things			
	Sun.–Thurs.	Fri.–Sat.	Matinee (Sun., Wed.)
Orchestra	$20	$22	$18
1st Balcony	$19	$20	$17
2nd Balcony	$17	$18	$15

1. Selma is buying 6 orchestra tickets for Saturday night. How much will they cost?

2. The Thespian Club bought 7 tickets for 1st balcony on Sunday evening and 5 tickets for the 1st balcony for the Sunday matinee. How much did they spend?

3. Suppose you received a $50 gift certificate for theater tickets. What is the greatest number of tickets you could buy? What seats and what days would they be good for? How much money would you have left?

1-6 Division

There are several different ways to show division. In the divisions below, the 6 is the **divisor**, the 24 is the **dividend**, and the 4 is the **quotient**.

$$6)\overline{24} \qquad 24 \div 6 = 4 \qquad \frac{24}{6} = 4$$

Multiplication and division are inverse operations.

In Arithmetic
If $4 \cdot 3 = 12$, then $12 \div 3 = 4$

In Algebra
If $ab = c$, then $c \div b = a$, $b \neq 0$

Example 1   Write a related equation: $3 \times 37 = 111$.

Solution $111 \div 37 = 3$

Your Turn   Write a related equation.
- $3 \times 92 = 276$ • $696 \div 8 = 87$ • $679 \times 9 = 6111$

Since multiplication and division are opposites, you can use multiplication to check division.

Example 2   Solve the equation $n = \dfrac{45}{3}$. Use multiplication to check your quotient.

Solution

$n = \dfrac{45}{3}$

$n = 15$

Check: $15 \times 3 = 45$

Remember how to divide whole numbers.

$$\begin{array}{r} 15 \\ 3)\overline{45} \\ -3 \\ \hline 15 \\ -15 \\ \hline \end{array}$$

Your Turn   Write and solve an equation for the problem. Check your answer.
- The quotient 68 divided by 5 is n.
- Mark traveled 160 mi on 12 gal of gas. How many miles per gallon does he get?

Class Exercises

Divide.

1. $3\overline{)39}$ **2.** $4\overline{)57}$ **3.** $4\overline{)577}$ **4.** $7\overline{)928}$

5. $72 \div 9$ **6.** $60 \div 12$ **7.** $221 \div 17$ **8.** $777 \div 37$

9. $\dfrac{56}{7}$ **10.** $\dfrac{121}{11}$ **11.** $\dfrac{48}{6}$ **12.** $\dfrac{182}{26}$

Exercises

Divide.

A   **1.** $8\overline{)96}$ **2.** $6\overline{)768}$ **3.** $7\overline{)896}$ **4.** $6\overline{)942}$

    5. $3\overline{)4678}$ **6.** $15\overline{)6380}$ **7.** $35\overline{)3254}$ **8.** $72\overline{)7609}$

Solve.

B   **9.** $a = 64 \div 4$ **10.** $b = \dfrac{88}{4}$ **11.** $m = 81 \div 9$

    12. $1102 \div 29 = x$ **13.** $12{,}750 \div 35 = m$ **14.** $585 \div 9 = t$

    15. $p = \dfrac{121}{11}$ **16.** $r = 144 \div 12$ **17.** $w = \dfrac{225}{15}$

Evaluate the expression using the given value of the variable.

18. $\dfrac{x}{7}$; $x = 49$ **19.** $121 \div a$; $a = 11$ **20.** $\dfrac{986}{y}$; $y = 17$

21. $m \div 50$; $m = 750$ **22.** $\dfrac{b}{78}$; $b = 632$ **23.** $5162 \div d$; $d = 89$

Write a variable expression for the phrase.

24. The quotient of 1335 and *n*.

25. 468 divided by *x*

26. *d* divided by 782

27. 76 more than the number *r*

28. *n* increased by 1580

29. the quotient of *w* and 339

Complete.

30. George took 120 pictures on his vacation.
Let *r* = the number of pictures on one roll of film.
Then __?__ = the number of rolls of film used.

31. Rhonda has 21 oz of cheese to divide among friends.
Let *n* = the number of friends.
Then __?__ = the amount of cheese for each friend.

C 32. The pups in a new litter registered the following weights:

Butch 15 oz	Sundance 17 oz	Foxy 14 oz
Wishbone 14 oz	Redeye 18 oz	

What is the average weight of a new pup from the litter?

33. Is division commutative? Give examples to explain your answer.

34. Is division associative? Give examples to explain your answer.

Problems

Solve.

1. When the Masons had telephone service installed in their new house, they were charged for the following services: Service order charge—$5.00, Home visit charge—$5.50, Telephone handling charge—$5.00 per phone. If 3 phones were installed, what was the total bill?

2. It takes about 8 to 10 oranges to make a quart of freshly squeezed orange juice. About how many oranges are needed for 5 quarts?

1-7 Identity Properties

The numbers 0 and 1 have some special properties. When 0 is added to any number the resulting sum is **identical** to the original number.

Identity Number for Addition
The identity number for addition is zero.
In Arithmetic   In Algebra
$0 + 7 = 7$ $0 + a = a$

When 1 is multiplied by any number, the resulting product is **identical** to the original number.

Identity Number for Multiplication
The identity number for multiplication is one.
In Arithmetic   In Algebra
$1 \cdot 3 = 3$ $a \cdot 1 = a$

Division by zero is impossible. To see why this is true consider the inverse relationship of multiplication and division.

$$\text{If } 8 \div 0 = n \text{ then } n \cdot 0 = 8$$

Since 0 multiplied by any number is 0, there is no value of n that will be a solution for the equation $n \cdot 0 = 8$.

Class Exercises

Add or multiply.

1. $6 + 0$	**2.** $84 \cdot 1$	**3.** $6842 + 0$	**4.** $398 \cdot 1$
5. $1 \cdot 9306$	**6.** $0 + 438$	**7.** $278 \cdot 1$	**8.** $4775 + 0$

Exercises

Add or multiply.

A **1.** $0 + 1$ **2.** $10 + 0$ **3.** $23 + 0$ **4.** $0 + 125$

    5. $1 \cdot 76$ **6.** $478 \cdot 1$ **7.** $0 + 0 + 0$ **8.** $999 \cdot 1$

    9. $0 + 9865$ **10.** $0 + 125$ **11.** $675 \cdot 1$ **12.** $1 \cdot 1$

Solve.

B **13.** $a = 96 \cdot 1$ **14.** $k = 36 + 0$ **15.** $p = 0 + 9$ **16.** $b = 58 \div 0$

    17. $z = 49 + 0 + 1$ **18.** $y = 19 \cdot 1$ **19.** $m = 0 + 999 + 1$ **20.** $y = 417 \cdot 1$

    21. $36 = 36 + m$ **22.** $46p = 46$ **23.** $56 + r = 56$ **24.** $n + 9 = 9$

    25. $9r = 9$ **26.** $p \cdot 46 = 46$ **27.** $(a)(102) = 102$ **28.** $n + 68 = 68$

What is the value of n?

C **29.** $na = a$ **30.** $n + b = b$ **31.** $n(3a) = 3a$ **32.** $n + 2b = 2b$

Solve the equation in your head.

33. $x = 37(468 - 467)$ **34.** $n = 725 + (95 - 95)$

35. $(229 - 228)76 = y$ **36.** $a = 400 + (637 - 637)$

Solve.

37. Is there an identity number for subtraction? Show why or why not.

Using the Calculator

Use your calculator to evaluate the following expressions for the given value of the variable.

1. $375x - 52$, if $x = 1953$ **2.** $4287n + 3781$, if $n = 274$

3. $11{,}597 - 13b$, if $b = 372$ **4.** $\dfrac{a}{359} + 23{,}281$, if $a = 1{,}447{,}488$

5. $\dfrac{162n}{156}$, if $n = 21{,}918$ **6.** $\dfrac{(328)(1404)}{y}$, if $y = 108$

1-8 Order of Operations

If you put on your shoes before your socks you'll achieve a different result than if you perform the activities in the reverse order. When you calculate mathematical expressions that have more than one operation the order of the operations also makes a difference.

Example 1 Morgan earns $15 per day. During one week he earned a $2 bonus for the week. During another week he earned a $2 bonus each day. How much did he earn for each 5-day week?

Solution

Week 1
$$(5)(15) + 2 = n$$
$$75 + 2 = n$$
$$77 = n$$
Morgan earned $77.

Week 2
$$5(15 + 2) = y$$
$$(5)(17) = y$$
$$85 = y$$
Morgan earned $85.

As you can see, computing with the same numbers can lead to different answers. Use the following rules to insure the correct order of operations.

1. Do all work in parentheses first.
2. When there are no parentheses, do all multiplications and divisions first, working from left to right. Next do all additions and subtractions working from left to right.

Example 2 Solve the equation $a = 3(5 + 2)4$.

Solution

$a = 3(5 + 2)4$ ⟵ *Work inside the parentheses first.*
$a = 3 \cdot 7 \cdot 4$
$a = 84$

23

Example 3   Solve the equation $a = 3^2 \cdot 5 + 2 \cdot 4$.

Solution
$a = 3^2 \cdot 5 + 2 \cdot 4$ ⟵ *First raise 3 to the second power.*
$a = 9 \cdot 5 + 2 \cdot 4$ ⟵ *Do the multiplications next.*
$a = 45 + 8$
$a = 53$

Your Turn   Solve the equation.
- $a = 3 \cdot 4 + 2$ $b = 3(4 + 2)$
- $m = 5^2 \cdot 6 - 2$ $n = (6 - 2)5^2$

The expression $7f + 9s + 2f$ contains three **terms,** $7f$, $9s$ and $2f$. The numbers 7, 9, and 2 are called the **coefficients** of the terms. When the variables in more than one term of an expression are the same, they are called **like terms.** You can combine like terms by addition or subtraction.

Example 4   Combine the like terms in the expression.
$6a + 4b + a$ ⟵
_____
| *The expression* $6a + a$
| *is the same as* $6a + 1a$.

Solution $6a + 4b + a = 7a + 4b$
| *When the terms are com-*
| *bined, the result is* $7a$.
_____

Your Turn   Add or subtract to combine the like terms.
- $7a + 3a$ • $13n + 9n + 2x$ • $14z - 8z + y$

Class Exercises

Solve the equation.

1. $6 + 2 \cdot 3 = a$
2. $(6 + 2)3 = b$
3. $14 \div 2 + 5 = x$
4. $14 \div (2 + 5) = y$
5. $4 + 3 \cdot 14 = m$
6. $(4 + 3)14 = n$

7. $4 \div 2 - 1 = a$ 8. $4 \div (2 - 1) = b$ 9. $4 \cdot 4 + 4 = q$

10. $4(4 + 4) = r$ 11. $0 \cdot 1 + 1 = w$ 12. $0(1 + 1) = z$

Exercises

Solve the equation.

A 1. **a.** $4 + 9 \cdot 2 = a$ 2. **a.** $5 \cdot 3^2 + 6 = y$ 3. **a.** $8 + 6 \cdot 3 = t$
   b. $(4 + 9)2 = r$ **b.** $5(3^2 + 6) = w$ **b.** $(8 + 6)3 = b$

4. **a.** $14 \cdot 3 + 9 = q$ 5. **a.** $15 + 4 \cdot 6 = m$ 6. **a.** $19 - 1 \cdot 8 = n$
   b. $14(3 + 9) = d$ **b.** $(15 + 4)6 = s$ **b.** $(19 - 1)8 = d$

7. **a.** $3 + 6 \cdot 4 + 4 = x$ 8. **a.** $17 - 8 \div 4 - 1 = k$ 9. **a.** $4^3 \cdot 3 + 6 \cdot 3 = h$
   b. $(3 + 6)(4 + 4) = g$ **b.** $(17 - 8) \div (4 - 1) = y$ **b.** $4^3(6 + 3)3 = j$

Name the coefficients and the variables in the expression.

B 10. $3a$ 11. $4b$ 12. $16n$

13. $87c$ 14. $47x$ 15. $963z$

16. $6n + 7m$ 17. $4p + 7q$ 18. $18g + 9g$

Combine like terms.

19. $n + 3n$ 20. $1n + 3n$ 21. $15x + x$

22. $13z + 17z + 4a$ 23. $66r - 19r - 17x$ 24. $75t + 92t - 6r$

25. $46s - 16s - 5t$ 26. $19j + 17k + j$ 27. $27e - 27b - 16e$

Let $a = 2$, $b = 8$, and $c = 6$. Evaluate the expression.

C 28. $a + bc$ 29. $(a + b)c$ 30. $3a + 8c$

31. $a(a + b)$ 32. $4 + 6b$ 33. $(b - c) \div a$

34. $ab - ac$ 35. $a(b - a)c$ 36. $a + b \div a + 4$

37. $\dfrac{(a + c)}{(c - a)}$ 38. $a^3 + ab$ 39. $2c^3 + 6$

Combine like terms.

40. $14x + 12y + 3x - 8y$ 41. $33a - 14a + 16b - 10b$

42. $48 + 3n - 4 + n$ 43. $63 - 4y + 12 + 6b$

Using the Computer

This program demonstrates that a computer using BASIC follows the rules for the Order of Operations. After trying this program, you may wish to try some examples of your own.

```
  5  REM: ORDER OF OPERATIONS
 10  PRINT "3*4+5*2 =";
 20  PRINT 3*4+5*2
 30  PRINT "3+4*5+2 =";
 40  PRINT 3+4*5+2
 50  PRINT "(3+4)*(5+2) =";
 60  PRINT (3+4)*(5+2)
 70  PRINT "6+8/2 =";
 80  PRINT 6+8/2
 90  PRINT "(6+8)/2 =";
100  PRINT (6+8)/2
110  PRINT "2↑3+5 =";
120  PRINT 2↑3+5
130  PRINT "2↑(3+5) =";
140  PRINT 2↑(3+5)
150  PRINT "4+2↑3 =";
160  PRINT 4+2↑3
170  PRINT "(4+2)↑3 =";
180  PRINT (4+2)↑3
190  END
```

Self Test 2

Solve.

1. Mr. Rice drove 360 mi on 25 gal of gas. How many miles did he travel per gallon of gas? **1-4**

2. $n = 17 \cdot 84$ 3. $46 \cdot 55 = x$ 4. $a = 27 \cdot 15$ **1-5**

5. $x = \dfrac{325}{25}$ 6. $1248 \div 16 = a$ 7. $y = \dfrac{1053}{39}$ **1-6**

8. $r = 872 \cdot 1$ 9. $q = 475 + 0$ 10. $z = 1006 + 0$ **1-7**

11. $3 + 7 \cdot 6 = n$ 12. $(14 \cdot 2) + 6 = a$ 13. $4 \cdot 3 + 6 \cdot 8 = n$ **1-8**

Algebra Practice

Evaluate the expression using the given value for the variable.

1. $(3w) + 2$; $w = 4$ **2.** $64 - (4y)$; $y = 7$ **3.** $r \div 5$; $r = 45$

4. $16 + (3b)$; $b = 2$ **5.** $100 - (10g)$; $g = 4$ **6.** $42 \div j$; $j = 6$

Write an equation for the sentence. Solve the equation.

7. The total of 406 and 97 is s.

8. The difference between 2001 and 1906 is a.

9. 89 increased by 19 is l.

10. 84,013 minus 47,219 is w.

11. The numbers 623 and 94 add to a sum of f.

12. 19 greater than 28 is p.

Write a variable expression for the phrase.

13. the product of h and 42   **14.** 89 divided by x

15. the quotient of 6273 and n **16.** j used as a factor 5 times

17. c increased by 50   **18.** the quotient of d and 12

Solve.

19. $b = 42 + 0$ **20.** $r = 17 \cdot 1$ **21.** $p = 0 + 683$

22. $71 = 71 + n$ **23.** $29c = 29$ **24.** $(65)(a) = 65$

25. $q = 14 \div 2 - 3$ **26.** $k = 8 \times 2 - 4 \div 2$ **27.** $3 + (3 \times 3) \div 3 = e$

Let $x = 4$, $y = 5$, $z = 1$. Evaluate the expression.

28. $x + y - z$ **29.** $(x + z)y$ **30.** $(2z + 4y)x$

31. $x(y - z)$ **32.** $2x(y + z)$ **33.** $y - z - x$

34. $4z - x + y$ **35.** $y(x - 4z)$ **36.** $z(y - x)$

Combine like terms.

37. $3n - 2n + n$ **38.** $4f - 3f + 1$ **39.** $4s + s + t$

40. $2a - 3b - a$ **41.** $4x + y + 4x$ **42.** $6p - 7 + 4p$

43. $5a + 6a - a + 8$ **44.** $3t + 7 + 8t - 1$ **45.** $4x + 2y - y - x$

Problem Solving on the Job

Construction Careers

Most construction jobs require the services of carpenters to transform wood into parts of a building.

CARPENTER

DESCRIPTION: Work on construction projects. Measure and mark room layouts.

QUALIFICATIONS: 4-5 years on-the-job training as an apprentice plus about 144 h per year of classroom work.

JOB OUTLOOK: Openings expected to be as good as other occupations; Outlook may change in response to reduced building activity.

	Year	Rate	Hours Worked	Total Pay
1.	1	$ 6.00	38	?
2.	1	?	35	$210.00
3.	1	$ 7.00	?	$294.00
4.	2	$ 8.50	40	?
5.	2	?	37	$323.75
6.	2	$ 9.35	?	$383.35

At various stages of a training program, an apprentice earns different rates of pay.

Complete the table.

Solve.

7. Molly is in the first year of her apprenticeship. She earns $6.00 per hour for the first 40 hours and time and a half for anything over 40 hours. How much will she earn in a week if she works 45 hours?

8. Ned is an apprentice earning $9.35 per hour. He earns double pay if he works on holidays. How much will he earn in a week that he works 38 regular hours and 7 holiday hours?

A contractor is building a housing development that will include 15 new houses.

9. Each new house will have a brick front. The bricklayer estimates that each house will require 500 bricks. How many bricks should be purchased for all the houses?

10. Each new house will need 8 doors for rooms, 5 doors for large closets, and 3 doors for small closets. How many doors of each size will be needed for the entire development?

11. The contractor estimates it will take 2 plumbers 4 days to do the major pipe installation in each house. If each plumber works 7 hours per day and earns $17 per hour, how much should the contractor estimate for the entire plumbing costs?

12. To wallpaper the living room of each new house requires 24 rolls of wallpaper. Each roll costs $9. How much will be spent for living room wallpaper for all the houses in the development?

13. Each house has a master bedroom 23 yd² in size. Wall-to-wall carpeting that costs $12 per yd² is being used for the master bedrooms for all the houses. How much will it cost?

14. Nine of the houses have patios 24 yd² in size. The size of the patios for the other 6 houses is 21 yd². Indoor-outdoor carpeting that costs $14 per yd² is being used to cover all the patios. How much will it cost?

15. The landscaping for each lot of the development includes the following trees and shrubs.

> 1 tree at $25
> 5 juniper bushes at $16 each
> 6 rose bushes at $13 each

How much will be spent on trees and shrubs for all the lots in the development?

Enrichment

Fibonacci Numbers

The Fibonacci Sequence consists of a series of numbers in which each term is the sum of the preceding two terms.

1, 1, 2, 3, 5, 8, 13, . . .

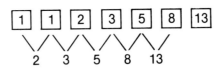

This sequence can be noted in many natural phenomena. The centers of most sunflowers have 21 spirals in one direction and 34 spirals in the other. The stalks of many plants have new leaves growing in a pattern such that starting at the base of the stalk and counting upward, each new leaf growing directly above another leaf is in a position corresponding to a Fibonacci number. If the first leaf is 1 the next leaf directly above it will be 8 and the next 13, etc.

1. Write the first 20 terms of the Fibonacci Sequence.
2. Find the sum of the first five terms of the sequence. Compare this to the seventh term. Repeat this for the sum of the first six terms and the first seven terms. What do you discover?
3. Use the results of Exercise 2 to write the sums of the first eight terms, ten terms, and fifteen terms without actually doing the addition.
4. It will take two months for a newborn pair of rabbits to produce their first pair of rabbits. After that they produce one new pair each month. If you start with one pair of rabbits producing according to this pattern and each succeeding pair of rabbits produces according to the same pattern, how many pairs of rabbits will you have at the end of each of the next twelve months?

Chapter Review

Match the word phrase to the expression.

1. n decreased by 56
2. the product of 56 and n
3. the sum of n and 56
4. a number n divided by 56
5. 56 reduced by a number n

a. $n + 56$
b. $n \div 56$
c. $n - 56$
d. $56 - n$
e. $56 \div n$
f. $56n$

Choose the correct answer. Write a, b, or c.

6. $479 + 378 = n$
  a. $n = 101$
  b. $n = 857$
  c. $n = 747$

7. $y = 3000 - 55$
  a. $y = 3055$
  b. $y = 2955$
  c. $y = 2945$

8. $t = 680 \cdot 79$
  a. $t = 53,720$
  b. $t = 48,020$
  c. $t = 43,720$

9. $14,874 \div 6 = x$
  a. $x = 2479$
  b. $x = 2110$
  c. $x = 3479$

10. $4 \cdot 8 + 6 = y$
  a. $y = 18$
  b. $y = 56$
  c. $y = 38$

11. $(33 \div 11) + (7 \cdot 6) = v$
  a. $v = 45$
  b. $v = 136$
  c. $v = 3$

12. $3(7 + 6) - 4 = j$
  a. $j = 23$
  b. $j = 27$
  c. $j = 35$

13. $(5 + 7)(10 - 2) = w$
  a. $w = 73$
  b. $w = 96$
  c. $w = 61$

True or false? Write *T* or *F*.

14. $12n + 5n = 17n$

15. $3b + 4a + b = 4b + 4a$

16. $16 + 3a = 19a$

17. $39y - 12 - 6y = 33y - 12$

18. $100a + 100b = 100ab$

19. $46x - 10x + 3 = 39x$

20. $45x - 3 - 10x = 35x - 3$

21. $6n + 2n - 5n = 3$

22. $16 + 4a + 3a = 7a - 16$

23. $5x + 2y + 3x + y = 8x + 3y$

24. $x + 4y - x - 4y = 2x + 8y$

25. $n + 8m + 20n = 21n + 8m$

26. $8a + 6b - 2a = 6a + 6b$

27. $50n - 49n = n$

28. $65r - 8r + 17r - 2 = 72r$

29. $25t - 15t - 10t = 0$

Chapter Test

Let $a = 5$, $b = 10$, and $c = 25$. Evaluate the expression.

1. $a + b$ **2.** ab **3.** $c - a$ **4.** $b \div a$ 1-1

5. $3a + c$ **6.** $5b \div c$ **7.** $2a + 2c$ **8.** $10b - 4c$

Solve the equation.

9. $n = 6872 + 439$ **10.** $6759 + 775 = y$ **11.** $d = 766 + 389$ 1-2

12. $a = 600 + 251$ **13.** $4007 + 398 = b$ **14.** $r = 773 + 86$

15. $h = 761 - 7$ **16.** $1256 - 87 = c$ **17.** $5306 - 93 = k$ 1-3

18. $y = 5002 - 111$ **19.** $4533 - 2779 = q$ **20.** $9000 - 625 = v$

Solve.

21. In June, Gordon's Fishcake Company spent $6215 for a TV ad, $488 for a newspaper ad, and $2715 for a direct-mail ad. What is the difference between the most and least expensive types of advertising purchased in June? 1-4

Solve the equation.

22. $(43)(16) = y$ **23.** $r = 750 \cdot 27$ **24.** $(7093)(86) = t$ 1-5

25. $3891 \cdot 54 = m$ **26.** $a = (973)(355)$ **27.** $466 \cdot 784 = n$

28. $1585 \div 5 = x$ **29.** $y = 6750 \div 27$ **30.** $d = 3036 \div 138$ 1-6

31. $5676 \div 22 = d$ **32.** $17,136 \div 48 = p$ **33.** $f = 3397 \div 79$

34. $r = 4862 \times 1$ **35.** $76,842 + 0 = n$ **36.** $q = 0 + 88,615$ 1-7

37. $t = 3541 \cdot 1$ **38.** $m = 0 + 15,987$ **39.** $(1)(3422) = k$

40. $b = 17 + (4 \cdot 8)$ **41.** $9 \cdot 8 + 18 \div 6 = a$ 1-8

42. $f = (16 + 4)(4 - 2)$ **43.** $(11 - 5) \div (2 \cdot 3) = n$

Combine like terms.

44. $2a - a + b$ **45.** $3x + 2y - 2x - x$

46. $m + 8m + 4t - 4t$ **47.** $16x + 9y + 2x - x - y$

Cumulative Review (Chapter 1)

Add, subtract, multiply, or divide.

1. $470 + 18 - 2$ **2.** $356 - 24 - 6$ **3.** $7305 - 38$ **4.** $4000 - 458$

5. $(279)(8)$ **6.** $(56)(18)$ **7.** $(56 \div 8) \cdot 5$ **8.** $(11{,}028)(0)$

9. $16 \cdot 5$ **10.** $230 + 7 - 49$ **11.** $(17 \cdot 8) - 3$ **12.** $6 + (8 \cdot 6)$

13. $45 - 5 \cdot 6$ **14.** $16 + 5 \cdot 8$ **15.** $36 \div 9 \cdot 4$ **16.** $25 - 4 \cdot 6$

Evaluate the expression.

17. $a + b$; $a = 2$, $b = 7$ **18.** $4x - 1$; $x = 12$ **19.** $x \div y$; $x = 80$, $y = 5$

20. $5n - 2x$; $n = 6$, $x = 1$ **21.** $7 + 3k$; $k = 12$ **22.** $2p + 4t$; $p = 1$, $t = 2$

23. $10m^2$; $m = 2$ **24.** $5y^2$; $y = 9$ **25.** $2 - 3x^2$; $x = 0$

26. $a^2 - 6$; $a = 4$ **27.** $4 - x^2$; $x = 1$ **28.** $(2m)^2$; $m = 4$

Simplify.

29. $4(x + 1)$ **30.** $a(6 - a)$ **31.** $5x + 2x$

32. $7a - a$ **33.** $4y - 2y - y$ **34.** $3(a - b)$

35. $12p + 6 - 6p$ **36.** $9n - 4n - 1$ **37.** $a + 7b - a$

38. $3w + 6 - 3w$ **39.** $15c - c + 21$ **40.** $8j - 6 - 6j$

41. $16e + 2f - 16e - 2f$ **42.** $12s + s - s$ **43.** $8g + 8h - g + h$

Write an expression for the phrase.

44. the product of some number n and 38

45. x decreased by 16

46. 14 greater than g

47. the sum of 54 and 16 added to w

48. c increased by n

Solve.

49. A laboratory technician ordered 12 boxes of microscope slides.
Let $c =$ cost of each box.
Then __?__ = the total cost.

50. Melba is taking a 7-day vacation.
Let $m =$ amount of spending money she has to spend.
Then __?__ = average amount of money she can spend each day.

2
Decimals

A picture is worth a thousand words. You can control the amount of light passing through the camera by adjusting the f-stop. Change it from 16 to 11 to increase the light, and from 5.6 to 8 to decrease the light.

2-1 Reading and Writing Decimals

Decimals represent a whole unit or part of a whole unit.

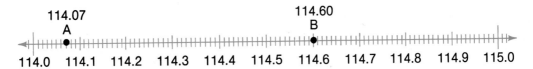

Point *A* shows the decimal one hundred fourteen and seven hundredths. Point *B* shows the decimals one hundred fourteen and six tenths, and one hundred fourteen and sixty hundredths. The decimals 114.6 and 114.60 are equal. **Equal decimals** name the same number.

Example 1   Read the decimal 13.0405.

Solution

millions	hundred-thousands	ten-thousands	thousands	hundreds	tens	ones	.	tenths	hundredths	thousandths	ten-thousandths	hundred-thousandths	millionths
					1	3	.	0	4	0	5		

Thirteen and four hundred five ten-thousandths

Example 2   Write the decimal sixty and four hundred thirteen thousandths.

Solution

millions	hundred-thousands	ten-thousands	thousands	hundreds	tens	ones	.	tenths	hundredths	thousandths	ten-thousandths	hundred-thousandths	millionths
					6	0	.	4	1	3			

Answer: 60.413

Your Turn
- Read the decimal 574.405.
- Write the decimal seven thousand five and two thousand two ten-thousandths.

Class Exercises

Complete.

1. $6.8 = \underline{?}$ and $\underline{?}$ tenths

2. $46.09 = \underline{?}$ and $\underline{?}$ hundredths

3. $9.031 = \underline{?}$ and $\underline{?}$ thousandths

4. $17.707 = \underline{?}$ and $\underline{?}$ thousandths

5. $0.88051 = \underline{?}$ hundred-thousandths

6. $0.0004 = \underline{?}$ ten-thousandths

7. $5.500613 = \underline{?}$ and $\underline{?}$ millionths

8. $0.234006 = \underline{?}$ millionths

Exercises

The number line below is divided into tenths and hundredths. Write the decimal for the given point.

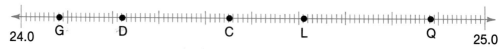

A **1.** Point C **2.** Point Q **3.** Point D **4.** Point G **5.** Point L

The number line below is divided into hundredths and thousandths. Write two equal decimals for the given point.

6. Point B **7.** Point A **8.** Point E **9.** Point C **10.** Point M

Write a, b, or c to show how to read the decimal.

B **11.** 0.12
    a. twelve thousandths
    b. twelve tenths
    c. twelve hundredths

12. 17.6
    a. seventeen and six tenths
    b. seventeen and six hundredths
    c. six and seventeen hundredths

13. 0.655
   a. six hundred fifty-five hundreds
   b. six hundred fifty-five thousandths
   c. six hundred five thousandths

14. 0.6505
   a. six hundred fifty-five ten-thousandths
   b. six thousand five hundred five ten-thousandths
   c. six thousand five hundred five thousandths

15. 4.00255
   a. two hundred fifty-five thousandths
   b. four and two hundred fifty-five hundred-thousandths
   c. four and two hundred fifty-five millionths

16. 10.005001
   a. ten and five thousand one millionths
   b. ten and five thousand one hundred-thousandths
   c. ten and five thousand one ten-thousandths

Write the decimal.

17. 7 tenths

18. 12 hundredths

19. 2401 ten-thousandths

20. 98 millionths

21. 17 and 6 thousandths

22. 500 and 5472 hundred-thousandths

23. 4001 and 23 ten-thousandths

24. 80 and 7 millionths

25. nine thousandths

26. twenty-two hundredths

27. four and six ten-thousandths

28. thirty and two hundred twenty-five thousandths

29. six hundred forty and fifty thousand two hundred one millionths

30. seven thousand thirteen and eight thousand eleven hundred-thousandths

Write a decimal for the given information.

C **31.** the part of a dollar represented by a dime

32. the part of a dollar represented by a nickel

33. the part of a decade represented by a year

34. the part of a century represented by a decade

35. the part of a centimeter represented by a millimeter

36. the part of a liter represented by a milliliter

37. the part of a meter represented by a centimeter

2-2 *Ordering Decimals*

The decimal coding system shown below is used to store merchandise at a store. Notice how the number of decimal places in the code increases as more information is given in the other charts.

RST Warehouse	3 Hair product	3.2 Shampoos	3.23 Natural
1 Stationery	3.1 Conditioners	3.21 Imperial	3.231 for normal hair
2 Cosmetics	3.2 Shampoos	3.22 Luster	3.232 for dry hair
3 Hair Products	3.3 Dyes/Tints	3.23 Natural	3.233 for oily hair
4 Soap Products	3.4 Perm Kits	3.24 Universal	3.234 for tinted hair
5 Miscellaneous	3.5 Combs/Brushes	3.25 Golden	3.235 for damaged hair

When an order needs to be filled, the code numbers from the order form are arranged in numerical order. To arrange decimals in numerical order, first recall how to compare decimals.

Example 1 Compare 5.46 and 5.41.

Solution Start from the left and compare the digits in corresponding places.

5.46 __?__ 5.41

$6 > 1$ ⟵ *The symbol $>$ means*

so $5.46 > 5.41$ *"is greater than."*

Example 2 Compare 27.4 and 27.456.

Solution It may be helpful to write zeros after 27.4 so that both decimals have the same number of decimal places.

27.400 __?__ 27.456.

$0 < 5$ ⟵ *The symbol $<$ means*

so $27.4 < 27.456$ *"is less than".*

Your Turn • Compare 4.1 and 4.16. • Compare 30.72 and 30.62.

Now use what you know about comparing decimals to arrange decimals in numerical order.

Example 3   Write the code numbers in
Form 6712 in order from
least to greatest.

Solution   3.246, 3.247, 3.31, 3.4

ORDER FORM 6712	
Code Number	Quantity
3.246	24
3.31	12
3.247	24
3.4	18

Class Exercises

Write > or < to compare the decimals.

1. 4.6 __?__ 5.2

2. 6.22 __?__ 6.32

3. 74.1 __?__ 74.4

4. 18.9 __?__ 19.4

5. 31.47 __?__ 31.48

6. 2.234 __?__ 2.214

7. 7.6 __?__ 7.63

8. 81.51 __?__ 81.5

9. 3.12 __?__ 31.2

10. 31.18 __?__ 31.187

11. 94.65 __?__ 946.5

12. 7.652 __?__ 7.65

13. 904.55 __?__ 904.5

14. 821.34 __?__ 821.3

15. 746.23 __?__ 74.623

Exercises

Write the decimals in order from least to greatest.

A **1.** 3.7, 6.8, 5.2

2. 9.6, 9.4, 9.1

3. 3.2, 4.2, 3.3

4. 4.16, 3.16, 4.26

5. 52.3, 53.3, 52.2

6. 86.15, 96.15, 86.17

7. 87.78, 87.77, 87.87

8. 641.3, 641.4, 642.3

9. 9.103, 9.301, 9.013

10. 57.62, 37.72, 57.63, 57.61

11. 413.32, 431.32, 413.23, 413.12

12. 5.6642, 566.42, 56.642, 5664.2

13. 2134.9, 21.349, 213.49, 2.1349

39

Write the decimals in order from greatest to least.

B **14.** 4.36, 4.3, 4.365 **15.** 0.008, 0.08, 0.8 **16.** 1.52, 1.522, 1.5

17. 39.954, 29.9, 29.95 **18.** 16.2, 16.23, 16.22 **19.** 71.45, 71.4, 71.455

20. 5.6432, 5.643, 5.64, 5.642 **21.** 7.2913, 7.291, 7.29, 7.28

22. 81.18, 81.187, 81.188, 81.178 **23.** 190.9, 190.09, 190.099, 19.009

24. 7.752, 7.7521, 7.75, 7.725 **25.** 61.008, 61.081, 610.8, 61.0813

Write $<$ or $>$ to compare the given numbers.

Example $b = 4.13$; $c = 4.1$ **Solution** $b > c$

C **26.** $a = 2.7$; $b = 2.4$ **27.** $x = 30.4$; $y = 3.04$ **28.** $d = 5.16$; $c = 5.17$
$a \underline{\ ?\ } b$ $x \underline{\ ?\ } y$ $d \underline{\ ?\ } c$

29. $h = 304.1$; $q = 301.4$ **30.** $c = 0.09$; $t = 0.08$ **31.** $f = 61.16$; $z = 16.61$
$h \underline{\ ?\ } q$ $c \underline{\ ?\ } t$ $f \underline{\ ?\ } z$

32. $g = 7.13$; $s = 71.34$ **33.** $m = 8.64$; $w = 8.6$ **34.** $e = 32.7$; $y = 32.78$
$g \underline{\ ?\ } s$ $m \underline{\ ?\ } w$ $e \underline{\ ?\ } y$

35. $k = 50.05$; $r = 50.032$ **36.** $d = 16.62$; $y = 16.624$ **37.** $p = 7.5$; $b = 7.543$
$k \underline{\ ?\ } r$ $d \underline{\ ?\ } y$ $p \underline{\ ?\ } b$

Problems

Write an expression for the given sentence.

1. a is greater than 4.1. **2.** p is less than 18.2.
3. s is equal to 5.21. **4.** 56.3 is greater than b.
5. 7.45 is less than c. **6.** 9.01 is equal to r.
7. h is greater than 7.04. **8.** w is equal to 32.6.
9. 7.6 is less than f. **10.** 9.211 is greater than q.

Challenge

You have six United States coins totaling $1.15 but no combination of the six coins totals $1.00. What are the six coins?

2-3 *The Rounding Process*

Power and light companies sometimes add a fuel adjustment charge
to the monthly bills of their customers. This charge helps to cover any
increases in the cost of the fuel
needed to produce electricity.

One month a charge of $0.146 per
kilowatt hour was added to each
customer's bill. The number 0.146 is
shown on the number line below.

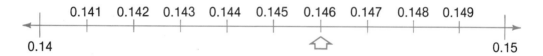

| | 0.141 | 0.142 | 0.143 | 0.144 | 0.145 | 0.146 | 0.147 | 0.148 | 0.149 | |
| 0.14 | | | | | | | | | | 0.15 |

Example 1 Use the number line to round $0.146 to the nearest
cent (hundredth of a dollar).

Solution 0.146 is closer to 0.15 than to 0.14, so $0.146 rounds to
$0.15.

Example 2 Round 26.158 to the nearest whole number.

Solution When the digit to the right of the place to which you
are rounding is less than 5, always round down.

Given Decimal	Digit to the Right	Is it 5 or More?	Round the Number
26.158	1	no	down to 26

Example 3   Round 7.526 to the nearest hundredth.

Solution   When the digit to the right of the place to which you are rounding is 5 or more, always round up.

Given Decimal	Digit to the Right	Is it 5 or More?	Round the Number
7.526	6	yes	up to 7.53

Your Turn   • Round 7.03 to the nearest tenth.
• Round 18.7855 to the nearest thousandth.

Example 4   Round 46.96 to the nearest tenth.

Solution   When the number in the place to which you are rounding is 9, and the digit to the right of that place is 5 or more, you'll need to round the number to its next greater place value.

Given Decimal	Digit to the Right	Is it 5 or More?	Round the Number
46.96	6	yes	up to 47.0

Your Turn   • Round 0.596 to the nearest hundredth.

Class Exercises

Use the number line to round the given decimal to the nearest tenth.

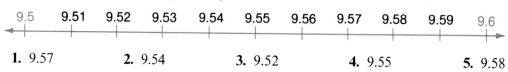

9.5   9.51   9.52   9.53   9.54   9.55   9.56   9.57   9.58   9.59   9.6

1. 9.57   **2.** 9.54   **3.** 9.52   **4.** 9.55   **5.** 9.58

Complete the chart.

	Round to the Nearest	Given Decimal	Digit to the Right	Is it 5 or More?	Round the Number
6.	whole number	7.38625	?	?	?
7.	tenth	7.38625	?	?	?
8.	hundredth	7.38625	?	?	?
9.	thousandth	7.38625	?	?	?

Exercises

Measure the segment to the nearest centimeter.

Example 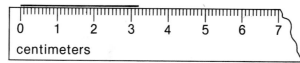 **Solution**   3 cm

A **1.** _____

2. _____

3. _____

4. _____

5. _____

Round to the nearest whole number.

B **6.** 0.6   **7.** 0.3   **8.** 1.5   **9.** 47.2   **10.** 45.5

11. 89.42   **12.** 538.81   **13.** 492.74   **14.** 389.554   **15.** 469.772

Round to the nearest tenth.

16. 0.45   **17.** 0.31   **18.** 4.76   **19.** 38.07   **20.** 42.03

21. 68.974   **22.** 72.213   **23.** 862.958   **24.** 497.0684   **25.** 766.3298

Round to the nearest hundredth.

26. 0.316   **27.** 0.567   **28.** 2.304   **29.** 31.209   **30.** 57.698

31. 38.4931   **32.** 757.6234   **33.** 840.3975   **34.** 624.8361   **35.** 590.0071

Round to the nearest thousandth.

36. 0.4567   **37.** 0.1306   **38.** 5.3496   **39.** 18.5792   **40.** 23.4575

41. 86.0052   **42.** 344.7214   **43.** 877.0555   **44.** 362.2397   **45.** 440.3205

Electrical energy is measured in kilowatt hours (kW · h). The chart at the right shows the amount of electrical energy used by some common electrical appliances in one year. Each amount has been rounded to the nearest whole number. What could the minimum and maximum amounts of electrical energy have been before they were rounded, if they had been measured to the nearest tenth?

	Electrical Appliance	Energy Used Per Year
C 46.	toaster	38 kW · h
47.	washer	100 kW · h
48.	dryer	993 kW · h
49.	stereo	109 kW · h
50.	fan	43 kW · h
51.	color TV	502 kW · h

Example   radio: 86 kW · h

Solution   minimum: 85.5 kW · h
maximum: 86.4 kW · h

Self Test 1

Write the decimal.

1. 28 and 7 hundredths

2. 7013 and 62 ten-thousandths   **2-1**

3. six hundred and four thousandths

4. five and fourteen millionths

Write the decimals in order from least to greatest.

5. 7.66, 8.66, 7.56, 7.67

6. 92.29, 82.29, 92.92, 92.28   **2-2**

7. 0.75, 0.7, 0.756, 0.76

8. 8.341, 8.3, 8.34, 8.44

Round to the nearest whole number.

9. 0.7   **10.** 4.28   **11.** 35.55   **12.** 129.86   **2-3**

Round to the nearest tenth.

13. 0.51   **14.** 8.39   **15.** 40.97   **16.** 112.65

2-4 *Adding Decimals*

The distance traveled each day on a biking trip was recorded as follows: Thursday 28.8 km, Friday 17.0 km, and Saturday 24.4 km.

You can find the total distance of the trip by adding the daily distances. Follow the same procedure for adding decimals as for adding whole numbers.

Let r = the total distance of the trip.
$r = 28.8 + 17.0 + 24.4$ ◄─────

$r = 70.2$

The total distance of the trip is 70.2 km.

Remember to line up the decimal points.

$$\begin{array}{r} 28.8 \\ 17.0 \\ +24.4 \\ \hline 70.2 \end{array}$$

To be sure that an answer is reasonable, estimate the sum.

Example 1 Estimate the total distance of the trip by rounding each addend to its greatest place value.

Solution

$$\begin{array}{rcl} 28.8 & \longrightarrow & 30 \\ 17.0 & \longrightarrow & 20 \\ +24.4 & \longrightarrow & +20 \\ \hline & & 70 \end{array}$$

The estimated distance of the trip is 70 km.
Since the estimated sum is close to the actual sum, 70.2 km is a reasonable answer.

Your Turn • Estimate the sum of 13.17 and 25.56 by rounding each addend to the greatest place value.

Example 2   Solve the equation $g = 5.6 + 7.529$.

Solution $g = 5.6 + 7.529$ ←———————

5.600	*Sometimes it's helpful*
+7.529	*to add zeros.*
13.129	

$g = 13.129$

Your Turn   • Solve the equation $y = 7.2 + 3.16 + 15.548$.

Class Exercises

Add.

1. 3.81	**2.** 6	**3.** 5.13	**4.** 26.6	**5.** 8.24
+2.7	+47.2	+0.652	5	2.075
			+117.8	+0.6

Exercises

Add.

A
1. $37.8 + 9.26$ **2.** $41.4 + 13.2$ **3.** $5.76 + 78.29$

4. $4.658 + 24.1$ **5.** $53.44 + 4$ **6.** $378.62 + 2.1985$

7. $47 + 654.72$ **8.** $85.111 + 26.81$ **9.** $532.11 + 275.6$

Round each addend to its greatest place value to estimate the sum.

10. $5.6 + 9.2$ **11.** $13.42 + 27.8$ **12.** $0.94 + 71.7$

13. $652.1 + 71.5$ **14.** $3.64 + 21.05$ **15.** $61.5 + 20.4$

16. $8.719 + 21.36$ **17.** $86.2 + 9.4$ **18.** $3.165 + 40.45$

Evaluate the expression using the given value for the variable.

B **19.** $3.29 + n$; $n = 4.6$ **20.** $7.31 + x$; $x = 26.5$

21. $0.743 + y$; $y = 31.6$ **22.** $0.81 + w$; $w = 31.7$

23. $z + 36.257$; $z = 81$ **24.** $b + 2134$; $b = 7.65$

25. $d + 31.66$; $d = 4.72$ **26.** $h + 3.007$; $h = 81.51$

27. $25.316 + m$; $m = 5.962$ **28.** $w + 18.934$; $w = 21.106$

29. $14 + t$; $t = 3.742$ **30.** $741.6 + r$; $r = 8.12$

Combine like terms.

C
31. $8.4b + 3.6b$ **32.** $0.7d + 4.6d$ **33.** $32.4a + 1.44a$

34. $1.75x + 12.1x$ **35.** $21.14g + 4.657g$ **36.** $0.321s + 52.1s$

37. $31.24m + 5.0006m$ **38.** $17.012c + 3.04c$ **39.** $162.34t + 31.17t$

40. $4.7b + 0.81a + 31.4b$ **41.** $0.91g + 21.4g + 3.77h$ **42.** $8.8p + 3.07r + 24.1r$

43. $8.3s + 57t + 16.61s$ **44.** $4.8w + 3.1w + 0.07z$ **45.** $0.24j + 3.9k + 7.7j$

Problems

Write an expression for the given phrase.

1. 3.7 added to x **2.** y plus 71.04   **3.** b increased by 0.92

4. 19.6 greater than p **5.** a more than 5.61   **6.** t added to 1.964

Write an expression to complete.

7. The distance to the school is 2.3 km further than the distance to the park.
Let b = the distance to the park.
Then, __?__ = the distance to the school.

8. Science class enrollment is 47 more at Jefferson High School than at Washington High School.
Let w = science class enrollment at Washington High School.
Then, __?__ = science class enrollment at Jefferson High School.

Write an equation for the given sentence. Solve the equation.

9. The sum of 4.85 and 31.7 is equal to b.
10. 8.4 increased by 0.45 is equal to a.
11. 2.13 added to 71.6 is equal to g.
12. y is equal to the sum of 21.1 and 34.62.
13. n is equal to 12.62 more than 3.074.
14. 76.81 greater than 2.7 is equal to w.

2-5 Subtracting Decimals

The qualifying time for entrants in a six-mile race is 51.6 min. One runner's time is 55.5 min.

Example 1   How much does the runner's time need to improve in order to qualify for the race?

Solution   Let n = how much the runner's time needs to improve.

$n = 55.5 - 51.6$ ◄——————— | *Remember to keep the*
$n = 3.9$ | *decimal points in line.*
The runner's time needs to | 55.5
improve 3.9 min. | $\underline{-51.6}$
| 3.9

Sometimes all you need is an estimated answer.

Example 2   Estimate about how much the runner's time needs to improve in order to qualify for the race.

Solution
$$55.6 \longrightarrow 56 \quad \longleftarrow \text{ Round each number.}$$
$$\underline{-51.6} \longrightarrow \underline{-52}$$
$$ 4$$

The runner's time needs to improve about 4 min.

Your Turn • Estimate the difference between 44.6 and 7.2.

Example 3 Solve the equation $t = 17.1 - 5.62$.

Solution $t = 17.1 - 5.52$ ←

17.10	*If you wish, add zeros to help*
− 5.62	*keep the decimal points*
11.48	*aligned.*

$t = 11.48$

Your Turn Solve the equation.
• $m = 12.6 - 4.71$ • $d = 25 - 9.34$

Class Exercises

Subtract.

1. 0.78 −0.32	**2.** 0.758 −0.317	**3.** 6.73 −0.26	**4.** 6.306 −4.728	**5.** 6.306 −0.72
6. 47.28 − 1.5	**7.** 0.7 −0.36	**8.** 37.6 − 7.746	**9.** 9 −6.3	**10.** 58 − 13.45

Exercises

Subtract.

A **1.** $0.963 - 0.721$ **2.** $42.376 - 11.153$ **3.** $14.76 - 10.59$

4. $196.381 - 78.265$ **5.** $321.15 - 76.28$ **6.** $0.4765 - 0.24$

7. $78.1 - 59.35$ **8.** $176 - 11.548$ **9.** $476 - 0.2486$

Round each number to the nearest whole number to estimate the difference.

10. $8.1 - 3.97$ **11.** $14.36 - 11.5$ **12.** $38.514 - 19.01$

13. $586.5 - 49.7$ **14.** $32.058 - 19.67$ **15.** $819.001 - 572.6$

16. $31.64 - 2.247$ **17.** $7.0892 - 0.99$ **18.** $567.55 - 76.51$

Solve the equation.

B **19.** $4.97 - 3.06 = g$ **20.** $567.8 - 324.5 = v$ **21.** $0.963 - 0.302 = z$

22. $1007.58 - 896.7 = d$ **23.** $119.76 - 89.84 = p$ **24.** $103.016 - 96.3 = h$

25. $d = 0.83 - 0.543$ **26.** $m = 96.13 - 3.257$ **27.** $h = 7.138 - 2.4$

28. $q = 42 - 36.245$ **29.** $f = 781 - 560.31$ **30.** $a = 13 - 9.1785$

Combine like terms.

31. $8.4a - 6.2a$ **32.** $13.7d - 9.4d$ **33.** $8.65x - 0.97x$

34. $16.43s - 8.4s$ **35.** $71.44b - 19.7b$ **36.** $364.2p - 1.0734p$

37. $9v - 5.4v$ **38.** $48t - 17.93t$ **39.** $245z - 76.892z$

Evaluate the expression using the given value for the variable.

40. $a - 4.3; a = 7.02$ **41.** $18.83 - d; d = 4.9$

42. $47.4 - d; d = 41.82$ **43.** $z - 21.3; z = 45.23$

44. $g - 222.2; g = 232.41$ **45.** $81.05 - k; k = 7.9$

46. $21.3 - w; w = 3.91$ **47.** $y - 77.7; y = 904.231$

Add or subtract to write the missing digits.

C **48.**
$$\begin{array}{r} 2.?317 \\ +8.6?2 \\ \hline 1?.28?7 \end{array}$$
49.
$$\begin{array}{r} ?88.5?6 \\ -\ 68.?79 \\ \hline 32?.51? \end{array}$$
50.
$$\begin{array}{r} 2?1.3?02 \\ -17?.893? \\ \hline 114.?671 \end{array}$$
51.
$$\begin{array}{r} ??.?03? \\ -\ 8.2?62 \\ \hline ?.75?4 \end{array}$$

Problems

Write an addition or subtraction expression for the given phrase.

1. 8.3 minus t **2.** 4.08 subtracted from x **3.** x decreased by 18.4

4. 27.91 less than y **5.** b increased by 7.41 **6.** c more than 97.62

Write an addition or subtraction equation for the given sentence. Solve the equation.

7. The difference between 21.073 and 4.68 is equal to r.

8. 67.9 decreased by 13.74 is equal to d.

9. The sum of 40.61 and 321.7 is equal to a.

10. 7.71 subtracted from 8.1 is equal to r.

2-6 *Problem Solving*

The answers to some problems must fall within a given set of limits. Quite often there will be key words such as *the greatest, the least, within,* and *at most,* to help determine if your answer is reasonable.

Example 1 A product planner spends from 2.75 to 4.9 hours per day making telephone calls. The planner estimates that about 23 h are spent on telephone calls during a 5-day week. Is that a reasonable estimate?

Solution Use the limits for the least and the greatest number of hours.

$$(2.75)(5) = 13.75 \qquad (4.9)(5) = 24.5$$

$13.75 < 23 < 24.5 \longleftarrow$ *The estimate falls within the limits.*

23 h is a reasonable estimate.

Example 2 At a certain bakery, a whole quiche sells for $7.95. A quiche cut into 8 slices is priced at $1.35 per slice. What is the greatest amount of money the bakery can get from selling 36 quiches?

Solution $(1.35)(8) = \$10.80 \longleftarrow$ *price of one quiche when sold in slices*

$\$10.80 > \$7.95$
$(\$10.80)(36) = \$388.80$

Your Turn • A clerk in a credit department spends from 3.5 to 5.25 hours per day recording payments. The clerk estimates that about 26.5 h are spent recording payments during a 6-day week. Is the estimate reasonable?
• Suppose the bakery in Example 2 cuts each quiche into 6 slices and sells each slice for $1.40. What is the greatest amount of money it can get from selling 25 quiches?

Class Exercises

Appointment Schedule	
Morning	Afternoon
9:00 A.M. to 10:00 A.M.	3:00 P.M. to 4:00 P.M.

Listed below are the times when calls were made for appointments. Were the calls made according to the appointment schedule?

1. 9:10 A.M.   **2.** 9:15 A.M.   **3.** 8:45 A.M.   **4.** 3:10 P.M.

5. 2:50 P.M.   **6.** 11:00 A.M.   **7.** 3:20 P.M.   **8.** 3:49 P.M.

Exercises

Use the table at the right to solve Exercises 1-5.

0.60	0.9	0.85	0.3
0.72	0.38	0.1	0.18
0.45	0.16	0.28	0.91
0.53	0.67	0.41	0.26

A
1. Find the greatest sum using three numbers.
2. Find the least sum using three numbers.
3. Which two numbers will produce the greatest product?
4. Which two numbers will produce the least product?
5. Two numbers are chosen so that the second number subtracted from the first produces a difference of 0.81. What are the two numbers?

B
6. Alan has a roll of crepe paper 24 ft long. He is cutting the paper in strips measuring 9 in., 1.5 ft, or 2.4 ft each. What is the greatest and the least number of strips that Alan can cut from the roll?
7. Three sets of tennis are scheduled for the championship play-off. It is estimated that each set will take 35 to 50 minutes. If the play-off begins at 2:30 P.M., what is the earliest and latest it can be expected to be completed?
8. The label on an extension-type curtain rod states that the rod can be extended from 22 in. to 48 in. Will the rod be long enough for a window that measures 2 ft 6 in. across? Will it be too long?

2-7 *Multiplying Decimals*

A part-time job at a deli pays $4.85 per hour. How much would you earn in 4 hours?

Let d = the total amount
$d = 4 \times \$4.85$
$d = \$19.40$

$\$\ 4.85$ ⟵ *2 decimal places*
$\underline{\times 4}$ ⟵ *+0 decimal places*
$\$19.40$ ⟵ *2 decimal places*

You would earn $19.40 in 4 hours.

   Sometimes it may be necessary to write zeros in the product so you have the correct number of decimal places.

Example 1   Solve the equation $s = 0.03 \times 0.21$.

Solution $s = 0.03 \times 0.21$
$s = 0.0063$

0.21 ⟵ *2 decimal places*
$\underline{\times 0.03}$ ⟵ *+2 decimal places*
0.0063 ⟵ *4 decimal places*

Your Turn   Solve the equation.
   • $y = 0.02 \times 0.42$ $x = 0.6 \times 0.031$

Example 2   Solve the equations $h = 10 \times 0.764$ and $p = 100 \times 4.3$.

Solution   Mentally move the decimal point one place to the right for each power of 10.

$h = 10 \times 0.764$ $p = 100 \times 4.3$
$h = 7.64$ $p = 430$

Your Turn   Solve the equation.
   • $100 \times 48.921 = a$ • $1000 \times 3.7006 = s$

Sometimes you may use a calculator to multiply decimals. Even with a calculator, mistakes can occur since you might press a wrong button.

Example 3 A calculator showed the product of 75.6 and 3.09 to be 78.69. To check that answer estimate the product.

Solution Round each factor.

$$
\begin{array}{ccc}
75.6 & \longrightarrow & 80 \\
\times 3.09 & \longrightarrow & \times 3 \\
\hline
& & 240
\end{array}
$$

Since 240 is nowhere near 78.69, we know we made a mistake. We pressed the ⊞ button instead of the ⊠ button. The correct product is 233.604.

Your Turn • Estimate the product of 26.731 and 481.2 by rounding each factor to the greatest place value.

Class Exercises

Multiply.

1.	0.3	**2.**	2.1	**3.**	0.7	**4.**	2.3	**5.**	4.5
	×6		×3		×0.5		×0.4		×0.2

6.	6.5	**7.**	1.47	**8.**	18.4	**9.**	0.4	**10.**	0.1
	×4.4		×2.3		×3.7		×0.02		×0.06

Multiply the given number by 10, 100, 1000, and 10,000.

	×	10	100	1000	10,000
11.	7.8234	?	?	?	?
12.	47.236	?	?	?	?
13.	625.0035	?	?	?	?

Exercises

Multiply.

A 1. 4×5.2

2. 13×17.6

3. 7.4×11.4

4. 3.2×4.1

5. 7.6×0.9

6. 8.42×3.2

7. 0.07×0.1

8. 8.32×4.21

9. 4.01×0.09

Round each factor to its greatest place value to estimate the product.

10. 47×36.2

11. 52×2.79

12. 345×41.4

13. 7.83×4.4

14. 3.19×21.6

15. 58.76×3.68

16. 8.99×876.1

17. 76.5×4.321

18. 852.9×45.3

Solve the equation.

B 19. $7 \times 6.13 = x$

20. $24 \times 0.15 = a$

21. $27.9 \times 0.8 = c$

22. $3.15 \times 2.6 = b$

23. $621.4 \times 3.7 = m$

24. $0.003 \times 0.2 = g$

25. $z = 8 \times 73.9$

26. $n = 41 \times 0.66$

27. $p = 52.6 \times 0.7$

28. $r = 0.473 \times 6.8$

29. $f = 0.06 \times 0.1$

30. $t = 0.007 \times 0.4$

Find the value of the expression if $a = 10$.

31. $0.4a$

32. $2.7a$

33. $3.68a$

34. $41.5a$

35. $74.63a$

36. $81.632a$

Find the value of the expression if $y = 100$.

37. $0.63y$

38. $7.85y$

39. $9.634y$

40. $24.31y$

41. $6.4y$

42. $8.3y$

Find the value of the expression if $z = 1000$.

43. $0.571z$

44. $7.634z$

45. $9.8124z$

46. $27.6312z$

47. $29.34z$

48. $171.5z$

Solve the equation.

C 49. $a = 2.1 + (0.4 \times 0.02)$

50. $4.21 \times 3.2 - 6.41 = c$

51. $z = 84.7 - (3.4 + 2.11)$

52. $(8.7 \times 100) + 6.62 = g$

53. $1.03 + 2.1 \times 10 = r$

54. $s = 8.84 - 2.1 \times 0.03$

Problems

Write an addition, subtraction, or multiplication expression for the given phrase.

1. 3.7 times *b*

2. 80.64 multiplied by *w*

3. *d* multiplied by 4.82

4. *s* minus 4.6

5. *m* added to 3.712

6. *a* multiplied by 0.08

Write an addition, subtraction, or multiplication expression to complete.

7. The length of the field is two times the width of the field. Let *h* = the width of the field. Then, __?__ = the length of the field.

8. Rob's height is 2.4 cm less than Joe's height. Let *r* = Joe's height. Then, __?__ = Rob's height.

Add, subtract, or multiply to solve.

What is the cost of the given amount of Swiss cheese?

9. 0.89 lb   **10.** 0.45 lb   **11.** 1.12 lb

How much does the given amount of cole slaw cost?

12. 0.38 lb   **13.** 0.66 lb   **14.** 1.09 lb

What is the cost of the given amount of sliced roast beef?

15. 0.54 lb   **16.** 1.78 lb   **17.** 2.42 lb

Today's Special

Swiss cheese $3.39/lb
Cole slaw $.99/lb
Roast beef $4.98/lb

Using the Computer

This program will round a decimal to the nearest tenth.

```
 5  REM: ROUNDING DECIMALS
10  PRINT "WHAT IS YOUR DECIMAL";
20  INPUT D            ←——————— This line prints a ?
30  PRINT D;" ROUNDED TO TENTHS IS";
40  PRINT INT(D*10+.5)/10
50  END
```

RUN this program for D = 2.34 and D = 2.36. Now try some other examples of your own.

2-8 Dividing Decimals

We follow the same procedure for dividing a decimal by a whole number as for dividing a whole number by a whole number.

Example 1   Solve the equation $y = 49.5 \div 9$.

Solution

$y = 49.5 \div 9$ ←
$y = 5.5$

$$
\begin{array}{r}
5.5 \\
9{\overline{)\,49.5}} \\
-45 \\
\hline
4\,5 \\
-4\,5 \\
\hline
\end{array}
$$

Notice that the decimal point in the quotient is placed to align with the decimal point in the dividend.

Example 2   Solve the equation $d = \dfrac{0.0048}{0.15}$.

Solution   Simplify the equation by multiplying the divisor and dividend by the power of 10 that makes the divisor a whole number.

$$\frac{0.0048 \times 100}{0.15 \times 100} = \frac{0.48}{15}$$

$d = \dfrac{0.48}{15}$ ←
$d = 0.032$

$$
\begin{array}{r}
0.032 \\
15{\overline{)\,0.480}} \\
-45 \\
\hline
30 \\
-30 \\
\hline
\end{array}
$$

Write an extra zero after the 8 in the dividend so that you can complete the division.

Your Turn • Solve the equation $t = \dfrac{3.71}{0.5}$.

To make sure that you put the decimal point in the correct place in a quotient, you can estimate the quotient.

Example 3   Estimate the quotient: $18.9 \div 35$.

Solution   Round the dividend and divisor to the nearest numbers that can be divided evenly.

$$35\overline{)18.9} \longrightarrow 40\overline{)20.0}^{\,0.5}$$

The estimated quotient is 0.5.

Your Turn   • Estimate the quotient of 3.71 divided by 0.5.

We can divide a decimal by a power of 10 mentally.

Example 4   Solve the equation $f = 4.1 \div 10$.

Solution   Move the decimal point one place to the left for each power of 10.
$f = 4.1 \div 10$
$f = 0.41$

Your Turn   • Solve the equation $d = 23.78 \div 100$.

Class Exercises

Divide.

1. $8\overline{)4.24}$ **2.** $6\overline{)0.45}$ **3.** $51\overline{)9.18}$ **4.** $23\overline{)89.7}$

5. $2.9\overline{)185.6}$ **6.** $0.38\overline{)24.32}$ **7.** $0.4\overline{)632.16}$ **8.** $0.9\overline{)80.55}$

Divide the given number by 10, 100, 1000, or 10,000.

÷	10	100	1000	10,000
9. 13,276.48	?	?	?	?
10. 1.764	?	?	?	?

Exercises

Divide.

A
1. $22.5 \div 5$ **2.** $5.32 \div 4$ **3.** $14.4 \div 72$ **4.** $1.35 \div 27$

5. $72.8 \div 0.4$ **6.** $46.72 \div 0.32$ **7.** $8.16 \div 12$ **8.** $19.3816 \div 3.461$

9. $0.02 \div 0.5$ **10.** $19.5 \div 0.6$ **11.** $4.03 \div 0.325$ **12.** $2.26 \div 0.452$

Estimate the quotient.

B
13. $23.8 \div 6.1$ **14.** $30.2 \div 4.5$ **15.** $41.6 \div .83$ **16.** $64.29 \div 7.89$

17. $177.6 \div 6.6$ **18.** $235 \div 3.8$ **19.** $419.1 \div 6.77$ **20.** $387.2 \div 7.71$

21. $36.2 \div 22.1$ **22.** $77.1 \div 36.62$ **23.** $119.8 \div 60.04$ **24.** $897.65 \div 258.6$

Find the value of the expression if $x = 100$.

25. $4523.6 \div x$ **26.** $196.41 \div x$ **27.** $62.3 \div x$ **28.** $0.16 \div x$

29. $5.5 \div x$ **30.** $0.01 \div x$ **31.** $6.486 \div x$ **32.** $41.13 \div x$

Find the value of the expression if $y = 1000$.

33. $5886.1 \div y$ **34.** $7330.4 \div y$ **35.** $821.6 \div y$ **36.** $75.58 \div y$

37. $4.312 \div y$ **38.** $8.8 \div y$ **39.** $0.921 \div y$ **40.** $0.65 \div y$

Evaluate the expression using the given value for the variable.

41. $\dfrac{7.654}{s}$; $s = 5$ **42.** $\dfrac{41.16}{c}$; $c = 7$ **43.** $\dfrac{46.72}{a}$; $a = 32$

44. $\dfrac{4.24}{m}$; $m = 0.8$ **45.** $\dfrac{1.962}{p}$; $p = 0.9$ **46.** $\dfrac{0.08}{t}$; $t = 0.016$

47. $\dfrac{62.3}{r}$; $r = 17.8$ **48.** $\dfrac{4.7}{q}$; $q = 2.5$ **49.** $\dfrac{142.48}{z}$; $z = 3.562$

To find the miles per gallon (MPG) rating of an automobile, divide the distance traveled by the amount of gasoline used.

Example Find the MPG rating of an automobile which used 19.1 gal of gasoline when traveling 496.6 mi.

Solution $496.6 \div 19.1 = 26$
The MPG rating is 26.

Find the MPG rating of each car.

	Car	Distance Traveled	Gasoline Used	Miles per Gallon
C 50.	A	362.7 mi	15.5 gal	?
51.	B	462.5 mi	18.5 gal	?
52.	C	452.4 mi	17.4 gal	?
53.	D	537.6 mi	16.8 gal	?
54.	E	452.2 mi	13.3 gal	?
55.	F	519.1 mi	17.9 gal	?

Problems

Write an expression for the given phrase.

1. 7.3 times a
2. 84.2 divided by z
3. 81.1 multiplied by b
4. h divided by 3.1
5. s multiplied by 3.8
6. f times 3.7
7. 47.2 decreased by d
8. r divided by 0.21
9. 5.18 divided by b

Write an equation for the given sentence. Solve the equation.

10. The quotient of 157.6 divided by 2 is a.
11. The product of 2.13 and 1.5 is z.
12. 4.116 divided by 2.1 is equal to s.
13. 80.08 decreased by 14.7 is equal to v.
14. a is equal to the quotient of 82.0 divided by 4.1.

2-9 Metric System

The metric system of measurement has three basic units of measure. Length is measured in **meters** (m), mass in **grams** (g), and capacity in **liters** (L). The same prefixes are used for units of length, mass, and capacity. Each prefix represents a power of 10. To change a metric unit to a smaller metric unit, multiply by a power of 10. To change a metric unit to a larger metric unit, divide by a power of 10.

Example   Change 0.48 m to centimeters and 7642 mg to grams.

Solution $0.48 \text{ m} \times 100 = 48 \text{ cm}$
$7642 \text{ mg} \div 1000 = 7.642 \text{ g}$

larger	**Metric Units of Length**				smaller	
kilometer	hectometer	dekameter	meter	decimeter	centimeter	millimeter
(km)	(hm)	(dam)	(m)	(dm)	(cm)	(mm)

$\times 10 \qquad \times 10$

$0.48 \text{ m} \times 100 = 48 \text{ cm}$

larger	**Metric Units of Mass**				smaller	
kilogram	hectogram	dekagram	gram	decigram	centigram	milligram
(kg)	(hg)	(dag)	(g)	(dg)	(cg)	(mg)

$\div 10 \qquad \div 10 \qquad \div 10$

$7642 \text{ mg} \div 1000 = 7.642 \text{ g}$

Your Turn   • Change 7 L to milliliters.   • Change 955 mL to hectoliters.

larger	**Metric Units of Capacity**				smaller	
kiloliter	hectoliter	dekaliter	liter	deciliter	centiliter	milliliter
(kL)	(hL)	(daL)	(L)	(dL)	(cL)	(mL)

Class Exercises

Would you multiply or divide to change metric units with the given prefixes? Write × or ÷.

1. centi to deka   **2.** hecto to deci   **3.** centi to milli   **4.** deci to kilo

Exercises

Complete.

A **1.** 1 km = _?_ m   **2.** 1 mm = _?_ cm   **3.** 1 m = _?_ dm

   4. 1 g = _?_ mg   **5.** 1 dag = _?_ kg   **6.** 1 cg = _?_ hg

B **7.** 6 m = _?_ cm   **8.** 0.456 km = _?_ m   **9.** 24 mm = _?_ cm

   10. 7 hm = _?_ dm   **11.** 8.6 cm = _?_ mm   **12.** 549 m = _?_ km

   13. 0.85g = _?_ cg   **14.** 0.936 kg = _?_ g   **15.** 743 dg = _?_ kg

   16. 16.2 dag = _?_ dg   **17.** 304 mg = _?_ g   **18.** 91 cg = _?_ mg

   19. 8342 mL = _?_ L   **20.** 4.72 hL = _?_ L   **21.** 0.45 daL = _?_ dL

   22. 3.4 cL = _?_ mL   **23.** 7.243 L = _?_ cL   **24.** 68 daL = _?_ kL

Self Test 2

Solve the equation.

1. $41.6 + 32.7 = f$ **2.** $8.75 + 21.9 = d$ **3.** $118.33 + 4.765 = c$ 2-4

4. $k = 8.96 - 4.38$ **5.** $s = 17.86 - 9.9$ **6.** $h = 325.8 - 27.65$ 2-5

Solve.

7. It usually takes a typist from 15 to 35 minutes to type a letter. He finished 12 letters in 5 hours. Is that time within his usual range?   2-6

Solve the equation.

8. $74 \times 2.36 = t$ **9.** $0.02 \times 0.43 = a$ **10.** $1000 \times 8.65 = c$ 2-7

11. $p = 220.8 \div 96$ **12.** $w = 8.4 \div 3.5$ **13.** $y = 36.2 \div 100$ 2-8

Complete.

14. 4562 mL = _?_ L   **15.** 0.841 kg = _?_ g   **16.** 724 m = _?_ km   2-9

Algebra Practice

Evaluate the expression using the given value for the variable.

1. $n(8 - n)$; $n = 2$ **2.** $81 - (6x)$; $x = 6$ **3.** $s \div 17$; $s = 51$

Solve.

4. $15 = 15 + r$ **5.** $22j = 22$ **6.** $6 + 4 - 8 \div 2 = w$

Simplify the expression.

7. $3.5r + 12.68r$ **8.** $4.3a + 9.7w + 8.12a$ **9.** $4.9c + 6.2f + 8.1c$

10. $15m - 12.78m$ **11.** $20.05b - 6.873b$ **12.** $31.05g - 0.982g$

13. $(6.6 - 3.97) + 4.806$ **14.** $18.3 - 4.6 \times 0.9$ **15.** $a(a + 4)$

Write an expression for the phrase.

16. the quotient of 263 and n **17.** h used as a factor 4 times

18. 14.2 greater than y **19.** g decreased by 2.6

20. c multiplied by 8.63   **21.** 7.5 divided by v

Write an equation for the sentence. Solve the equation.

22. 8090 minus 5989 is a.
23. 196 increased by 505 is n.
24. The sum of 8.93 and 12.7 is equal to t.
25. The difference between 36.409 and 19.7 is equal to h.
26. 3.2 times 0.6 is equal to j.
27. 18 divided by 3.6 is equal to b.
28. The product of 5.216 and 0.05 is f.
29. The numbers 11.097, 230.28, and 1.973 add to a sum of s.
30. The quotient of 14.04 and 1.8 is z.
31. 21 increased by 804 is y.
32. The number n is the quotient of 16 divided by 0.4.
33. 27.4 greater than 37 is q.
34. The number t is the product of 0.16 and 0.042.
35. The number j is 6 used as a factor twice.
36. The factors 2.01 and 0.07 have a product m.
37. 86 decreased by 2.71 is h.
38. The sum of 4 greater than 18 and 25 is x.
39. The number n is the product of 2.01 and 0.009.
40. 4.01 decreased by itself is z.

Problem Solving for the Consumer

Buying a Car

The amount of money a dealer pays a factory for a car is called the dealer cost. To make a profit a dealer must charge a customer more than the dealer cost. To give you an idea of a fair price for that charge, consumer magazines will estimate a cost factor. This cost factor is often expressed as a decimal. To determine dealer cost, just multiply the list price the dealer tells you by the cost factor.

Complete the chart to find the approximate dealer cost.

	Model	List Price	Cost Factor	Dealer Cost
1.	Fire fly	$10,370	0.88	?
2.	Cruiser	$ 7095	0.86	?
3.	Wanderer	$ 6714	0.90	?
4.	Stallion	$ 8979	0.84	?

Options are not included in the basic price of a car, so you have to add those extra charges onto the cost. Here's a list of some options on one car model and the list prices. Answer the questions.

5. What's the total cost of these options?

6. Which options total $881?

7. The cost factor on each of these options is 0.84. What is the dealer cost on the air conditioning? on the radio/tape?

OPTIONS	LIST PRICE
POWER BRAKES	100.00
POWER STEERING	315.95
TINTED GLASS	99.95
AIR CONDITIONING	685.50
RADIO/TAPE	195.50

Often you can bargain with the car salesperson over the price you are willing to pay. The difference between dealer cost and the total list price is roughly the amount of bargaining room you have. Complete the chart on the next page.

	Basic Car			Options			(a) Total List Price	(b) Total Dealer Cost	Difference (a − b)
	List Price	Cost Factor	Dealer Cost	List Price	Cost Factor	Dealer Cost			
8.	$6278	0.88	?	$1375	0.85	?	?	?	?
9.	$7762	0.93	?	$ 495	0.84	?	?	?	?
10.	$8449	0.87	?	$1036	0.87	?	?	?	?

It's smart to compare the gas mileage ratings of cars. The average driver might drive 12,000 mi per year. Suppose gas costs $1.08 a gallon. Complete the chart. Round the answers to the nearest hundredth.

	Model	Miles Per Gallon (Highway driving)	Number of Gallons Per Year	Cost of Gas Per Year
11.	Darter	42.6	281.69	?
12.	XM-Hawk	33.4	?	?
13.	Winslow	25.7	?	?

Solve.

14. A Cobra gets 39 mi per gallon. When air conditioning is running it gets 4.2 fewer miles per gallon. How many gallons of gas would be used for running the air conditioner constantly on a 420-mile trip?

15. Suppose you have the following options, but you decide to buy only two of them. How many different combinations of two options do you have?
automatic transmission   air conditioning   power brakes
stereo with AM-FM system   fog lights

Enrichment

Binary Digits and the Computer

You know that the number system we use is a decimal system. The digits 0–9 are used in this system. The only digits used in the binary system are 0 and 1. These binary digits are important to the operation of digital computers. If a circuit in a computer is open, the computer reads that as 1. If the circuit is closed, the computer reads that as 0. A listing of a code system called Binary Coded Decimal (BCD) is shown below.

010001 A	010010 B	010011 C	010100 D	010101 E	010110 F	010111 G
011000 H	011001 I	100001 J	100010 K	100011 L	100100 M	100101 N
100110 O	100111 P	101000 Q	101001 R	110010 S	110011 T	110100 U
110101 V	110110 W	110111 X	111000 Y	111001 Z		

Notice that different combinations of the binary digits are used as codes for the letters of the alphabet.

The code for the word *math* is 100100 010001 110011 011000.
The word 010001 010100 010100 is decoded as *add*.

Write the code for each word using the BCD system.

1. number **2.** addition **3.** binary **4.** computer

5. multiply **6.** divide **7.** homework **8.** decimal

Use the BCD listing to decode the name of a famous mathematician who worked with binary digits.

9. 100001 100110 011000 100101   110101 100110 100101
100101 010101 110100 100100 010001 100101 100101

Chapter Review

Match.

1. the standard form of 4 and 6 ten-thousandths
2. 4000.551 rounded to the nearest tenth
3. the product when you multiply 20.003 by 2
4. the sum when you add 1.453 and 2.54706
5. the standard form of four and six tenths
6. 400.00618 rounded to the nearest thousandth
7. the difference between 92.438 and 52.378
8. the quotient when you divide 1001.5 by 2.5

A. 40.006
B. 40.06
C. 4.0006
D. 400.6
E. 400.006
F. 4.6
G. 4.00006
H. 4000.6

Are the decimals in order? Write *Yes* or *No*.

9. 31.18, 31.1, 31.19, 32.19
10. 8.8, 8.89, 8.94, 8.9, 8.98
11. 34.14, 34.146, 34.15, 34.162
12. 184.481, 18.4481, 18.48, 1848.1
13. 7.006, 7.007, 7.706, 7.707
14. 21.01, 21.1, 21.12, 21.121, 21.122

Choose the equivalent measure. Write *a*, *b*, or *c*.

15. 6.4 cm   **a.** 64 m   **b.** 64 mm   **c.** 640 mm
16. 984 mg   **a.** 9.84 g   **b.** 0.984 dg   **c.** 0.984 g
17. 7.4 km   **a.** 7400 m   **b.** 740 m   **c.** 740 hm
18. 0.5 L   **a.** 50 mL   **b.** 500 mL   **c.** 500 cL
19. 51.3 dag   **a.** 5.13 g   **b.** 513 mg   **c.** 513 g

Solve the equation.

20. $4.002 + 31.6 = y$
21. $48.9 - 21.75 = d$
22. $1000 \times 6.75 = m$
23. $1.512 \div 3.6 = a$
24. $0.84 \times 3.6 = z$
25. $27.85 + 191.9 = b$
26. $d = 93.2 \div 100$
27. $s = 771.1 - 39.865$
28. $m = 2.07 \div 4.6$
29. $x = 4.7 \times 0.09$
30. $n = 62 - 0.63$
31. $10 \times 7.62 = n$
32. $0.06(4.3 - 1.02) = m$
33. $(0.001)(5.8) = a$
34. $4.1(6.2 + 3.5) = c$

Chapter Test

Write the decimal.

1. 36 and 4 thousandths

2. 424 and 307 ten-thousandths

3. seventeen and four hundred five hundred-thousandths

4. two and six thousand twelve millionths

2-1

Write the decimals in order from least to greatest.

5. 6.12, 6.11, 6.22, 6.21

6. 9.83, 9.8, 9.8347, 9.834

7. 766, 7.16, 7.1, 7.61

8. 41.14, 41.1, 41.141, 41.41

2-2

Round to the nearest tenth.

9. 0.2

10. 5.753

11. 16.545

12. 245.8364

2-3

Round to the nearest hundredth.

13. 2.476

14. 31.555

15. 0.8327

16. 7.5426

Solve the equation.

17. $8.62 + 4.09 = a$

18. $74.5 + 2.13 = b$

19. $684.2 + 39.754 = h$

2-4

Solve the equation.

20. $24.36 - 9.75 = w$

21. $45.2 - 3.16 = s$

22. $745.3 - 74.536 = g$

2-5

Solve.

23. A certain computer game ranges in price from $29.95 to $34.50, depending upon different store prices. What is the minimum you could pay for 4 of those games? The maximum?

2-6

Solve the equation.

24. $87 \times 42.2 = y$

25. $0.04 \times 0.22 = v$

26. $100 \times 78.6 = m$

2-7

Solve the equation.

27. $15.51 \div 47 = n$

28. $9.5 \div 0.25 = t$

29. $8.45 \div 1000 = r$

2-8

Complete.

30. $45 \text{ cm} = \underline{\ ?\ } \text{ mm}$

31. $0.64 \text{ daL} = \underline{\ ?\ } \text{ L}$

32. $7.5 \text{ L} = \underline{\ ?\ } \text{ mL}$

2-9

Cumulative Review (Chapters 1–2)

Add, subtract, multiply, or divide.

1. $(1)(87,019)$
2. $4.7 - 2.35$
3. $(83.5)(0)$
4. $16 \div 4 - 1.5$
5. $3.5(2 + 8)$
6. $14.08 - 2.4 \div 2$
7. 10×7.4
8. $(10)(10)(16.83)$
9. $(100)(4.807)$
10. $(10)(4.639)(0)$
11. $(4.7)(0) + 84.213$
12. $6.4 - (9.01)(0)$
13. 0.004×1000
14. $2 \cdot 2 \cdot 2 \cdot 2$
15. $3 \cdot 3 \cdot 3 \cdot 3$

Round to the place of the underlined digit.

16. 43<u>8</u>1
17. 52<u>9</u>4
18. 0.7<u>11</u>
19. 32<u>1</u>.6
20. <u>0</u>.9
21. <u>2</u>8,300
22. 0.4<u>66</u>
23. 52<u>95</u>
24. 2<u>9</u>.9
25. 3.<u>98</u>

Choose the best estimate.

26. $82 + 19$ **a.** 95   **b.** 100   **c.** 110   **d.** 115
27. $162 \div 43$ **a.** 4   **b.** 3   **c.** 40   **d.** 30
28. $78.8 \div 5.1$ **a.** 15   **b.** 25   **c.** 16   **d.** 18
29. $\$1.95 \times 9$ **a.** \$20   **b.** \$18   **c.** \$9   **d.** \$17

Simplify.

30. $a \cdot b$
31. $a \cdot a$
32. $c \cdot c \cdot c \cdot c$
33. $5 \cdot a \cdot a \cdot a$
34. $2 \cdot 2 \cdot n \cdot n$
35. $5 \cdot x \cdot y \cdot z$
36. $6 \cdot k^2$
37. $4 \cdot t \cdot t \cdot w$
38. $x \cdot y \cdot z \cdot x$
39. $s \div s$
40. $n(n + 1)$
41. $j(7 + j)$

Write an equation for the sentence. Solve the equation.

42. When 4.2 is increased by 18.9, the result is x.
43. 83 is greater than 47.1 by g.
44. 160 reduced by 45.5 is n.
45. He traveled d by driving 55 mi per hour for 2 hours.

Write an equation for the problem. Solve it.

46. The Norwalk High School band has twice as many members as the 45-member Drama Club. How many members does the band have?

3
Number Theory

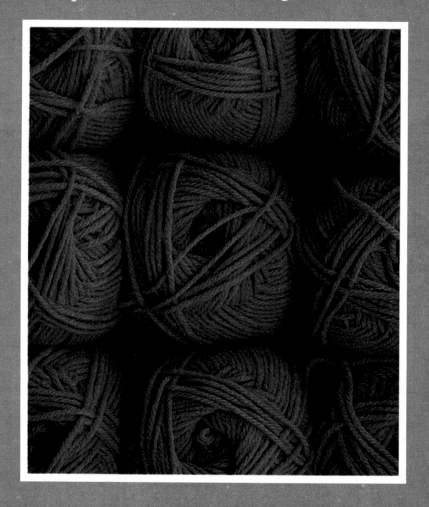

Imagine knitting 20,160 squares! That's the number of 6-in. squares needed to make the largest blanket in the world. The length and width had to be multiples of 6 in., 68 ft by 100 ft to be exact.

3-1 Divisibility Tests

Some division problems have remainders. Others do not. On a calculator a remainder is indicated by decimal places in the quotient.

$650 \div 5 = 130$

650 is **divisible** by 5.

$650 \div 4 = 162.25$

650 is not divisible by 4.

A number a is divisible by another number b if the quotient $a \div b$ is a whole number.

Every number is divisible by 1. $a \div 1 = a$

No number is divisible by 0. $\cancel{a \div 0 = ?}$

To decide if a number is divisible by a number other than 1, you can divide the first number by the second number. At times you may wish to use a calculator to do the division. For certain divisors, however, it is often easier to use a few simple **divisibility tests.** In some cases you can determine divisibility just by looking at the number!

Divisor	Divisibility Tests
2	The last digit is 0, 2, 4, 6, or 8.
3	The sum of the digits is divisible by 3.
4	The number named by the last two digits is divisible by 4.
5	The last digit is 0 or 5.
6	The number is divisible by both 2 and 3.
8	The number named by the last three digits is divisible by 8.
9	The sum of the digits is divisible by 9.
10	The last digit is 0.

Notice that a number is divisible by 6 if it is divisible by its factors 2 and 3. In the same way a number is divisible by 12 if it is divisible by 3 and 4, a number is divisible by 15 if it is divisible by 3 and 5, and so on.

Example 1   Is 24,570 divisible by 3?

Solution   Sum of digits: $2 + 4 + 5 + 7 + 0 = 18$
Is 18 divisible by 3? ←———— $18 \div 3 = 6$
Then 24,570 is also divisible by 3.

Example 2   Is 1286 divisible by 4?

Solution   Look at last two digits: 1286
Is 86 divisible by 4? No. ←———— $86 \div 4 = 21.5$
Then 1286 is not divisible by 4.

Your Turn   • Is 92,424 divisible by the number?
• 2   • 3   • 4   • 5   • 6   • 10

The numbers 1, 2, 3, 4, 5, . . . are called **consecutive whole numbers.**
A number that is divisible by 2 is called an **even number.** A number
that is not divisible by 2 is called an **odd number.**

Consecutive even numbers: 2, 4, 6, 8, 10, . . .
Consecutive odd numbers: 1, 3, 5, 7, 9, . . .

Example 3   Is 117,594 an even number, or an odd number?

Solution   Look at the last digit: 117,594
The last digit is divisible by 2.
So 117,594 is divisible by 2 and 117,594 is an even
number.

Your Turn   • Is 20,371 an even number or an odd number?
• What is the next consecutive whole number after 16?
• What is the next consecutive even number after 16?

Class Exercises

Look at the last digit of each number. Is the number divisible by 2? By 5? By 10?

1. 528	**2.** 1730	**3.** 384	**4.** 3992	**5.** 2000
6. 75	**7.** 5556	**8.** 328	**9.** 975	**10.** 850

11–20. Look at the last two digits of each number in Exercises 1–10. Tell whether the number is divisible by 4.

21–30. Add the digits of each number in Exercises 1–10. Tell whether the number is divisible by 3.

Exercises

Is the first number divisible by the second number? Write *Yes* or *No*.

A
1. 526 by 2	**2.** 1723 by 3	**3.** 804 by 1	**4.** 157 by 0
5. 630 by 5	**6.** 726 by 4	**7.** 1938 by 6	**8.** 621 by 3
9. 478 by 4	**10.** 445 by 2	**11.** 895 by 10	**12.** 970 by 5
13. 96 by 4	**14.** 490 by 0	**15.** 836 by 6	**16.** 802 by 2

Is the number divisible by 2? By 3? By 4? By 5? By 6? By 8? By 9? By 10?

B
17. 480	**18.** 373	**19.** 1980	**20.** 2000	**21.** 882
22. 98	**23.** 138	**24.** 575	**25.** 324	**26.** 2670
27. 582	**28.** 777	**29.** 360	**30.** 517	**31.** 184
32. 51,840	**33.** 68,672	**34.** 41,847	**35.** 502,164	**36.** 881,280

True or False? Write *T* or *F*.

37. All even numbers are divisible by 2.

38. All odd numbers are divisible by 3.

39. Some even numbers are divisible by 5.

40. The sum of two odd numbers is an odd number.

41. The sum of two even numbers is an even number.

By which of the numbers is 17,844 divisible?

42. 3; 4; 12   **43.** 2; 5; 10   **44.** 2; 3; 6   **45.** 4; 6; 24

Let *n* be a whole number. Write an expression for the next consecutive number.

46. n **47.** $2n$ **48.** $n - 1$ **49.** $n + 1$ **50.** $2n + 1$

Let *n* be a whole number. Does the expression represent an even number?

C   **51.** $2n$ **52.** $2n + 1$ **53.** $2n - 1$ **54.** $2n + 2$ **55.** $4n + 2$

Problems

Solve.

1. For graduation, 96 students must be seated on stage.
   Each row must have the same number of seats.
   Is it possible to accomplish this with rows of 12 seats?
2. A teacher wishes to arrange 30 desks in a classroom.
   Each row is to have the same number of seats.
   List all the ways that this can be done.
3. For a relay race 72 students are to be divided into teams of equal size. List all the possible ways that teams of more than 6 students may be formed.
4. A "square" for a square dance is made up of 4 couples. If there are 156 people at a square dance, is it possible for all of them to dance in "squares" at the same time?

Using the Calculator

An exclamation point is used in the following way to show multiplication of consecutive whole numbers.

$$2! = 1 \cdot 2, \quad 3! = 1 \cdot 2 \cdot 3, \quad 4! = 1 \cdot 2 \cdot 3 \cdot 4, \text{ and so on.}$$

The symbol 2! is read *2 factorial.*

Using a calculator find the value of the factorial expression.

  1. 3!   **2.** 6!   **3.** 10!   **4.** 5! $\cdot$ 3!   **5.** 8! $\cdot$ 6!

3-2 *Factors and Prime Numbers*

Alpha Games plans to ship a new electronic game to stores in cartons containing 12 games each. The shipping department recommends cartons with 3 stacks of 4 games in each.

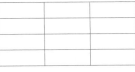

$3 \times 4 = 12$

The numbers 3 and 4 are called **factors** of 12. A number is divisible by each of its factors. Other factors of 12 are 1, 2, 6, and 12. The number 1 is a factor of every number.

Sometimes a number, like 7, has only two factors, itself and 1. Such a number is called a **prime number.** If a number has more than two factors, it is called a **composite number.** The number 1 is not considered to be a prime since we would have to show it as a factor in all examples such as Example 1 following. The number 0 is not considered to be a composite since, unlike any other composite number, it would have all numbers as its factors.

Factors of 7: 1 and 7   Factors of 6: 1, 2, 3, 6
7 is a prime number   6 is a composite number.

Composite numbers can be written as the product of prime factors, called the **prime factorization** of the number. The array of factors and connecting lines shown below is called a **factor tree.**

Example 1   Write the prime factorization of 42.

Solution   First write 42 as a product of two factors. Replace composite factors with a product of two factors until all numbers are prime.

$42 = 2 \times 3 \times 7$

Sometimes there is more than one choice for a set of factors. However, the end result is always the same.

Example 2   Show two different ways of finding the prime factorization of 48.

Solution

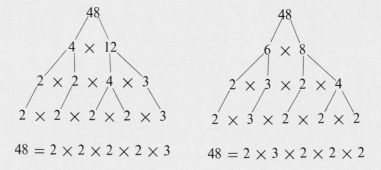

$$48 = 2 \times 2 \times 2 \times 2 \times 3 \qquad 48 = 2 \times 3 \times 2 \times 2 \times 2$$

Because of the commutative property the order of the factors does not matter. The two solutions are equal.

Sometimes it's convenient to use exponents to write repeated factors.

Example 3   Write the prime factorization of 48 using exponents.

Solution   Use the prime factorization of 48 in Example 2.

$$48 = \underbrace{2 \times 2 \times 2 \times 2}_{2^4} \times 3$$
$$= \qquad 2^4 \qquad \times 3$$

Your Turn   Complete.

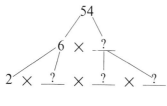

• $54 = 2 \times 3^?$

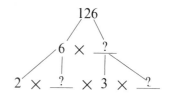

• $126 = 2 \times 3^? \times \underline{\,?\,}$

Divisibility tests can help you write the prime factorization of a number.

In Arithmetic

$12 \div 3 = 4$

So both 3 and 4 are factors of 12.

In Algebra

$a \div b = c$

So, b and c are factors of a.

Example 4   Write the prime factorization of 78.

Solution   The last digit of 78 is 8. So 78 is divisible by 2. There-fore 2 is a factor of 78 and $78 \div 2$, or 39, is also a factor.

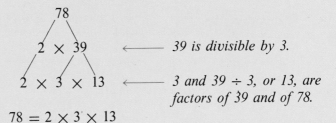

2×39 ←——— *39 is divisible by 3.*

$2 \times 3 \times 13$ ←——— *3 and $39 \div 3$, or 13, are factors of 39 and of 78.*

$78 = 2 \times 3 \times 13$

Your Turn   Write the prime factorization of the number.

• 104   • 220   • 1350

Class Exercises

Complete the factor tree. State the prime factorization of the first number.

1.

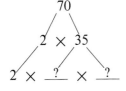

$70 = 2 \times \underline{\ ?\ } \times \underline{\ ?\ }$

2.

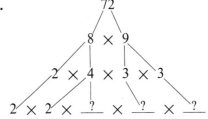

$72 = 2^? \times 3^?$

Exercises

Copy and complete the factor tree. Write the prime factorization of the first number.

A   **1.** 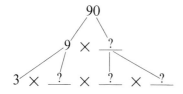   **2.**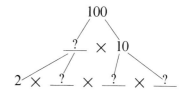

$$90 = 3 \times \underline{\ ?\ } \times \underline{\ ?\ } \times \underline{\ ?\ }$$
$$100 = 2 \times \underline{\ ?\ } \times \underline{\ ?\ } \times \underline{\ ?\ }$$

Draw a factor tree. Write the prime factorization of the number.

B   **3.** 24   **4.** 50   **5.** 63   **6.** 44   **7.** 30

   8. 140   **9.** 66   **10.** 112   **11.** 128   **12.** 200

   13. 130   **14.** 900   **15.** 6300   **16.** 512   **17.** 176

Use exponents to write the prime factorization of the number.

   18. 180   **19.** 378   **20.** 280   **21.** 660   **22.** 550

   23. 1600   **24.** 1360   **25.** 784   **26.** 1296   **27.** 1225

Solve.

C   **28.** One of the largest known prime numbers is $2^{3217} - 1$. Is it an even number or an odd number? Give a reason for your answer.

   29. How many prime numbers are there between 0 and 50? Between 50 and 75?

Problems

Twin primes are two prime numbers whose difference is 2. Here are some examples of twin primes.

      3 and 5   5 and 7   11 and 13

Solve.

   1. How many pairs of twin primes are there less than 50? Name them.

   2. Janice claims that she can name a pair of even numbers that are twin primes. Evelyn says that she must have made a mistake. Who is right? Why?

3-3 The Greatest Common Factor

Sometimes two numbers have some of the same factors. These are called **common factors.**

Factors of 30: 1, 2, 3, 5, 6, 10, 15, 30
Factors of 54: 1, 2, 3, 6, 9, 18, 27, 54

Common factors of 30 and 54 are: 1, 2, 3, and 6. The **greatest common factor (GCF)** of 30 and 54 is 6.

Example 1   What is the GCF of 48 and 64?

Solution   Factors of 48: 1, 2, 3, 4, 6, 8, 12, 16, 24, 48
Factors of 64: 1, 2, 4, 8, 16, 32, 64
Common factors: 1, 2, 4, 8, and 16
Greatest common factor: 16

Instead of listing all the factors as in Example 1, it is sometimes easier to use the prime factorization of each of the numbers.

Example 2   What is the GCF of 36 and 54?

Solution   The GCF is the product of all common factors.
$36 = 2 \times 2 \times 3 \times 3$
$54 = 2 \times 3 \times 3 \times 3$
The GCF is $2 \times 3 \times 3$, or 18.

Your Turn   List the factors and write the GCF of the numbers.
● 8 and 12   ● 15 and 50   ● 12 and 32
Use the prime factorization to find the GCF of the numbers.
● 12 and 18   ● 9 and 30   ● 16 and 24

Class Exercises

Tell what numbers are missing.

1. Factors of 8: 1, 2, 4, 8
Factors of 12: 1, 2, 3, 4, 6, 12
Common factors: 1, 2, _?_
GCF of 8 and 12: _?_

2. Factors of 6: 1, 2, 3, 6
Factors of 15: 1, 3, 5, 15
Common factors: 1, _?_
GCF of 6 and 15: _?_

3. $14 = 2 \times 7$
$42 = 2 \times 3 \times$ _?_
GCF of 14 and 42: $2 \times$ _?_ $=$ _?_

4. $16 = 2 \times 2 \times 2 \times 2$
$20 = 2 \times 2 \times$ _?_
GCF of 16 and 20: $2 \times$ _?_ $=$ _?_

Exercises

List all the factors of each number.

A **1.** 6   **2.** 12   **3.** 15   **4.** 30   **5.** 24

Use the answers to Exercises 1–5 to determine the GCF.

6. of 6 and 12   **7.** of 12 and 15   **8.** of 15 and 30   **9.** of 30 and 24

Copy and complete. Use the prime factorization of the numbers to find the GCF.

10. $4 = 2 \times$ _?_
$6 = 2 \times$ _?_
GCF: _?_

11. $28 = 2 \times 2 \times$ _?_
$70 = 2 \times 5 \times$ _?_
GCF: $2 \times$ _?_ $=$ _?_

12. $30 = 2 \times 3 \times$ _?_
$75 = 3 \times 5 \times$ _?_
GCF: _?_ $\times$ _?_ $=$ _?_

Write the GCF of the pair of numbers.

B **13.** 4 and 8   **14.** 5 and 15   **15.** 6 and 18   **16.** 7 and 70

17. 12 and 16   **18.** 9 and 15   **19.** 6 and 15   **20.** 8 and 20

21. 36 and 54   **22.** 27 and 45   **23.** 12 and 18   **24.** 15 and 24

25. 48 and 112   **26.** 84 and 66   **27.** 99 and 108   **28.** 65 and 66

29. 180 and 168   **30.** 176 and 192   **31.** 216 and 180   **32.** 210 and 231

Two numbers are *relatively prime* if their GCF is 1. Tell if the pair of numbers is relatively prime.

C **33.** 8 and 9   **34.** 6 and 8   **35.** 9 and 15   **36.** 12 and 35

3-4 Factoring in Algebra

To **factor** an expression means to write it as the product of factors.

$$6x^2 = 2 \cdot 3 \cdot x \cdot x \qquad\qquad 5a^2b = 5 \cdot a \cdot a \cdot b$$

We will call $2 \cdot 3 \cdot x \cdot x$ the **algebraic factorization** of $6x^2$.

In arithmetic, to find the greatest common factor (GCF), you first wrote the prime factorization of each number. Then the GCF was equal to the product of all the common prime factors.

In algebra, to find the GCF, you first write the algebraic factorization of each expression. The GCF is equal to the product of all the common factors of the algebraic factorizations.

Example 1   Write the GCF of $5a^2b$, $10a^3$, and $15a^2$.

Solution   Factor each expression.
$5a^2b = 5 \cdot a \cdot a \cdot b$
$10a^3 = 2 \cdot 5 \cdot a \cdot a \cdot a$
$15a^2 = 3 \cdot 5 \cdot a \cdot a$
GCF: $5 \cdot a \cdot a$, or $5a^2$

After a little practice you will be able to do this first step in your head.

Your Turn   Write the GCF of the set of expressions.
- $5m^2$ and $10m$
- $6r^2$, $3rs$, and $9s^2$

When an expression has two or more terms you can often use the distributive property to factor the expression.

Distributive Property

In Arithmetic
$2 \cdot 6 + 2 \cdot 5 = 2(6 + 5)$
$2 \cdot 6 - 2 \cdot 5 = 2(6 - 5)$

In Algebra
$ab + ac = a(b + c)$
$ab - ac = a(b - c)$

Example 2   Use the distributive property to factor $9a + 6$.

Solution   What is the GCF of $9a$ and 6? The GCF is 3.

$9a + 6 = (3 \cdot 3a) + (3 \cdot 2)$ ←——— *Factor both 9a and 6.*

$ = 3(3a + 2)$ ←——— *Use the distributive property.*

Check by multiplying.

$3(3a + 2) = 9a + 6$ ←——— *It checks.*

Example 3   Factor the expression $3p^2 - 6p$.

Solution   What is the GCF of $3p^2$ and $6p$? The GCF is $3p$.

$3p^2 - 6p = (3p \cdot p) - (3p \cdot 2)$ ←——— *Factor $3p^2$ and 6p.*

$ = 3p(p - 2)$ ←——— *Use the distributive property.*

Check by multiplying.

$3p(p - 2) = 3p^2 - 6p$ *It checks.*

Your Turn   Use the distributive property to factor the expression.

- $8x - 4$ • $5y^2 + y$ • $7a + 14b - 7c$

Class Exercises

Find the GCF.

1. a. 2 and 6 **2. a.** 5 and 7 **3. a.** 10, 12, and 20
   b. a^2 and a **b.** b^2 and b **b.** c, c^2, and $c\,d$
   c. $2a^2$ and $6a$ **c.** $5b^2$ and $7b$ **c.** $10c$, $12c^2$, and $20cd$

Complete.

4. GCF of $6d^2$ and $3d$: _?_
    $6d^2 + 3d = \underline{\ ?\ }(2d + \underline{\ ?\ })$

5. GCF of $30e$ and $20e^2$: _?_
    $30e - 20e^2 = \underline{\ ?\ }(3 - \underline{\ ?\ })$

6. GCF of $5f^2$, $10f$, and 15: _?_
    $5f^2 + 10f - 15 =$
    $\underline{\ ?\ }(\underline{\ ?\ } + \underline{\ ?\ } - 3)$

7. GCF of 8, $12g$, and $2g^2$: _?_
    $8 + 12g - 2g^2$
    $\underline{\ ?\ }(4 + \underline{\ ?\ } - \underline{\ ?\ })$

Exercises

Write the GCF.

A **1.** h^2 and $2h$ **2.** $5j$ and 5 **3.** $2k^3$ and 6 **4.** $4m^2$ and $8m$

     5. n and $6n^2$ **6.** $4p$ and $10p^2$ **7.** q^2 and q **8.** $6r^3$ and 18

Use the distributive property to complete.

9. $3s^2 - 3 = 3(\underline{\ ?\ } - \underline{\ ?\ })$ **10.** $4u + u^2 = u(\underline{\ ?\ } + \underline{\ ?\ })$

11. $5v^2 - 10 = \underline{\ ?\ }(v^2 - \underline{\ ?\ })$ **12.** $3w + w^2 = \underline{\ ?\ }(3 + \underline{\ ?\ })$

13. $12x^3 - 4 = \underline{\ ?\ }(3x^3 - \underline{\ ?\ })$ **14.** $10y + 15y^2 = \underline{\ ?\ }(2 + \underline{\ ?\ })$

Use the distributive property to factor the expression.

B **15.** $a^2 + 2a$ **16.** $30 + 10b^2$ **17.** $c^2 - 2c$

     18. $d + d^2$ **19.** $e - 2e^2$ **20.** $10f - 25$

     21. $6g^2 - 3g$ **22.** $12h - 4h^2$ **23.** $15j + 20j^2$

     24. $18k - 12k^2$ **25.** $30m + 25m^2$ **26.** $24n - 20n^2$

     27. $6pq + p^2q^2$ **28.** $4rs + 8r^2$ **29.** $12u^2v - 8uv^2$

     30. $3x^2 - 6x + 9$ **31.** $ay^2 + ay + a$ **32.** $12 + 4z - 8z^2$

     33. $12 + 8a - 6a^2$ **34.** $10b - 5b^2 - 15$ **35.** $12c^2 - 8c - 4cd$

Self Test 1

Is 12,550 divisible by the given number?

1. 2   **2.** 4   **3.** 3   **4.** 5   **3-1**

Write the prime factorization of each number.

5. 84   **6.** 920   **7.** 147   **8.** 1880   **3-2**

Write the GCF of each set of numbers.

9. 18 and 28   **10.** 20 and 45   **11.** 6, 10, and 15   **3-3**

Use the distributive property to factor the expression.

12. $3a - 12$ **13.** $5b^2 + 10b$ **14.** $2 + 6c - 4c^2$ **3-4**

Self Test answers and Extra Practice are at the end of the book.

3-5 *Problem Solving*

Finding and using a pattern can be the key to solving certain kinds of problems.

Example 1 A city planner needs to know the number of street intersections before ordering new traffic lights. If there are 20 North-South streets intersecting with 20 East-West streets, how many intersections are there?

Solution First we'll find how many intersections there are if there are fewer streets in each direction. We draw diagrams and count to determine the number of intersections if there are 1, 2, 3, or 4 streets in each direction.

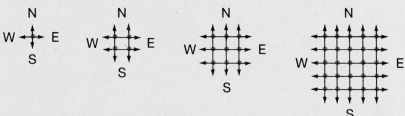

Next, record the results and look for a pattern.

Number of streets in each direction	Number of intersections
1	1
2	4
3	9
4	16
⋮	⋮
20	400

Pattern:
$1^2 = 1$
$2^2 = 4$
$3^2 = 9$
$4^2 = 16$

The last step is to extend the pattern:

$20^2 = 400$

If 20 North-South streets intersect 20 East-West streets, there are 400 intersections in all.

In Example 1, the numbers 1, 4, 9, 16, . . . form a **sequence.** Each number, or **term** of a sequence, is obtained by applying a rule. You are probably already familiar with many sequences that you can use in problem solving.

Example 2 If January 1 is a Sunday, what is the date for the last Sunday of January? Write a sequence and state the rule.

Solution You know that there are 7 days from one Sunday to the next. The Sundays of January are the following:

January 1,   8,   15,   22,   29   36

+7   +7   +7   +7   +7

Rule: The first term is 1. Each of the other terms is obtained by adding 7 to the term before it. No term is greater than 31, so the last Sunday of January is January 29.

Your Turn • Beginning at 7:45 A.M. buses leave Lawrence for Macon every 45 minutes. What is the departure time that is closest to 12 Noon? Write a sequence and state the rule.

Class Exercises

State the next three terms of the sequence.

1. 10, 20, 30, 40, . . .

2. 1, 3, 5, 7, . . .

3. 1980, 1984, 1988, 1992, . . .

4. A, C, E, G, I, K, M, . . .

5. A, Ɐ, B, ᙠ, C, Ɔ, . . .

6. Jan 31, Feb 28, Mar 31, . . .

7. 0, 1, 10, 11, 100, 101, . . .

8. 12:00, 12:20, 12:40, 1:00, . . .

9. 2, 4, 8, 16, . . .

10. Sun., Tues., Thurs., . . .

11. ZYX, YXW, XWV, . . .

12. 3, 9, 27, 81, . . .

Exercises

Solve.

A
1. Four people can be seated at a single card table. If two tables are placed end to end, 6 people can be seated. How many tables must be placed end to end to seat 22 people?

2. January 1, 2000 will be a Saturday. On what day of the week will July 4 be that year? (The year 2000 will be a leap year.)

3. A signature of a book resembles a small booklet of 32 pages. It is made by folding a large piece of paper several times and cutting the edges so that the pages will open. How many times must a piece of paper be folded to make a signature?

B
4. Buses that go up Temple Avenue stop at every even-numbered street (Second Street, Fourth Street, Sixth Street, etc.). Houses on Temple Avenue that are numbered 10 to 19 are between First Street and Second Street. Houses numbered 20–29 are between Second Street and Third Street, and so on. What is the closest bus stop to 178 Temple Avenue?

5. How many squares, of any size, are there on a standard, 8 × 8-square checkerboard?

Most comets travel around the sun in orbits that are oval shaped. The time it takes a comet to complete an orbit is called its period. We can calculate its approximate period from records of previous observations and then predict the approximate date of the comet's return. Solve.

C
6. Halley's comet is one of the most famous and brilliant of comets. It was observed in 1531, 1607, 1682, 1759, 1835, and 1910.
   a. What is the average period (number of years from one observation to the next)?
   b. Use the average period to predict when Halley's comet will next be visible.

3-6 *The Least Common Multiple*

Tennis balls are sold in cans containing 3 balls each. There are 3 balls in 1 can, 6 balls in 2 cans, 9 balls in 3 cans, and so on. The numbers 3, 6, 9, . . . are **multiples** of 3.

A multiple of a number is any product of that number and a natural number. Multiples of a number, *n*, are *n*, 2*n*, 3*n*, 4*n*, ...

A pair of numbers may have **common multiples.**

Multiples of 4: 4, 8, 12, 16, 20, 24, . . . , 36, . . . , 48, . . .
Multiples of 6: 6, 12, 18, 24, . . . , 36, . . . , 48, . . .

The numbers 12, 24, 36, 48, . . . are common multiples of 4 and 6.
The **least common multiple (LCM)** of 4 and 6 is 12.

Example 1 Write the LCM of 4, 6, and 10.

Solution Multiples of 4: 4, 8, 12, . . . , 60, . . . , 120, . . .
Multiples of 6: 6, 12, 18, . . . , 60, . . . , 120, . . .
Multiples of 10: 10, 20, 30, . . . , 60, . . . , 120, . . .
Common multiples: 60, 120, . . .
LCM: 60

Your Turn Write the LCM of the numbers.
• 3, 8, and 12
• 4, 9, and 15

The LCM of two or more numbers contains as factors all the factors of each number. You can use the prime factorization of a set of numbers to write the LCM. The highest power of each factor is used in the LCM.

Example 2   Use the prime factorizations of 100, 80, and 54 to write the LCM.

Solution   Write the prime factorizations.
$$100 = 2 \cdot 2 \cdot 5 \cdot 5 \qquad 80 = 2 \cdot 2 \cdot 2 \cdot 2 \cdot 5 \qquad 54 = 2 \cdot 3 \cdot 3 \cdot 3$$
$$ = 2^2 \cdot 5^2 \qquad = 2^4 \cdot 5 \qquad = 2 \cdot 3^3$$
The highest power of 2 in any factor is 2^4.
The highest power of 3 in any factor is 3^3.
The highest power of 5 in any factor is 5^2.
The LCM is $2^4 \cdot 3^3 \cdot 5^2$, or 10,800

Example 3   Write the LCM of $10a^2$, $5a$, and 40.

Solution   Factor each expression.
$$10a^2 = 2 \cdot 5 \cdot a^2 \qquad 5a = 5 \cdot a \qquad 40 = 2^3 \cdot 5$$
Use the highest power of each factor to form the LCM.
The LCM is $2^3 \cdot 5 \cdot a^2$, or $40a^2$

Your Turn   Use prime factorization to write the LCM.
- 6 and 8
- 9, 15, and 75
- $2x^2$, $4x$, and 16

Class Exercises

State the first five multiples of each number.

1. 3   **2.** 10   **3.** 5   **4.** 9   **5.** m **6.** $2p$

State the common multiples less than 30. What is the LCM?

7. 2 and 3   **8.** 2 and 5   **9.** 3 and 4   **10.** 3 and 5

Exercises

Write the numbers.

A **1.** Multiples of 8 less than 40   **2.** Multiples of 12 less than 50

3. Multiples of 4 between 10 and 20   **4.** Multiples of 5 between 20 and 40

Write the common multiples less than 100. Ring the LCM.

5. 5 and 6	**6.** 6 and 9	**7.** 2 and 7	**8.** 8 and 12
9. 6 and 15	**10.** 5 and 7	**11.** 10 and 12	**12.** 8 and 10

Write the LCM of the numbers.

B **13.** 2 and 4	**14.** 3 and 7	**15.** 3 and 9	**16.** 4 and 10
17. 4 and 9	**18.** 5 and 9	**19.** 4 and 10	**20.** 3 and 12
21. 6 and 10	**22.** 8 and 20	**23.** 10 and 15	**24.** 7 and 8
25. 18 and 30	**26.** 42 and 56	**27.** 48 and 60	**28.** 64 and 72
29. a^2 and $2a$	**30.** $3b$ and $6b^3$	**31.** $4c$ and $8c^2$	**32.** $5d^2$ and $6d$
33. $10e^3$ and $15e^2$	**34.** $4f$ and $2f^2$	**35.** h and j	**36.** $2k$ and $3m$

Use prime factorizations to write the LCM.

37. 10, 20, and 25	**38.** 20, 60, and 45	**39.** 24, 30, and 36
40. 12, 18, and 20	**41.** 15, 20, 25, and 30	**42.** 6, 9, 15, 18, and 24
43. $25d$, $15d^2$, and $5d^3$	**44.** $3n^2$, $6n^3$, and $4n$	**45.** $12x^2$, $6xy$, and $8y^2$

Using the Computer

This program will find pairs of factors of a whole number. RUN this program for 500, 600, and 700. Then try some examples of your own.

```
 5  REM: FACTORING A WHOLE NUMBER
10  PRINT "WHAT IS YOUR WHOLE NUMBER";
20  INPUT N
30  FOR F=1 TO N
40  LET Q=N/F
50  IF Q <> INT(Q) THEN 70
60  PRINT F;Q,
70  NEXT F
80  END
```

3-7 *Scientific Notation*

Over 2.1×10^5 cubic feet of water flow over Niagara Falls each second. The number 2.1×10^5 is written in **scientific notation.**

Scientific notation is especially useful when you are working with very large or very small numbers. Many calculators and computers use scientific notation for these numbers to save space. In a future course you will learn how to write numbers less than 1 using scientific notation.

A number written in scientific notation is the product of two factors.

The first factor is a number between 1 and 10.	2.1×10^5	*The second factor is a power of 10.*

Example 1   Are these numbers written in scientific notation?
36×10^5; 7.2×10^3; 4.0×5^2

Solution 36×10^5 ———— No, 36 is not between 1 and 10.
7.2×10^3 ———— Yes.
4.0×5^2 ———— No, 5^2 is not a power of 10.

Any number in scientific notation can be written in standard notation. To do this you multiply. You can think of moving the decimal point as many places as the power of 10.

Example 2   Write 8.6×10^9 in standard form.

Solution $8.6 \times 10^9 = 8,600,000,000$
                    9 places

Your Turn   Is the number written in scientific notation?
  • 8.7×10^3 • 19×10^7 • 1.2×100^2
  Write the number in standard form.
  • 1.2×10^5 • 6.1×10^{10} • 7.5×10^8

If a number is given in standard form, you can write it in scientific notation.

Example 3   Write the number 361,000,000,000 in scientific notation.

Solution   First, move the decimal point to the left to obtain a number between 1 and 10. The number of places the decimal moves is the exponent of 10.

$$361,000,000,000. = 3.61 \times 10^{11}$$

11 places

Your Turn   Write the number in scientific notation.
- 54,000
- 2,038,000,000
- 3,100,000

Class Exercises

Is the number in scientific notation? Answer *Yes* or *No*.

1. 2.0×10^7 **2.** 1.2×5^5 **3.** 86×10^9 **4.** 7.0×10^{32}

What is the standard form of the number?

5. 3.1×10^1 **6.** 8.6×10^3 **7.** 5.4×10^2 **8.** 1.2×10^3

Tell what number completes the equation.

9. $679,000 = 6.79 \times 10^?$ **10.** $5,800,000,000 = 5.8 \times 10^?$

11. $20,000 = \underline{\,?\,} \times 10^4$ **12.** $15,000,000,000 = \underline{\,?\,} \times 10^{10}$

Exercises

Write the number in standard form.

A   **1. a.** 10^3 **2. a.** 10^6 **3. a.** 10^5 **4. a.** 10^8
    b. 3×10^3 **b.** 4×10^6 **b.** 2×10^5 **b.** 9×10^8
    c. 3.1×10^3 **c.** 4.9×10^6 **c.** 2.1×10^5 **c.** 9.3×10^8

Write the number in scientific notation.

5. a. 10,000   **6. a.** 1000   **7. a.** 1,000,000   **8. a.** 100,000
   b. 50,000   **b.** 6000   **b.** 7,000,000   **b.** 400,000
   c. 51,000   **c.** 6100   **c.** 7,200,000   **c.** 432,000

Write the number in standard form.

B **9.** 9.3×10^3 **10.** 8.1×10^5 **11.** 1.7×10^4 **12.** 3.0×10^6

13. 1.8×10^2 **14.** 7.6×10^1 **15.** 3.8×10^5 **16.** 6.9×10^8

17. 3.9×10^{12} **18.** 9.2×10^{10} **19.** 7.1×10^9 **20.** 8.5×10^{15}

Write the underlined number in standard form.

21. An ocean sunfish may lay as many as 3.0×10^8 eggs at one time.
22. There are about 3.9×10^5 known types of plants in existence.
23. There are over 1.23×10^6 known types of animals in existence.

Write the underlined number in scientific notation.

24. The deepest mine on Earth is the Western Deep Levels Mine in the Transvaal, South Africa. It is 12,600 ft deep.
25. The temperature of Earth at its core is about 76000°F.
26. The maximum take-off weight of a "Jumbo Jet" is 775,000 lb.

Self Test 2

Solve by using a pattern.

1. January 1, 2000 will be a Saturday. On what day of the week will   **3-5**
July 3 be that year? (The year 2000 will be a leap year.)

Write the LCM of each set of expressions.

2. 60 and 72   **3.** $3a$ and $2a$ **4.** $2b$, $6b^2$, and 8   **3-6**

Write the number in standard form.

5. 8.0×10^3 **6.** 1.7×10^5 **7.** 9.4×10^8 **3-7**

Write the number in scientific notation.

8. 2,800,000   **9.** 96,000   **10.** 54,000,000

Algebra Practice

Write an equation for the sentence. Solve the equation.

1. 2398 increased by 9746 is r.

2. The difference between 5000 and 2509 is a.

3. The product of 780 and 67 is m.

4. 2 used as a factor 3 times is w.

5. 2784 divided by 87 is equal to f.

6. 1000 minus 19.65 is equal to t.

7. The numbers 1986.21, 8.005, and 497.6 add to a sum of d.

8. The quotient of 1.424 and 1.6 is n.

9. 0.41 times 0.25 is equal to y.

Simplify the expression.

10. $6.8h + 2.5h$ **11.** $7.5a + 5.2b + 3.6a$ **12.** $12k - 8.53k$

13. $9.2p - 4.8q - 2.3p$ **14.** $8.63 - (2.08 + 3.5)$ **15.** $14.2 - (0.33 \times 26.8)$

Write the GCF of each set of expressions.

16. $3c^2$ and 6   **17.** $2r^3$ and $10r^2$ **18.** $3h^2$, $12h$, and $9h^2$

19. $12m$ and $4m^2$ **20.** 14 and $2b^2$ **21.** $4y^2$, $8y$, and $2xy$

Use the distributive property to factor each expression.

22. $30k^3 + 24k^2$ **23.** $2s^3 - 6s^2t + 10s$ **24.** $3m^2 + 6n$

25. $5b^3 - 15b$ **26.** $4j^2k - (jk)^2 + jk^3$ **27.** $(fg)^2 + f^2g - fg^2$

28. $6mn^2 - 18m^2$ **29.** $8x^2y - (2x)^2 + 4xy^2$ **30.** $9a^2t^2 + 12at^2 - 15a^2t$

Write the LCM of each set of expressions.

31. w^2 and $3w$ **32.** $2d$ and $8d^3$ **33.** $3q$ and $5q^2$

34. 15 and a **35.** $3k^2$ and n **36.** $5h^2$ and $7h$

37. 12, $5d^2$, and $3d$ **38.** $8y$, $6x$, and $3y$ **39.** $10s$, r^2, and $4r$

40. $3m^2$, $6mn$, and m^2n^2 **41.** $4a^2b$, $12b^2$, and $16a^2$ **42.** $5d$, $20h$, $4d^2h$

Problem Solving On The Job

Administrative Assistant

Judy Alison is administrative assistant to Mrs. Eastland, a senior vice-president at Hampton Associates.

ADMINISTRATIVE ASSISTANT

DESCRIPTION: Coordinate secretarial services and handle details so that the head person in an office can attend to other matters.

QUALIFICATION: High school; college or business school desirable; on-the-job training

JOB OUTLOOK: Excellent opportunities for individuals who want to work their way up

One of Judy's duties is to keep attendance records for all the staff in the office. Use the table below to do Exercises 1–9.

	Years of Service Completed	Days Absent This Year	
		Sickness	Vacation
Mike	2	0	5
Judy	4	3	6
Frank	5	1	3
Kim	1	5	1

Up to 10 days a year are allowed each person for sickness. How many more days are allowed each person for sickness this year?

1. Mike   **2.** Judy   **3.** Frank   **4.** Kim

5. If a person is absent from work for 2–4 hours, one-half day is deducted from the 10 days allowed for sickness. Because of a doctor's appointment. Mike was out of the office for 3 hours. How many more days is Mike allowed for sickness this year?

During the first two years of service, each person has 10 vacation days. After two years, 15 days are allowed. How many more days are allowed each person for vacation this year?

6. Mike   **7.** Judy   **8.** Frank   **9.** Kim

Mrs. Eastland has asked Judy to have copies of these documents made.

Document A: 2 pages, 100 copies   Document B: 32 pages, 15 copies
Document C: 10 pages, 125 copies   Document D: 60 pages, 10 copies

Sometimes Judy sends material to a copy center to be copied. For each page of a document, the copy center charges $.05 per page for the first 100 copies and $.03 per page for additional copies. The copy center will also assemble and staple the documents for $.03 per staple.

How much will it cost the company to have the documents copied, assembled and stapled at the copy center?

10. Document A   **11.** Document B   **12.** Document C   **13.** Document D

Judy calls The Temporary Agency for assistance when there is a work overload. The agency charges $7.10 an hour for a general typist and $11.50 an hour for a word processor operator (WP operator). How much does The Temporary Agency charge Hampton Associates each week?

14. Week 1:   **15.** Week 2:   **16.** Week 3:
   Typist: 18 hours   Typist: 30 hours   Typist: 8 hours
   WP operator: 24 hours   WP operator: 20 hours   WP operator: 12 hours

Mrs. Eastland is planning a conference for 50 people. It becomes Judy's job to make reservations for a meeting room and a catered lunch. Solve the problem.

17. The Bay View Hotel charges $8.95 per person for lunch. There is no additional charge for the meeting room if lunch is served. If the conference is held at the Bay View Hotel, how much will it cost?

18. Renting a hall for one day costs $225. A caterer will serve lunch for $5.50 per person. If Judy rents the hall for the conference and has the lunch catered, how much will it cost the company?

Enrichment

Arithmetic Sequences

In the sequence 3, 7, 11, 15, 19, . . . each term is obtained by adding 4 to the previous term. The difference between any two consecutive terms is 4.

$$3 \;,\; 7 \;,\; 11 \;,\; 15 \;,\; 19$$
$$+4 \quad +4 \quad +4 \quad +4$$

Such a sequence is called an **arithmetic sequence** and 4 is called the **common difference.**

An arithmetic sequence is a sequence in which the difference obtained when any term is subtracted from the next is always the same number.

We can use a pattern to find the nth term of an arithmetic sequence. Suppose a is the first term of an arithmetic sequence. Then $a + d$ is the second term, and $a + d + d$ or $a + 2d$ is the third term, and so on. The coefficient of d is always 1 less than the number of the term. The sequence looks like this.

Term 1	Term 2	Term 3	Term 4	. . .	Term n
a	$a + 1d$	$a + 2d$	$a + 3d$	. . .	$a + (n-1)d$

$$+d \qquad +d \qquad +d$$

We can find the 100th term of the sequence 3, 7, 11, 15, 19, ... The first term, a, is 3. The common difference, d, is 4.

nth term $= a + (n - 1)d$
100th term $= 3 + (100 - 1)\,4$ ← *Since we are looking for the*
$\qquad\quad = 3 + (99)\,4 \qquad$ *100th term, n is 100.*
$\qquad\quad = 3 + 396,$ or 399

Find the specified term of the sequence 5, 10, 15, 20, ...

1. 10th term
2. 50th term
3. 80th term
4. 200th term

Find the 100th term of each sequence.

5. 1, 3, 5, 7, 9, . . .
6. $1.50, $3.00, $4.50, $6.00, . . .

Chapter Review

Match. An answer may be used more than once.

1. A number that is divisible by 6 and 10
2. A number that is divisible by 2 but not by 4
3. A number that is divisible by 3 and 5
4. An odd number that is not divisible by 3
5. An even number that is divisible by 5

a. 132
b. 91
c. 780
d. 98

Match. An answer may be used more than once.

6. The prime factorization of 60
7. The prime factorization of 108
8. The LCM of 12 and 15
9. The LCM of 36 and 27
10. The GCF of 54 and 162
11. The GCF of 48 and 144
12. Scientific notation for 2000

a. $2 \cdot 3^3$
b. $2 \cdot 10^3$
c. $2^2 \cdot 3 \cdot 5$
d. $2^2 \cdot 3^3$
e. $2^4 \cdot 3$

Use the distributive property to factor each expression.

13. $6a + 2$

14. $5b^2 - b$

15. $2c - 4c^2 + 6$

True or False? Write *T* or *F*.

16. The LCM of $4x^2$ and $6x$ is $12x^2$.

17. Standard notation for 2.8×10^5 is 2,800,000.

18. Scientific notation for 57,000 is 57×10^3.

19. The GCF of $32a^2$ and $18a$ is $2a$.

20. An odd number is always 1 more than an even number.

Solve by using a pattern.

21. A movie which lasts for 1 h 45 min starts at 1:15 P.M. and runs for 5 shows. There are 20 min between showings. At what time does the last show start?

Chapter Test

Is the first number divisible by the second number? Answer *Yes* or *No*.

1. 528 by 4
2. 729 by 3
3. 1001 by 2   **3-1**
4. 940 by 5
5. 765 by 10
6. 94 by 6

Use exponents to write the prime factorization of each number.

7. 840
8. 98
9. 416   **3-2**

Write the GCF of each set of expressions.

10. 12 and 20
11. 16 and 36
12. 15, 30, and 40   **3-3**
13. $2n$ and 8
14. a^2 and $2a$
15. $3b^3$, $6b^2$, and 96   **3-4**

Use the distributive property to factor each expression.

16. $5x - 10$
17. $y^2 + 2y$
18. $4s^2 + 2s + 10$
19. $3m - 12m^2$
20. $5h + 25h^2$
21. $6d^2 - 18$

Solve by using a pattern.

22. Subway trains leave Central Station for Uptown Station every   **3-5**
    4 min during rush hour. The trip takes 22 min. How many trips
    are completed during a two-hour period?

Write the LCM of each set of expressions.

23. 15 and 20
24. $2n$ and n^2
25. 6, $2a$, and $4a^2$ **3-6**
26. $3m$, $12m^2$, 4
27. d^2, $5d$, 20
28. 7, $14h^2$, $2h$

Write the number in scientific notation.

29. 8,000,000
30. 1200
31. 93,000,000   **3-7**
32. 51,200
33. 806,000
34. 4,570,000

Write the number in standard form.

35. 2.0×10^3
36. 4.9×10^5
37. 1.8×10^6
38. 6.2×10^4
39. 8.4×10^6
40. 1.1×10^7

Cumulative Review (Chapters 1–3)

Add, subtract, multiply, or divide.

1. 100×6.7

2. $356 \div 10$

3. $2 \cdot 2 \cdot 2$

4. $10 \cdot 10 \cdot 10 \cdot 10$

5. $10 \cdot 10 \cdot 10 \cdot 10 \cdot 10$

6. $(4.1 + 5)(2.01 - 1)$

7. 5^3

8. $2^3 + 7$

9. $10^2 + 7^2 - 6$

10. $3 \times 12 \div 9$

11. $4^2 \div 8 + 8$

12. $21 - 3^2 \cdot 2$

Simplify.

13. $4(a + 1) + a(7 - 3)$

14. $r(12 - 3) + 5(5 - r)$

15. $2p + 8y - p + 6y - 2$

16. $(15 - 14)x + (8 + 9)x$

17. $a \cdot 1 + 4a - b$

18. $23s - (5)(s) + s(6 + 1)$

Factor.

19. $3x + 3$

20. $9a - 9$

21. $8 - 2c$

22. $n^2 + n$

23. $5y^2 + 10$

24. $6j^2 + 3j$

25. $6s^2 - 3s + 15$

26. $8p^2 - 8$

27. $10f^2 + 5f - 25$

Write an expression for the phrase.

28. $2x$ increased by 9

29. 35 decreased by n

30. m greater than the sum of n and 17

31. the sum of 12 and the product of 5 and y

Solve.

32. Chris is 5 years older than her sister Martha.
Let y = Chris' age.
Then __?__ = Martha's age.

33. It took a carpenter 3 weeks to complete a job.
Let d = the number of days worked per week.
Then __?__ = the number of days spent on the job.

34. For each hour of classroom time in a college course, students are expected to spend from 2 to $2\frac{1}{2}$ hours on homework. Libby spent 15 hours on homework for 8 hours of classroom time. Was her homework time within the expected range?

4
Fractions

You can vary the length of a column of air inside a flute by opening and closing holes at different locations. One-half of the column will give a sound with pitch higher than the whole column.

4-1 Fraction Concept

On the number line below, the distance from 0 to 1 is divided into 8 equal parts. The distance from 0 to point P is five of those eight parts, or $\frac{5}{8}$ of the distance from 0 to 1.

Every fraction has a **numerator** and a **denominator**.

$$\frac{5}{8} \xleftarrow{\hspace{1cm}} \textit{numerator} \qquad \frac{x}{y} \xleftarrow{\hspace{1cm}} \textit{numerator}$$
$$\phantom{\frac{5}{8}} \xleftarrow{\hspace{1cm}} \textit{denominator} \qquad \phantom{\frac{x}{y}} \xleftarrow{\hspace{1cm}} \textit{denominator}$$

The numerator of a fraction can be any whole number, while the denominator can be any whole number except zero. For the fraction $\frac{x}{y}$, y cannot be 0. We write $y \neq 0$ to mean y is not equal to zero.

You can see from the number line that $\frac{5}{8}$ is less than 1. Not all fractions are less than 1. Study the number line and notice that the fraction $\frac{8}{8}$ could replace the whole number 1. The fraction $\frac{8}{8}$ is an example of a fraction that is equal to 1.

The fact that $a \div b = \frac{a}{b}$ enables us to write other whole numbers as fractions.

$$\text{Since } 18 \div 6 = \frac{18}{6}, \frac{18}{6} = 3$$

$$\text{Since } 10 \div 1 = \frac{10}{1}, \frac{10}{1} = 10$$

$$\text{Since } 6 \div 6 = \frac{6}{6}, \frac{6}{6} = 1$$

In Arithmetic	In Algebra
$8 \div 8 = \dfrac{8}{8} = 1$	$x \div x = \dfrac{x}{x} = 1 \qquad (x \neq 0)$
$10 \div 1 = \dfrac{10}{1} = 10$	$y \div 1 = \dfrac{y}{1} = y$

101

The number line below is divided into sixths. We know that $\frac{6}{6}$ is 1.
The fraction $\frac{7}{6}$, therefore, can be written as the mixed number $1\frac{1}{6}$.

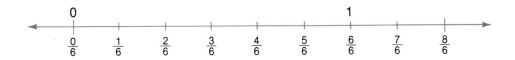

The examples below show how to write a fraction as a mixed number
and a mixed number as a fraction.

Example 1 Write $\dfrac{28}{9}$ as a mixed number.

Solution $\dfrac{28}{9}$ ⇨ $9\overline{)\,28\,}$ with $\dfrac{3}{-27}\,\dfrac{}{1}$ ⇨ $3\dfrac{1}{9}$ ← *Write the remainder as a fractional part of the divisor 9.*

Your Turn • Write $\dfrac{35}{8}$ and $\dfrac{26}{9}$ as mixed numbers.

Example 2 Write $3\dfrac{4}{5}$ as a fraction.

Solution $5 \times 3 = 15$ ← *Multiply the denominator and the whole number.*

$15 + 4 = 19$ ← *Add the numerator to the product.*

$\dfrac{19}{5}$ ← *Write the sum over the denominator.*

Shortcut: $\dfrac{(5 \times 3) + 4}{5} = \dfrac{19}{5}$

Your Turn • Write $6\dfrac{2}{7}$ and $8\dfrac{1}{6}$ as fractions.

Class Exercises

Name the fraction at point *A*.

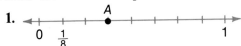

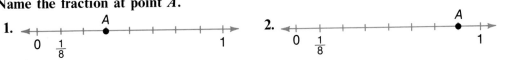

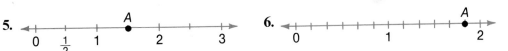

5. number line from 0, 1/2, 1, 2, 3 with A between 1 and 2

6. number line from 0, 1, 2 with A near 2

Exercises

Write the whole number as a fraction. Use 1 as the denominator.

A **1.** 6   **2.** 7   **3.** 8   **4.** 9   **5.** 10   **6.** 11

7. 12   **8.** 13   **9.** 14   **10.** 15   **11.** 16   **12.** 17

Write as a fraction.

13. $4 \div 1$ **14.** $5 \div 3$ **15.** $7 \div 2$ **16.** $8 \div 5$ **17.** $9 \div 4$

18. $6 \div 7$ **19.** $8 \div 11$ **20.** $12 \div 11$ **21.** $5 \div 13$ **22.** $14 \div 6$

Write the mixed number as a fraction.

B **23.** $1\frac{1}{2}$ **24.** $1\frac{1}{3}$ **25.** $2\frac{1}{2}$ **26.** $2\frac{1}{3}$ **27.** $1\frac{1}{4}$

28. $2\frac{1}{4}$ **29.** $1\frac{3}{4}$ **30.** $2\frac{2}{3}$ **31.** $3\frac{1}{2}$ **32.** $2\frac{2}{5}$

Write a mixed number or a whole number for the variable.

33. $\frac{3}{2} = a$ **34.** $\frac{4}{3} = b$ **35.** $\frac{6}{5} = c$ **36.** $\frac{7}{3} = d$ **37.** $\frac{8}{4} = e$

38. $\frac{9}{4} = f$ **39.** $\frac{7}{1} = g$ **40.** $\frac{9}{2} = h$ **41.** $\frac{8}{1} = j$ **42.** $\frac{10}{7} = k$

103

Write a fraction for the variable.

43. $5\dfrac{1}{4} = l$ **44.** $3\dfrac{2}{5} = m$ **45.** $8\dfrac{1}{6} = n$ **46.** $7\dfrac{3}{5} = p$ **47.** $6\dfrac{2}{7} = q$

48. $9\dfrac{6}{7} = r$ **49.** $10\dfrac{3}{5} = s$ **50.** $11\dfrac{5}{8} = t$ **51.** $12\dfrac{9}{10} = v$ **52.** $13\dfrac{5}{9} = w$

Substitute. Write the fraction as a whole number or as a mixed number.

53. $\dfrac{a}{b}$ when $a = 5$ and $b = 4$ **54.** $\dfrac{c}{d}$ when $c = 11$ and $d = 8$

55. $\dfrac{e}{f}$ when $e = 12$ and $f = 5$ **56.** $\dfrac{c}{h}$ when $c = 18$ and $h = 6$

57. $\dfrac{a}{p}$ when $a = 24$ and $p = 7$ **58.** $\dfrac{e}{s}$ when $e = 36$ and $s = 1$

59. $\dfrac{t}{u}$ when $t = 32$ and $u = 16$ **60.** $\dfrac{l}{m}$ when $l = 14$ and $m = 13$

61. $\dfrac{e}{d}$ when $e = 31$ and $d = 20$ **62.** $\dfrac{c}{g}$ when $c = 35$ and $g = 35$

Problems

Three groups of people were asked if they would use Brite-White tooth-paste. Write a fraction for the given response.

1. In a group of 21 people, n people answered yes.
2. In a group of x people, 15 people answered no.
3. In a group of 30 people, m people were undecided.

The first year class at the Taft School voted for a class president. The results were posted on the bulletin board as shown at the right.

4. What part of the class voted for Peter?
5. What part of the class voted for Helen?
6. What part of the class did not vote for Joel?
7. What part of the class did not vote for Marie?
8. Who got the larger fraction of the votes, Helen or Joel?
9. Suppose every one voted for Helen. What fraction of the votes did she get?

Election Results	
Candidate	Votes
Helen Ling	28
Peter Allard	20
Joel Lopes	17
Marie Stevens	16

4-2 *Renaming Fractions*

When you rename a fraction you can either multiply or divide the numerator and the denominator by the same nonzero number. By doing so you do not actually change the value of the fraction. You simply find a fraction that is equal to the given fraction.

Example 1   Write a fraction that is equal to $\dfrac{7}{10}$.

Solution $\dfrac{7}{10} = \dfrac{7 \times 3}{10 \times 3}$ ←———— *Multiply the numerator and the denominator by 3.*

$\qquad\quad = \dfrac{21}{30}$

Here are other examples of fractions that are equal to $\frac{7}{10}$: $\frac{14}{20}$, $\frac{28}{40}$, and $\frac{35}{50}$.

When you divide to find an equal fraction, you rename the original fraction with a different numerator and denominator. If you divide by the greatest common factor (GCF), you rename the fraction in **lowest terms.**

Example 2   Write $\dfrac{16}{20}$ in the lowest terms.

Solution $\dfrac{16}{20} = \dfrac{16 \div 4}{20 \div 4}$ ←———— *4 is the GCF of 16 and 20.*

$\qquad\quad = \dfrac{4}{5}$

Here is a shortcut you may have learned before.

$\dfrac{\overset{4}{\cancel{16}}}{\underset{5}{\cancel{20}}} = \dfrac{4}{5}$ ←———— *If you cannot think of the GCF right away, you can do the division in several steps.*

105

You can also use the factors of the numerator and denominator to help you find a fraction in lowest terms.

Example 3   Write $\dfrac{18}{24}$ in lowest terms.

Solution $\dfrac{18}{24} = \dfrac{3 \cdot 2 \cdot 3}{2 \cdot 2 \cdot 2 \cdot 3}$ ⟵ *Notice that the GCF is 2 · 3.*

$$= \dfrac{3 \cdot \overset{1}{\cancel{2}} \cdot \overset{1}{\cancel{3}}}{2 \cdot 2 \cdot \underset{1}{\cancel{2}} \cdot \underset{1}{\cancel{3}}} = \dfrac{3}{4}$$

Example 4   Write $\dfrac{6x^2}{9x}$ in lowest terms.

Solution $\dfrac{6x^2}{9x} = \dfrac{\overset{2}{\cancel{6}} \cdot \overset{1}{\cancel{x}} \cdot x}{\underset{3}{\cancel{9}} \cdot \underset{1}{\cancel{x}}}$

$$= \dfrac{2x}{3}$$

Your Turn   Write the fraction in lowest terms.

- $\dfrac{2xy}{2x}$
- $\dfrac{4a}{8ab}$
- $\dfrac{5ac^2}{10ac}$

Class Exercises

Complete.

1. $\dfrac{3 \times 2}{4 \times 2} = \dfrac{?}{8}$

2. $\dfrac{1 \times ?}{3 \times 3} = \dfrac{3}{9}$

3. $\dfrac{4 \times ?}{5 \times ?} = \dfrac{16}{20}$

4. $\dfrac{8 \div 2}{10 \div 2} = \dfrac{4}{?}$

5. $\dfrac{7 \div ?}{14 \div ?} = \dfrac{1}{2}$

6. $\dfrac{16 \div ?}{20 \div ?} = \dfrac{4}{5}$

Exercises

Complete to find an equal fraction.

A **1.** $\dfrac{3}{7} = \dfrac{?}{14}$ **2.** $\dfrac{4}{5} = \dfrac{?}{35}$ **3.** $\dfrac{7}{9} = \dfrac{?}{63}$ **4.** $\dfrac{5}{8} = \dfrac{?}{24}$

5. $\dfrac{1}{14} = \dfrac{3}{?}$ **6.** $\dfrac{7}{19} = \dfrac{14}{?}$ **7.** $\dfrac{24}{36} = \dfrac{2}{?}$ **8.** $\dfrac{56}{64} = \dfrac{?}{8}$

Write in lowest terms.

9. $\dfrac{6}{9}$ **10.** $\dfrac{10}{12}$ **11.** $\dfrac{15}{30}$ **12.** $\dfrac{8}{20}$

13. $\dfrac{8}{32}$ **14.** $\dfrac{9}{24}$ **15.** $\dfrac{18}{12}$ **16.** $\dfrac{35}{21}$

B **17.** $\dfrac{x}{x}$ **18.** $\dfrac{3y}{y}$ **19.** $\dfrac{2b}{2}$ **20.** $\dfrac{2a}{a}$

21. $\dfrac{3}{6c}$ **22.** $\dfrac{4}{4m}$ **23.** $\dfrac{2e}{4e}$ **24.** $\dfrac{10n}{5n}$

25. $\dfrac{3xy}{3x}$ **26.** $\dfrac{4st}{4s}$ **27.** $\dfrac{2ab}{6b}$ **28.** $\dfrac{9cd}{3d}$

29. $\dfrac{6mc}{9mc}$ **30.** $\dfrac{4nb}{12b}$ **31.** $\dfrac{10cv}{15v}$ **32.** $\dfrac{7xu}{21x}$

33. $\dfrac{21t}{28t^2}$ **34.** $\dfrac{8ax}{4a^2x}$ **35.** $\dfrac{9m^2b}{6m}$ **36.** $\dfrac{15cd^2}{10cd}$

Simplify.

Example $\dfrac{6a + 12}{6} = \dfrac{6(a + 2)}{6} = a + 2$

C **37.** $\dfrac{3x + 3}{3}$ **38.** $\dfrac{5n + n}{2n}$ **39.** $\dfrac{7b + b}{b}$ **40.** $\dfrac{2x + x}{x}$

41. $\dfrac{8c + 4c}{4c}$ **42.** $\dfrac{6a + 3a}{3a}$ **43.** $\dfrac{2x}{4x + 2x}$ **44.** $\dfrac{10b}{5b + 5b}$

45. $\dfrac{6r}{10r + 4}$ **46.** $\dfrac{5d + 10}{15d}$ **47.** $\dfrac{12e + 4}{8}$ **48.** $\dfrac{16a}{4a + 8}$

4-3 Adding Fractions

Students at Eastdale High School are sponsoring concerts to buy band uniforms. Profits from the first concert paid for $\frac{1}{8}$ of the cost of the uniforms. Profits from the second concert paid for another $\frac{3}{8}$. What part of the cost of the uniforms is paid for by the two concerts?

To find the answer, you can add $\frac{1}{8}$ and $\frac{3}{8}$. We add fractions with like denominators by writing the sum of the numerators over the denominator.

In Arithmetic

$$\frac{2}{5} + \frac{1}{5} = \frac{2+1}{5} = \frac{3}{5}$$

In Algebra

$$\frac{a}{c} + \frac{b}{c} = \frac{a+b}{c} \qquad (c \neq 0)$$

Example 1   Solve $n = \frac{1}{8} + \frac{3}{8}$. Write the sum in lowest terms.

Solution $n = \dfrac{1+3}{8}$ ⟵——— *Add the numerators.*

$= \dfrac{4}{8} = \dfrac{1}{2}$

Example 2   Find the sum. $\dfrac{5}{a} + \dfrac{4}{a}$

Solution $\dfrac{5}{a} + \dfrac{4}{a} = \dfrac{5+4}{a} = \dfrac{9}{a}$ ⟵——— *Add the numerators.*

To add fractions with different denominators we must first rewrite the fractions with a common denominator and then add. Try to choose the least common multiple (LCM) as the common denominator. We call that number the **least common denominator** (LCD).

Example 3   Write the sum of $\frac{5}{12}$ and $\frac{3}{8}$ in lowest terms.

Solution   The LCM for 12 and 8 is 24, so the LCD is 24.

$$\frac{5}{12} = \frac{10}{24} \text{ and } \frac{3}{8} = \frac{9}{24}$$

$$\frac{10}{24} + \frac{9}{24} = \frac{19}{24} \longleftarrow \textit{Add the numerators.}$$

Example 4   Write the sum: $5\frac{3}{10} + 3\frac{2}{5}$.

Solution
$$5\frac{3}{10} = 5\frac{3}{10}$$
$$+3\frac{2}{5} = +3\frac{4}{10}$$
$$\overline{\phantom{+3\frac{2}{5}} \quad 8\frac{7}{10}}$$

Your Turn   Find the sum.

- $2\frac{3}{4} + 3\frac{2}{3}$
- $\frac{5}{d} + \frac{7}{d}$
- $\frac{e}{6} + \frac{e}{6}$

Class Exercises

Write the sum in lowest terms.

1. $\frac{2}{3} + \frac{4}{5}$ **2.** $\frac{2}{3} + \frac{5}{8}$ **3.** $\frac{3}{4} + \frac{7}{9}$ **4.** $9 + \frac{4}{7}$

5. $\frac{3}{4} + \frac{3}{5}$ **6.** $\frac{7}{10} + 10$ **7.** $\frac{11}{20} + \frac{3}{8}$ **8.** $\frac{4}{5} + 13$

Exercises

Simplify the expression using the given value for the variable.

A
1. $5\frac{2}{9} + e$; $e = 4\frac{3}{8}$

2. $6\frac{1}{7} + p$; $p = 3\frac{5}{8}$

3. $2\frac{1}{12} + r$; $r = 4\frac{5}{8}$

4. $3\frac{4}{11} + l$; $l = 5\frac{1}{4}$

5. $4\frac{5}{11} + w$; $w = 6\frac{3}{8}$

6. $8\frac{5}{6} + v$; $v = 4\frac{4}{9}$

7. $b + 9\frac{4}{7}$; $b = 3\frac{6}{7}$

8. $a + 2\frac{7}{8}$; $a = 7\frac{1}{12}$

9. $c + 6\frac{1}{11}$; $c = 4\frac{2}{3}$

Is the first expression greater than the second expression? Guess first, then check your guess. Assume the variable is greater than zero.

B
10. $\frac{1}{x} + \frac{5}{x}$ $\frac{2}{x} + \frac{3}{x}$

11. $\frac{3}{a} + \frac{5}{a}$ $\frac{1}{a} + \frac{3}{a} + \frac{2}{a}$

12. $\frac{5y}{7} + \frac{2y}{7}$ $\frac{3y}{7} + \frac{y}{7}$

13. $\frac{4e}{5} + \frac{e}{5}$ $\frac{2e}{5} + \frac{e}{5} + \frac{3e}{5}$

Find the sum.

14. $\frac{6}{a} + \frac{7}{b}$

15. $\frac{8}{c} + \frac{2}{w}$

16. $\frac{5}{m} + \frac{5}{r}$

17. $\frac{3}{b} + \frac{2}{c}$

18. $\frac{x}{2} + \frac{x}{3}$

19. $\frac{b}{4} + \frac{b}{8}$

20. $\frac{x}{2} + \frac{y}{3}$

21. $\frac{m}{10} + \frac{f}{5}$

Problems

Write an equation. Solve.

1. The sum of $4\frac{1}{2}$ and $\frac{3}{5}$ is some number. What is it?

2. Some number is the total of 15 and $6\frac{5}{8}$. What is it?

3. Brian bought $\frac{3}{4}$ lb of grapes and $\frac{1}{2}$ lb of plums. How many pounds of fruit did he buy?

4. A board $\frac{3}{4}$ in. thick is placed on top of a board $\frac{5}{8}$ in. thick. What is the total thickness of the two boards?

5. Beth cycled $\frac{5}{8}$ mi from home to school, $\frac{5}{6}$ mi from school to the library, $\frac{2}{3}$ mi from the library to work, and $\frac{1}{4}$ mi from work to her home. How many miles did Beth travel?

4-4 *Subtracting Fractions*

The rules for subtracting fractions are similar to the rules for adding fractions. We subtract fractions with like denominators by writing the difference of the numerators over the denominator.

In Arithmetic

$$\frac{5}{7} - \frac{3}{7} = \frac{5-3}{7} = \frac{2}{7}$$

In Algebra

$$\frac{a}{c} - \frac{b}{c} = \frac{a-b}{c} \qquad (c \neq 0)$$

Example 1   Find the difference. $\dfrac{25}{a} - \dfrac{16}{a}$

Solution $\dfrac{25}{a} - \dfrac{16}{a} = \dfrac{25-16}{a}$ ⟵——— *Subtract the numerators.*

$$= \frac{9}{a}$$

To subtract fractions with unlike denominators, we first rewrite the fractions with a common denominator, then subtract.

Example 2   Find the difference: $\dfrac{a}{4} - \dfrac{a}{6}$.

Solution   The LCD is 12.

$$\frac{a}{4} = \frac{?}{12}$$ ⟵——— *4 must be multiplied by 3 to get 12.*
$$= \frac{3a}{12}$$ ⟵——— *So multiply a by 3 to get 3a.*

$$\frac{a}{6} = \frac{2a}{12}$$

$$\frac{3a}{12} - \frac{2a}{12} = \frac{a}{12}$$

111

Your Turn   Find the difference.

- $\dfrac{12}{x} - \dfrac{3}{x}$
- $\dfrac{5}{xy} - \dfrac{2}{xy}$
- $\dfrac{2a}{4} - \dfrac{a}{5}$

When you subtract mixed numbers you may have to regroup. A mixed number is said to be in simplest form if the fraction part is in lowest terms.

Example 3   Write the difference in simplest form.

$$8\dfrac{1}{3} - 6\dfrac{5}{7}$$

Solution $8\dfrac{1}{3} - 6\dfrac{5}{7}$ $\Rightarrow$ $8\dfrac{7}{21} - 6\dfrac{15}{21}$ $\longleftarrow$ *Rewrite $\dfrac{1}{3}$ and $\dfrac{5}{7}$ with the LCD.*

Write $8\dfrac{7}{21}$ as $7\dfrac{28}{21}$ so that you can subtract $\dfrac{15}{21}$ from $\dfrac{28}{21}$.

$$7\dfrac{28}{21} - 6\dfrac{15}{21} = 1\dfrac{13}{21}$$

Your Turn   Write the difference in simplest form.

- $\dfrac{b}{3} - \dfrac{b}{6}$
- $4\dfrac{2}{5} - 3\dfrac{4}{5}$
- $6\dfrac{1}{2} - 4\dfrac{3}{5}$

Class Exercises

Rewrite the fraction with the given denominator.

1. $\dfrac{a}{5} = \dfrac{?}{10}$ 2. $\dfrac{b}{6} = \dfrac{?}{12}$ 3. $\dfrac{c}{4} = \dfrac{?}{12}$ 4. $\dfrac{d}{3} = \dfrac{?}{18}$

5. $2\dfrac{1}{4} = 1 + \dfrac{?}{4} + \dfrac{1}{4}$ 6. $3\dfrac{5}{6} = 2 + \dfrac{?}{6} + \dfrac{5}{6}$ 7. $4\dfrac{3}{8} = 3 + \dfrac{?}{8} + \dfrac{?}{}$

8. $5\dfrac{1}{12} = \dfrac{?}{} + \dfrac{12}{12} + \dfrac{1}{12}$ 9. $\dfrac{?}{} = 6 + \dfrac{11}{11} + \dfrac{3}{11}$ 10. $\dfrac{?}{} = 7 + \dfrac{12}{12} + \dfrac{7}{12}$

Exercises

Write the difference in lowest terms.

A 1. $\dfrac{7}{6} - \dfrac{5}{6}$ 2. $\dfrac{5}{8} - \dfrac{2}{8}$ 3. $\dfrac{17}{10} - \dfrac{16}{10}$ 4. $\dfrac{11}{12} - \dfrac{4}{12}$

5. $\dfrac{7}{10} - \dfrac{3}{10}$ 6. $\dfrac{11}{18} - \dfrac{5}{18}$ 7. $\dfrac{17}{15} - \dfrac{7}{15}$ 8. $\dfrac{23}{24} - \dfrac{5}{24}$

9. $\dfrac{1}{2} - \dfrac{1}{20}$ 10. $\dfrac{6}{10} - \dfrac{1}{3}$ 11. $\dfrac{4}{7} - \dfrac{3}{14}$ 12. $\dfrac{5}{8} - \dfrac{11}{32}$

B 13. $\dfrac{15}{y} - \dfrac{10}{y}$ 14. $\dfrac{31}{g} - \dfrac{29}{g}$ 15. $\dfrac{21}{s} - \dfrac{11}{s}$ 16. $\dfrac{32}{f} - \dfrac{29}{f}$

17. $\dfrac{3a}{n} - \dfrac{2a}{n}$ 18. $\dfrac{4e}{q} - \dfrac{3e}{q}$ 19. $\dfrac{10m}{3} - \dfrac{10m}{3}$ 20. $\dfrac{16r}{8} - \dfrac{9r}{8}$

21. $\dfrac{b}{3} - \dfrac{b}{9}$ 22. $\dfrac{2r}{3} - \dfrac{r}{6}$ 23. $\dfrac{m}{6} - \dfrac{m}{8}$ 24. $\dfrac{3d}{4} - \dfrac{d}{12}$

Solve the equation. Write the difference in lowest terms.

25. $\dfrac{7}{8} - \dfrac{3}{24} = a$ 26. $\dfrac{5}{6} - \dfrac{2}{30} = b$ 27. $\dfrac{7}{10} - \dfrac{1}{4} = c$ 28. $\dfrac{11}{12} - \dfrac{3}{8} = d$

29. $3\dfrac{1}{9} - 2\dfrac{2}{3} = j$ 30. $4\dfrac{1}{2} - 2\dfrac{3}{5} = k$ 31. $6\dfrac{3}{4} - 5\dfrac{1}{6} = l$ 32. $7\dfrac{1}{4} - 5\dfrac{5}{8} = m$

33. $4\dfrac{7}{9} - 2\dfrac{3}{4} = n$ 34. $5\dfrac{3}{7} - 3\dfrac{2}{3} = p$ 35. $6\dfrac{2}{3} - 2\dfrac{2}{15} = q$ 36. $4\dfrac{1}{10} - 2\dfrac{3}{4} = r$

Subtract.

Example $\dfrac{5n + 8}{6} - \dfrac{1}{6} = \dfrac{5n + 8 - 1}{6}$

$= \dfrac{5n - 7}{6}$

C 37. $\dfrac{7a + 5}{8} - \dfrac{1}{8}$ 38. $\dfrac{4b + 7}{9} - \dfrac{2}{9}$ 39. $\dfrac{8c + 5}{12} - \dfrac{c}{12}$

40. $\dfrac{6a - 3}{3}$ 41. $\dfrac{4a - 4b}{2}$ 42. $\dfrac{10x - 5m}{5}$

Problems

Write an equation. Solve.

1. $4\frac{2}{3}$ decreased by $1\frac{1}{3}$ is what number?
2. $6\frac{4}{5}$ increased by $\frac{1}{2}$ is some number. What is it?
3. How much greater is $\frac{5}{8}$ than $\frac{1}{3}$?
4. The librarian shelved the books in $\frac{5}{6}$ h yesterday and in $\frac{3}{5}$ h today. Find the decrease in time, expressed as a fraction.
5. Joan jogged $\frac{7}{8}$ of a mile, and Sarah jogged $\frac{7}{10}$ of a mile. How much farther did Joan jog than Sarah?
6. A nut has a center hole $\frac{21}{64}$ in. across. The hole must be enlarged to $\frac{3}{8}$ in. How much wider will the new hole be?

Self Test 1

Write a fraction for the variable.

1. $7\frac{2}{3} = t$ 2. $4\frac{5}{6} = m$ 3. $5\frac{4}{9} = x$ 4. $10\frac{1}{7} = b$ **4-1**

Substitute. Write the fraction or whole number as a mixed number.

5. $\frac{x}{y}$ when $x = 21$ and $y = 10$ 6. $\frac{c}{p}$ when $c = 24$ and $p = 8$

Write the fraction in lowest terms.

7. $\frac{12ab}{6ab}$ 8. $\frac{9dx^2}{3d}$ 9. $\frac{7c^2m}{14cm}$ 10. $\frac{10a}{16b}$ **4-2**

Find the sum.

11. $\frac{5x}{8} + \frac{x}{8}$ 12. $\frac{6}{b} + \frac{7}{b}$ 13. $\frac{10}{c} + \frac{1}{c}$ **4-3**

Simplify the expression using the given value for the variable.

14. $y + 7\frac{3}{5}; \ y = 2\frac{7}{15}$ 15. $8\frac{3}{7} + p; \ p = \frac{5}{28}$

Write the difference in lowest terms.

16. $\frac{12n}{r} - \frac{9n}{r}$ 17. $\frac{5s}{3} - \frac{3s}{8}$ 18. $10\frac{3}{16} - 8\frac{7}{12}$ **4-4**

4-5 *Multiplying Fractions*

The average weight of a pilot whale is $\frac{3}{4}$ t. You can find the weight of five pilot whales by multiplying $\frac{3}{4}$ by 5. To multiply a fraction and a whole number, you multiply the whole number and the numerator and write the product over the denominator.

Example 1 Solve $w = \frac{3}{4} \times 5$. Write the product in lowest terms.

Solution $w = \dfrac{3 \times 5}{4}$ ⟵ *Multiply the numerator and the whole number.*

$= \dfrac{15}{4}$

$= 3\dfrac{3}{4}$ ⟵ *Write $\dfrac{15}{4}$ in simplest form.*

You know that the whole number 5 can be written as a fraction. So, in Example 1 above, you could rewrite the equation as $w = \dfrac{3}{4} \times \dfrac{5}{1}$.

Example 2 Solve the equation. Write the product in lowest terms.
$$t = \frac{2}{3} \times \frac{7}{8}$$

Solution $t = \dfrac{2 \times 7}{3 \times 8}$ ⟵ *Multiply the numerators. Multiply the denominators.*

$= \dfrac{14}{24} = \dfrac{7}{12}$

Notice that in Example 2 we could have divided by common factors before multiplying. When you divide first, the multiplication is usually made easier since you will be multiplying smaller numbers.

$$\frac{\overset{1}{\cancel{2}}}{3} \times \frac{7}{\underset{4}{\cancel{8}}} = \frac{1 \times 7}{3 \times 4} = \frac{7}{12} \longleftarrow \boxed{\begin{array}{l}\textit{Divide the numerator 2 and} \\ \textit{the denominator 8 by 2.}\end{array}}$$

Example 3   Write the product in lowest terms.

$$2\frac{1}{2} \cdot \frac{x}{5}$$

Solution $$2\frac{1}{2} \cdot \frac{x}{5} = \frac{5}{2} \cdot \frac{x}{5} \longleftarrow \textit{Write } 2\frac{1}{2} \textit{ as the fraction } \frac{5}{2}.$$

$$= \frac{\overset{1}{\cancel{5}}}{2} \cdot \frac{x}{\underset{1}{\cancel{5}}} = \frac{x}{2}$$

Your Turn   Find the product in lowest terms.

- $\dfrac{4}{5} \times \dfrac{1}{2}$
- $3\dfrac{1}{2} \times \dfrac{a}{10}$

In Arithmetic	In Algebra
$\dfrac{1}{2} \times \dfrac{3}{5} = \dfrac{1 \times 3}{2 \times 5} = \dfrac{3}{10}$	$\left(\dfrac{a}{b}\right) \cdot \left(\dfrac{c}{d}\right) = \dfrac{ac}{bd} \qquad (b \neq 0,\ d \neq 0)$

Class Exercises

Write the mixed number or whole number as a fraction.

1. $1\dfrac{1}{2}$ 2. $2\dfrac{2}{3}$ 3. $2\dfrac{3}{4}$ 4. 3 5. $1\dfrac{4}{5}$

6. $2\dfrac{1}{4}$ 7. 9 8. $3\dfrac{5}{6}$ 9. $4\dfrac{2}{7}$ 10. $4\dfrac{3}{8}$

Exercises

Write the product in lowest terms.

A **1.** $\dfrac{1}{2} \cdot 5$ **2.** $\dfrac{1}{3} \cdot 8$ **3.** $\dfrac{7}{12} \cdot 6$ **4.** $3 \cdot \dfrac{2}{3}$

5. $\dfrac{4}{7} \cdot 7$ **6.** $\dfrac{3}{5} \cdot 15$ **7.** $16 \cdot \dfrac{1}{16}$ **8.** $6 \cdot \dfrac{7}{12}$

9. $\dfrac{1}{2} \cdot \dfrac{2}{3}$ **10.** $\dfrac{1}{2} \cdot \dfrac{1}{4}$ **11.** $\dfrac{1}{2} \cdot \dfrac{1}{2}$ **12.** $\dfrac{1}{3} \cdot 3$

Write the product in lowest terms.

B **13.** $\dfrac{1}{b} \cdot 8$ **14.** $\dfrac{9}{11} \cdot g$ **15.** $\dfrac{7}{13} \cdot f$ **16.** $\dfrac{x}{6} \cdot \dfrac{3}{x}$

17. $\dfrac{1}{b} \cdot \dfrac{b}{3}$ **18.** $\dfrac{2}{a} \cdot \dfrac{a}{8}$ **19.** $\dfrac{2c}{c} \cdot \dfrac{4e}{e}$ **20.** $\dfrac{5}{d} \cdot \dfrac{2d}{15}$

21. $3\dfrac{1}{7} \cdot \dfrac{m}{3}$ **22.** $2\dfrac{2}{3} \cdot \dfrac{3}{x}$ **23.** $3\dfrac{1}{3} \cdot \dfrac{b}{5}$ **24.** $4\dfrac{1}{2} \cdot \dfrac{2}{y}$

Write the given expression as shown below.

Example $\dfrac{1}{2} \cdot x = \dfrac{1 \cdot x}{2} = \dfrac{x}{2}$

25. $\dfrac{1}{3} \cdot b$ **26.** $\dfrac{1}{5} \cdot c$ **27.** $\dfrac{1}{6} \cdot g$ **28.** $\dfrac{1}{7} \cdot p$ **29.** $\dfrac{1}{8} \cdot q$

Problems

Write an equation. Solve.

1. Twice $\frac{3}{4}$ is some number. What is the number?
2. The product of $\frac{2}{3}$ and 20 is some number. What is the number?
3. 25 times $\frac{4}{5}$ is some number. What is the number?
4. A gas tank holds 18 gal of gasoline. When the tank is $\frac{3}{4}$ full, how many gallons does it contain?
5. Twenty-five students in an art class were each given $\frac{2}{3}$ lb of clay. What was the total weight of the clay distributed to the 25 students?

4-6 *Dividing Fractions*

Before you divide fractions and mixed numbers, it is useful to know about **reciprocals.** Two numbers are reciprocals if their product is 1. The numbers 4 and $\frac{1}{4}$ are reciprocals because $4 \times \frac{1}{4} = 1$. Since the product of 0 and another number is always 0 and never 1, the number 0 has no reciprocal.

Suppose you have to share $\frac{2}{3}$ of a lasagna equally among four people. You would write the problem as $\frac{2}{3} \div 4$.

Instead of dividing by 4, you could multiply by $\frac{1}{4}$, since dividing by a number is the same as multiplying by its reciprocal.

$$\frac{2}{3} \div 4 = \frac{2}{3} \times \frac{1}{4} = \frac{\overset{1}{\cancel{2}} \times 1}{3 \times \underset{2}{\cancel{4}}} = \frac{1}{6}$$

In Arithmetic	In Algebra
$\frac{2}{3} \div \frac{4}{1} = \frac{2}{3} \times \frac{1}{4} = \frac{1}{6}$	$\frac{a}{b} \div \frac{d}{c} = \frac{a}{b} \times \frac{c}{d} \qquad (b \neq 0,\ c \neq 0,\ d \neq 0)$

Example 1 Solve the equation. Write the quotient in lowest terms.

$$x = \frac{5}{16} \div 10$$

Solution $x = \frac{5}{16} \times \frac{1}{10} \quad \longleftarrow$ *Dividing by 10 is the same as multiplying by $\frac{1}{10}$.*

$$= \frac{\overset{1}{\cancel{5}}}{16} \times \frac{1}{\underset{2}{\cancel{10}}}$$

$$x = \frac{1}{32}$$

Example 2 Write the quotient in lowest terms: $\dfrac{x}{5} \div x$.

Solution $\dfrac{x}{5} \div x = \dfrac{x}{5} \times \dfrac{1}{x}$ ⟵ *Dividing by x is the same as multiplying by $\dfrac{1}{x}$.*

$$\dfrac{\overset{1}{\cancel{x}}}{5} \times \dfrac{1}{\underset{1}{\cancel{x}}} = \dfrac{1}{5}$$

Your Turn Write the quotient in lowest terms.

- $\dfrac{4}{7} \div 4$
- $\dfrac{d}{6} \div d$
- $\dfrac{5}{m} \div 10$

Class Exercises

Write the reciprocal.

1. 3
2. 5
3. 6
4. 7
5. 8
6. 10

7. $\dfrac{1}{2}$
8. $\dfrac{3}{2}$
9. $\dfrac{4}{5}$
10. $\dfrac{4}{9}$
11. $\dfrac{7}{2}$
12. $\dfrac{14}{1}$

Exercises

Write the quotient in lowest terms.

A

1. $\dfrac{1}{2} \div 3$
2. $\dfrac{1}{3} \div 3$
3. $\dfrac{1}{4} \div 5$
4. $\dfrac{3}{4} \div 2$

5. $\dfrac{2}{3} \div \dfrac{1}{5}$
6. $\dfrac{1}{4} \div \dfrac{3}{8}$
7. $\dfrac{7}{2} \div \dfrac{2}{3}$
8. $\dfrac{3}{5} \div \dfrac{1}{5}$

9. $\dfrac{3}{8} \div \dfrac{3}{8}$
10. $\dfrac{2}{5} \div \dfrac{4}{3}$
11. $\dfrac{3}{5} \div \dfrac{6}{15}$
12. $\dfrac{4}{7} \div \dfrac{1}{14}$

B 13. $4\dfrac{1}{2} \div \dfrac{9}{x}$
14. $3\dfrac{1}{4} \div \dfrac{4}{b}$
15. $2\dfrac{1}{8} \div \dfrac{8}{c}$
16. $4\dfrac{1}{6} \div \dfrac{m}{5}$

17. $2\dfrac{1}{2} \div \dfrac{3}{2e}$ **18.** $3\dfrac{1}{3} \div \dfrac{7f}{3}$ **19.** $4\dfrac{1}{4} \div \dfrac{13}{4g}$ **20.** $2\dfrac{1}{5} \div \dfrac{5}{6h}$

21. $\dfrac{11}{10} \div \dfrac{7r}{2}$ **22.** $2\dfrac{2}{5} \div \dfrac{3}{4n}$ **23.** $2\dfrac{1}{16} \div \dfrac{9e}{4}$ **24.** $1\dfrac{5}{6} \div \dfrac{11}{3b}$

Write the given expression with a fraction as the coefficient of the variable. Write the coefficient in lowest terms if possible.

Example $2a \div 3 = \dfrac{2a}{3} = \dfrac{2 \cdot a}{3} = \dfrac{2}{3}a$

25. $4b \div 5$ **26.** $6x \div 7$ **27.** $3m \div 8$ **28.** $2b \div 6$

29. $3m \div 9$ **30.** $7e \div 14$ **31.** $8r \div 14$ **32.** $10p \div 30$

Problems

Write an expression for the phrase.

1. The quotient of 24 divided by x.
2. The quotient of some number divided by 16.

Solve.

3. At the science fair, four students plan to display their exhibits on a table $18\frac{1}{2}$ ft long. How long will each section be if the students divide the space equally among themselves?
4. Joan plans to stake 5 tomato plants in her garden. How tall will each stake be if she cuts a pole $21\frac{1}{4}$ ft long into 5 stakes of equal length?
5. How many small pillows can be made from a piece of fabric that measures $3\frac{1}{3}$ yd?
6. How many large pillows can be made from $6\frac{2}{3}$ yd of fabric?
7. How many yards of fabric do you need for three large pillows and four small ones?

Decorative Pillows
Material needed

$\frac{5}{6}$ yd (small pillow)

$1\frac{2}{3}$ yd (large pillow)

4-7 *Problem Solving*

Suppose that you are in charge of ordering ingredients for the "Potato Pancake Fest" your class is sponsoring. You know that 230 tickets have been sold. You also know that the recipe for 20 pancakes includes the ingredients listed below.

> **Potato Pancakes (makes 20)**
>
> | | |
> |---|---|
> | 6 large potatoes | salt & pepper to taste |
> | 1 onion | 2 T flour |
> | 2 eggs | 2½ T vegetable oil |

Using the information you know, you can work backward to determine the amount of each ingredient to buy. First, decide how many pancakes 230 people might eat. Let's say each person might eat 5.

$$230 \times 5 = 1150$$

Next, how many recipes of pancakes are needed to make 1150.

$$\begin{array}{l} \textit{pancakes needed} \longrightarrow \\ \textit{pancakes in 1 recipe} \longrightarrow \end{array} \frac{1150}{20} = 57\frac{1}{2}$$

You can now multiply the amount of ingredient in one recipe by the number of recipes to find the amount of each ingredient to buy.

Example 1 How many pounds of potatoes should you order if 6 large potatoes weigh about 2 lb?

Solution Use the information you know:
6 potatoes are needed for 1 recipe.
6 potatoes weigh about 2 lb.
$57\frac{1}{2}$ recipes are needed.

$$2 \times 57\frac{1}{2} = 115$$

So, you should order 115 lb of potatoes.

When you work backward to solve a problem, you often have to reverse the operation.

Example 2   Each year an automobile is worth about $\frac{4}{5}$ of what its value was the previous year. If an automobile is now worth $6000, what was its value last year?

Solution   You know that last year's value was multiplied by $\frac{4}{5}$ to get $6000.
Reverse the operation to find the value you want.

$$\$6000 \div \frac{4}{5} = \overset{1500}{\cancel{6000}} \times \frac{5}{\underset{1}{\cancel{4}}} = \$7500$$

Your Turn   • This year a machine is worth $\frac{3}{5}$ of what it was worth last year. If the machine is worth $1600 this year, what was it worth last year?

Class Exercises

Write a new exercise by reversing the operation, then solve the new exercise.

1. $200 \times \frac{5}{3}$

2. 48×16

3. $4\frac{1}{8} \div \frac{3}{5}$

4. $24p \times 20p$

5. $15y \times 1\frac{1}{4}$

6. $\frac{9}{11} \div 3\frac{2}{3}$

Exercises

Solve. Use the recipe for potato pancakes to answer questions 1–2.

A   **1.** How many cups of vegetable oil do you need for 10 recipes, if 1 c contains 16 T?

     2. If one carton contains 12 eggs, how many cartons of eggs do you need for 60 recipes?

B 3. Each year a bicycle is worth about $\frac{3}{4}$ of its value the previous year. If a bicycle is now worth $45, what was its value one year ago? Two years ago?

4. A highway construction crew can build $\frac{3}{8}$ mi of highway per day. How many working days should be allowed to meet the time deadline for constructing 12 mi of highway?

5. Suppose that you have a minibike that can travel $24\frac{1}{2}$ mi on 1 qt of gas. You have $2\frac{1}{4}$ qt of gas and want to make a 52-mi trip without buying more gas. Can you make the trip?

6. The price of a Super Sedan is now $\frac{9}{5}$ times what the price was last year. If the sedan now sells for $6700, what was last year's price?

7. The price tag on a car stereo is now $\frac{5}{8}$ of last week's price. If the price is now $230, what was last week's price?

8. A magazine club has 3125 subscribers this month. This is $\frac{1}{4}$ more than the number of subscribers last month. How many subscribers were there last month?

9. City Cab charges a flat fee of $2.05 plus $.15 for every $\frac{1}{6}$ mile. A City Cab driver charged Jane Norton a total of $4.75 for a trip from her home to the airport. How many miles is Jane's home to the airport?

10. An architect estimates that the average person in a dense crowd requires a fraction of a square yard of space. Based on this estimate she decided that a 4000 yd^2 shopping mall will hold 10,000 people. Find the fraction.

4-8 Fractions and Decimals

The label on the package of ground beef that Jason bought reads 0.75 lb. Did he buy the amount of ground beef that is shown on the list?

To find out, we can change $\frac{3}{4}$ to a decimal and then compare.

$$\frac{3}{4} = 3 \div 4$$

$\frac{3}{4}$ lb is equal to 0.75 lb.

Jason's Shopping List
$\frac{3}{4}$ lb ground beef
$\frac{1}{2}$ gal milk
1 lb carrots

We say that $\frac{3}{4}$ is equal to the **terminating decimal** 0.75. Sometimes when we divide, the division process does not end.

Example 1   Write $\frac{3}{11}$ as a decimal.

Solution $\frac{3}{11} = 3 \div 11 = 0.2727\ldots$

You can see that $\frac{3}{11}$ is equal to the **repeating decimal** 0.2727... We may write this decimal as $0.\overline{27}$. The bar indicates that 27 repeats.

A terminating decimal can be written as a fraction or mixed number. Such a fraction will have a power of 10 as its denominator.

Example 2   Write 0.352 as a fraction in lowest terms.

Solution $0.352 = \dfrac{352}{1000}$ ⟵ *Remember that 0.352 is $\dfrac{352}{1000}$.*

$= \dfrac{352 \div 8}{1000 \div 8}$ ⟵ *The GCF of 352 and 1000 is 8.*

$= \dfrac{44}{125}$

Your Turn Write as a fraction or as a decimal.

- 0.25   • $\dfrac{2}{3}$ • $\dfrac{5}{6}$

Class Exercises

Is the decimal repeating or terminating? Write *R* for repeating and *T* for terminating.

1. 0.4   **2.** 0.7   **3.** 0.10   **4.** $0.1\overline{2}$ **5.** $018\overline{6}$

6. 2.18   **7.** $4.27\overline{33}$ **8.** 0.005   **9.** $7.19\overline{93}$ **10.** $6.\overline{6}$

Exercises

Write the fraction as a terminating or repeating decimal.

A   **1.** $\dfrac{5}{2}$ **2.** $\dfrac{3}{20}$ **3.** $\dfrac{6}{25}$ **4.** $\dfrac{7}{11}$ **5.** $\dfrac{5}{8}$

   6. $\dfrac{4}{3}$ **7.** $\dfrac{5}{9}$ **8.** $\dfrac{7}{2}$ **9.** $\dfrac{5}{4}$ **10.** $\dfrac{3}{8}$

   11. $\dfrac{7}{8}$ **12.** $\dfrac{5}{16}$ **13.** $\dfrac{4}{20}$ **14.** $\dfrac{5}{12}$ **15.** $\dfrac{7}{15}$

Write the decimal as a fraction or mixed number in lowest terms.

16. 0.34   **17.** 0.96   **18.** 0.100   **19.** 0.220   **20.** 0.165

21. 0.37   **22.** 0.009   **23.** 2.7   **24.** 6.12   **25.** 1.125 *prime*

26. 1.025   **27.** 1.625   **28.** 2.215   **29.** 7.024   **30.** 4.666 *— fact*

Write a terminating or repeating decimal for the variable.

B   **31.** $\dfrac{32}{10} = a$ **32.** $\dfrac{21}{16} = b$ **33.** $\dfrac{19}{11} = c$ **34.** $\dfrac{53}{30} = d$ **35.** $\dfrac{41}{12} = e$

   36. $\dfrac{49}{25} = f$ **37.** $\dfrac{87}{20} = g$ **38.** $\dfrac{65}{40} = h$ **39.** $\dfrac{77}{50} = j$ **40.** $\dfrac{95}{45} = k$

If $x = 10$, write the fraction as a decimal.

   41. $\dfrac{x}{15}$ **42.** $\dfrac{x}{25}$ **43.** $\dfrac{x}{30}$ **44.** $\dfrac{x}{45}$ **45.** $\dfrac{x}{50}$

Using the Computer

This program will change a fraction to a decimal. You will need to input the numerator and denominator separately. RUN it for 11/16 and 11/15 and then try some examples of your own.

```
 5  REM: CHANGING A FRACTION TO A DECIMAL
10  PRINT "WHAT IS THE NUMERATOR";
20  INPUT N
30  PRINT "WHAT IS THE DENOMINATOR";
40  INPUT D
50  PRINT N;" /";D;" =";N/D
60  END
```

Self Test 2

Write the product in lowest terms.

1. $\dfrac{7}{13} \cdot p$

2. $10 \cdot \dfrac{x}{5}$

3. $4\dfrac{4}{5} \cdot \dfrac{d}{12}$

4-5

Write the quotient in lowest terms.

4. $\dfrac{5n}{6} \div \dfrac{18}{n}$

5. $\dfrac{12b}{11} \div 7\dfrac{1}{3}$

6. $\dfrac{8}{9}p \div 8p$

4-6

Work backward to find the answer.

7. A carton contains enough cereal mix to make $8\frac{1}{3}$ c of cooked cereal. The $8\frac{1}{3}$ c is enough for $12\frac{1}{2}$ servings. How many cups of cereal is one serving?

4-7

Write a terminating or repeating decimal for the variable.

8. $\dfrac{2}{33} = w$

9. $\dfrac{7}{40} = y$

10. $\dfrac{9}{16} = h$

4-8

Write a fraction or mixed number for the variable.

11. $b = 0.59$

12. $d = 0.010$

13. $f = 5.65$

126 *Self Test answers and Extra Practice are at the back of the book.*

Algebra Practice

Solve.

1. $12 = 12r$
2. $63 = 63 + q$
3. $3 + 5 \times 2$
4. $4 + 4 \times 4 \div 4$
5. $4 + 4 \div 4 - 4$
6. $4 \times 4 + 4 \div 4$

Use the distributive property to factor each expression.

7. $3b^3 + 6bc + 15b$
8. $(rs)^2 + 2rs^2 + 5r^3s$
9. $3jk^2 + (jk)^2 + 2j^2k$

Write the LCM of each set of expressions.

10. 12 and $9x$
11. $3a$, $4b$, and $6a^2$
12. m^2n, $5m$, and $2n$

Substitute. Write the fraction as a whole number or as a mixed number.

13. $\dfrac{c}{r}$ when $c = 28$ and $r = 14$
14. $\dfrac{y}{m}$ when $y = 13$ and $m = 4$

Write the fraction in lowest terms.

15. $\dfrac{18}{3b}$
16. $\dfrac{17h}{2h}$
17. $\dfrac{d^2e}{de^2}$
18. $\dfrac{15c^3}{10cf}$

Write an equation. Solve.

19. The difference between 4073 and 2981 is a.

20. 7680 increased by 1979 is equal to g.

21. The quotient of 9632 and 112 is u.

22. 7.18 times 6.82 is equal to n.

23. The sum of 86.28, 100.5, and 1.097 is t.

24. 2.3018 divided by 0.17 is equal to r.

25. Some number is the sum of $2\dfrac{1}{3}$ and $1\dfrac{1}{4}$.

26. How much greater is $\dfrac{11}{16}$ than $\dfrac{17}{32}$?

27. The product of $\dfrac{3}{4}$ and 18 is some number.

28. The quotient of $4\dfrac{3}{8}$ divided by $1\dfrac{3}{4}$ is some number.

29. $2\dfrac{2}{5}$ multiplied by $1\dfrac{7}{8}$ is some number.

30. $4\dfrac{1}{5}$ divided by $1\dfrac{1}{3}$ is some number.

Problem Solving for the Consumer

Investing in Stocks

A shareholder, or stockholder, is a person who buys one or more shares of stock in a corporation. A share of stock represents part ownership in a corporation. Stock prices are published each day in the business sections of daily newspapers. Prices are usually quoted in eighths of a dollar. Below is a listing of stocks.

Stock	Open	High	Low	Close	Chg	
Balloongram	$64\frac{5}{8}$	$65\frac{1}{2}$	$64\frac{3}{8}$	$64\frac{1}{2}$	$-\frac{1}{8}$	⟵ *The price of stock went down.*
NJK	$98\frac{1}{2}$	$100\frac{1}{8}$	98	$98\frac{1}{8}$	$-\frac{3}{8}$	
ECO	$25\frac{1}{8}$	$26\frac{3}{8}$	25	$25\frac{3}{4}$	$+\frac{5}{8}$	⟵ *The price of stock went up.*
Winter Ice	$30\frac{3}{8}$	$31\frac{1}{8}$	$30\frac{3}{8}$	$30\frac{3}{4}$	$+\frac{3}{8}$	
Fructose	$36\frac{3}{4}$	$36\frac{3}{4}$	$35\frac{1}{2}$	$35\frac{1}{2}$	$-1\frac{1}{4}$	
Tots	49	$50\frac{1}{2}$	$48\frac{3}{4}$	49	0	
Techtron	$36\frac{7}{8}$	39	$36\frac{7}{8}$	$38\frac{1}{8}$	$+1\frac{1}{4}$	
Reliable Motors	$64\frac{1}{2}$	$65\frac{1}{2}$	$64\frac{3}{8}$	$64\frac{1}{2}$	0	
Windpower Electric	$42\frac{3}{4}$	$44\frac{7}{8}$	$42\frac{3}{4}$	$43\frac{3}{4}$	$+1$	
Top Notch Equipment	$96\frac{3}{8}$	$101\frac{1}{4}$	$95\frac{5}{8}$	99	$+2\frac{5}{8}$	

Notice that Winter Ice's highest and lowest prices for the day were $31\frac{1}{8}$ and $30\frac{3}{8}$. The stock opened at $30\frac{3}{8}$ and closed at $30\frac{3}{4}$. The closing price was $\frac{3}{8}$ higher than the closing price on the previous day. If you owned 100 shares of Winter Ice's stock, the value of your investment would have increased $\frac{3}{8}$ per share that day.

$$\$\frac{3}{8} \cdot 100 = \frac{300}{8} = \$37.50$$

Use the stock price listing to give the answer.

1. Write the closing price of NJK.
2. Write the high price of Tots.
3. Write the low price of Reliable Motors.
4. Write the opening price of Top Notch Equipment.

5. Find the difference between the high and low prices of Reliable Motors.

6. If you owned 30 shares of Techtron, how much would your investment have increased between closing times?

7. If you owned 40 shares of ECO, what was the highest value of your investment during the day?

8. What would it have cost you to purchase 10 shares of Tots at its low price?

9. If you had $500 to buy stock, how many shares of Windpower Electric could you have purchased at its opening price?

10. Fructose stock opened at $36\frac{3}{4}$ and closed at $35\frac{1}{2}$. Did the value increase or decrease during that day? By how much did it increase or decrease?

When a stock splits "n for 1", the number of shares are multiplied by n, and the price is divided by n, so that the stockholders investment keeps its value.

Use the low price from the listing to find the value for the stock split.

Example Suppose a stockholder owns 10 shares of a stock at $60 per share. The value is $60 · 10, or $600. If the stock splits 3 for 1, the stockholder now owns 10 · 3, or 30 shares. The value is $\frac{\$60}{3}$ · 30, or $600.

11. ECO: 4 for 1
10 shares

12. Tots: 2 for 1
10 shares

13. NJK: $1\frac{1}{2}$ for 1
10 shares

Enrichment

Repeating Decimals as Fractions

You have seen that every terminating decimal can be written as a fraction. Repeating decimals can also be written as fractions. One method for finding a fraction for a repeating decimal is shown below. Study each step carefully.

Write $7.\overline{82}$ as a fraction in lowest terms.

Step 1: Write an equation using n to represent the fraction you want to find.

$n = 7.\overline{82}$ or $n = 7.82\overline{82}$ ⟵——— *82 is called the repetend because it repeats.*

Step 2: Since there are two digits in the repetend, we multiply both sides of the equation by 10^2 or 100.

$n \cdot 100 = 7.82\overline{82} \cdot 100$

$100n = 782.\overline{82}$ ⟵——— *The decimal point moves two places to the right.*

Step 3: Subtract the original equation from the new equation.

$$100n = 782.\overline{82}$$
$$-n = 7.\overline{82}$$
$$99n = 775$$

Step 4: Solve $99n = 775$ ⟵——— *Divide by 99 to solve for n.*

$n = \dfrac{775}{99}$ ⟵——— $\dfrac{775}{99}$ *is already in lowest terms.*

Check: Divide 775 by 99.

$$
\begin{array}{r}
7.82 \\
99\overline{)775.00} \\
693 \\
\hline
82\,0 \\
79\,2 \\
\hline
2\,80 \\
1\,98 \\
\hline
82 \\
\end{array}
$$

Write each repeating decimal as a fraction in lowest terms.

1. $0.\overline{3}$ 2. $0.\overline{9}$ 3. $1.\overline{2}$ 4. $3.\overline{4}$ 5. $2.\overline{35}$

6. $7.\overline{19}$ 7. $0.\overline{62}$ 8. $8.\overline{07}$ 9. $2.1\overline{35}$ 10. $4.7\overline{31}$

Chapter Review

Match.

1. a fraction that equals 1

2. $2\frac{1}{9}$ expressed as a fraction

3. $\frac{21}{10}$ expressed as a mixed number

4. $\frac{7}{25}$ expressed as a terminating decimal

5. 3.125 expressed as a mixed number

6. 0.875 expressed as a fraction

7. $\frac{5}{6}$ expressed as a repeating decimal

8. the reciprocal of $\frac{11}{21}$

a. $\frac{21}{11}$

b. $\frac{7}{8}$

c. $3\frac{1}{8}$

d. $0.8\overline{3}$

e. $2\frac{1}{10}$

f. $\frac{871}{871}$

g. 0.28

h. $\frac{19}{9}$

Choose the answer that renames the expression. Write a, b, or c.

9. $\frac{12}{4x}$ a. $\frac{1}{3x}$ b. $\frac{3}{x}$ c. $\frac{3}{4}$

10. $\frac{17df}{d^2f}$ a. $\frac{17}{d}$ b. $\frac{17}{f}$ c. $\frac{17}{df}$

11. $\frac{18em^2}{24em}$ a. $\frac{18}{24}$ b. $\frac{2m}{3}$ c. $\frac{3m}{4}$

Solve.

12. $\frac{3}{x} + \frac{8}{x}$

13. $\frac{c}{8} + \frac{3c}{8}$

14. $\frac{9x}{n} - \frac{5x}{n}$

15. $\frac{7r}{8} - \frac{5r}{12}$

16. $5\frac{1}{2} \cdot \frac{2}{y}$

17. $\frac{4b}{b} \cdot \frac{4c}{c}$

18. $\frac{3n}{8} \div \frac{n}{16}$

19. $\frac{m^2r}{r} \div mr$

Simplify the expression using the given value for the variable.

20. $g + 4\frac{5}{9};\ g = 2\frac{3}{4}$

21. $m + \frac{14}{15};\ m = 1\frac{2}{5}$

22. $k + 4\frac{1}{6};\ k = \frac{6}{7}$

23. $8\frac{3}{7} - p;\ p = 7\frac{1}{14}$

24. $4\frac{5}{12} - y;\ y = 1\frac{2}{9}$

25. $10\frac{1}{4} - t;\ t = \frac{9}{10}$

131

Chapter Test

Write a fraction for the variable.

1. $8\frac{2}{3} = m$ **2.** $9\frac{1}{5} = t$ **3.** $7\frac{3}{4} = a$ **4.** $6\frac{10}{11} = b$ **4-1**

Write a mixed number or a whole number for the variable.

5. $c = \frac{7}{3}$ **6.** $d = \frac{8}{5}$ **7.** $f = \frac{10}{2}$ **8.** $g = \frac{16}{1}$

Write the fraction in lowest terms.

9. $\frac{10x}{12xy}$ **10.** $\frac{7ab}{14b}$ **11.** $\frac{18cd^2}{16cd}$ **12.** $\frac{9f^2}{27f}$ **4-2**

Find the sum.

13. $\frac{4}{c} + \frac{6}{c} + \frac{1}{c}$ **14.** $\frac{2f}{5} + \frac{3f}{5} + \frac{f}{5}$ **15.** $\frac{0}{a} + \frac{1}{a} + \frac{7}{a}$ **4-3**

Simplify the expression using the given value for the variable.

16. $4\frac{5}{7} + y; \ y = 1\frac{3}{14}$ **17.** $p + 3\frac{4}{9}; \ p = 2\frac{1}{6}$

Write the difference in lowest terms.

18. $\frac{9m}{x} - \frac{7m}{x}$ **19.** $\frac{3b}{4} - \frac{2b}{8}$ **20.** $9\frac{5}{6} - 7\frac{4}{5}$ **4-4**

Write the product in lowest terms.

21. $15 \cdot \frac{b}{3}$ **22.** $\frac{x}{14} \cdot \frac{7}{2x}$ **23.** $\frac{t}{20} \cdot 2\frac{2}{9}$ **4-5**

Write the quotient in lowest terms.

24. $\frac{3r}{8} \div 2\frac{1}{2}$ **25.** $4\frac{1}{6} \div \frac{5}{2x}$ **26.** $\frac{4}{7}p \div 4p$ **4-6**

Work backward to find the answer.

27. Susan Chu can drive 304 mi on a full tank of gasoline. On the average, she can drive 22 mi on a gallon of gasoline. How many gallons of gasoline does the gas tank of Susan's car hold?   **4-7**

Write a terminating or repeating decimal for the variable.

28. $\frac{12}{25} = y$ **29.** $\frac{3}{16} = p$ **30.** $\frac{5}{11} = j$ **4-8**

Cumulative Review (Chapters 1–4)

Solve the equation.

1. $17.4 \div 10 = y$ **2.** $x = (4.82)(4)$ **3.** $12 + 18 - 12 = m$

4. $17.1 \div 3.8 = s$ **5.** $5^2 - 2^2 = r$ **6.** $(3.2 + 4)(1.8 - 1) = y$

7. $\dfrac{1}{3} + \dfrac{2}{3} = n$ **8.** $w = \dfrac{3}{8} \times \dfrac{2}{3}$ **9.** $\dfrac{1}{2} \times \dfrac{2}{5} = k$

10. $j = \dfrac{3}{4} \div \dfrac{3}{5}$ **11.** $1\dfrac{1}{2} + 6\dfrac{3}{4} = m$ **12.** $x = 12 \div \dfrac{6}{7}$

13. $15 \cdot \dfrac{1}{15} + \dfrac{2}{5} = n$ **14.** $\dfrac{4}{5} \cdot \dfrac{10}{13} = c$ **15.** $y = 1\dfrac{2}{3} \div 3$

16. $\dfrac{3}{8} \times \dfrac{8}{3} = s$ **17.** $16.3 + \dfrac{1}{3} - 16.3 = p$ **18.** $25 \cdot \dfrac{1}{25} + 50 = a$

Simplify.

19. $8y - 2x - 6y + 13x$ **20.** $9m - 6 + 7m - 7m$ **21.** $12 - 6p + 6p$

22. $x \div x$ **23.** $5 + 6n - 5$ **24.** $2x + 3y + z - x$

25. $4x(x + 1)$ **26.** $12r - 8 + 8$ **27.** $9(x - 2)$

28. $4 \cdot \dfrac{1}{r}$ **29.** $\dfrac{1}{8} \cdot d$ **30.** $s \cdot \dfrac{1}{4}$

31. $\dfrac{24x^2}{x}$ **32.** $\dfrac{12n^2}{3n}$ **33.** $\dfrac{5ab}{20a^2b}$

Solve.

34. The temperature at 9:00 A.M. was 68°.
The temperature at 2:00 P.M. went up to t degrees.
Then __?__ = the number of degrees the temperature went up.

35. The average family throws out 17 lb of trash per day.
Let $t =$ time in days.
Then __?__ = amount of trash it throws out in t days.

36. A dime is about 1 mm thick. If a stack of dimes is 3 cm high, about how many dimes are in the stack?

37. Every two years Jo has to pay to register her car in her state. Last year the charge was $14. The cost has been increasing $2 every year. At this rate, how much will Jo be charged the next time she registers? the time after that?

5
Solving Equations

Most American cookbooks use volume measure for the ingredients of recipes. Did you know, however, that European cooks and professional chefs often use balance scales to measure their ingredients?

5-1 Equations

A statement that two expressions are equal is called an **equation.** A set of scales in balance can be used to picture an equation.

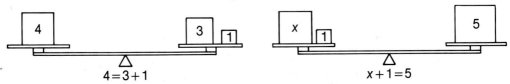

$$4 = 3 + 1 \qquad\qquad x + 1 = 5$$

Some equations, like $x + 1 = 5$, contain a **variable.** If you substitute 4 for x in the equation $x + 1 = 5$, the scales are in balance and the equation is true. The number 4 is called the **solution** of $x + 1 = 5$. We write $x = 4$.

Example 1 Is 3 a solution of $5 + y = 7$?

Solution Substitute 3 for y in the equation.

$$5 + y = 7$$

$$\begin{array}{c|c} 5 + 3 & 7 \\ 8 & 7 \end{array}$$

$8 \neq 7 \longleftarrow \neq$ *means is not equal to*

So 3 is not a solution and $y \neq 3$.

Example 2 Is 2 a solution of $5 + y = 7$?

Solution Substitute 2 for y in the equation.

$$5 + y = 7$$

$$\begin{array}{c|c} 5 + 2 & 7 \\ 7 & 7 \end{array}$$

$7 = 7 \; \checkmark$

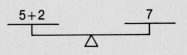

So 2 is a solution and $y = 2$.

Your Turn
- Is 2 a solution of $x + 8 = 10$?
- Is 10 a solution of $n - 5 = 7$?
- Is 3 a solution of $5x = 15$?

Example 3   Solve $c - 5 = 13$.

Solution   What number minus 5 equals 13?
$18 - 5 = 13$
So, $c = 18$.

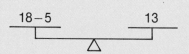

Example 4   Solve $12 = 2x$.

Solution   12 equals 2 times what number?
$12 = 2 \cdot 6$
So, $x = 6$.

Your Turn
- Solve $3h = 15$.
- Solve $10 = y - 4$.
- Solve $\frac{1}{2}z = 20$.

Class Exercises

Is 5 a solution of the equation?

1. $x + 1 = 6$ **2.** $y - 4 = 10$ **3.** $20 = 5z$ **4.** $5 - n = 3$

5. $2 + a = 3$ **6.** $3b = 15$ **7.** $\frac{1}{4}c = 20$ **8.** $p + 7 = 8$

Solve the equation.

9. $p - 2 = 8$ **10.** $4q = 20$ **11.** $3 + r = 9$ **12.** $q - 4 = 5$

13. $10u = 30$ **14.** $32 = 8v$ **15.** $\frac{w}{2} = 7$ **16.** $7s = 14$

Exercises

Is the given number a solution of the equation? Write *Yes* or *No*.

A **1.** $2a = 16$; 8   **2.** $b + 9 = 10$; 9   **3.** $g - 6 = 37$; 31

4. $81 = 3x$; 6   **5.** $39 = d - 12$; 27   **6.** $4h = 56$; 14

7. $4 + e = 28$; 32   **8.** $2f = 46$; 23   **9.** $3x = 21$; 24

Solve the equation.

10. $k + 9 = 15$ **11.** $m - 4 = 12$ **12.** $3n = 36$ **13.** $\frac{1}{2}p = 6$

14. $4 + q = 6$ **15.** $15 = r - 5$ **16.** $16 = 8s$ **17.** $u - 7 = 3$

Tell which of the numbers is a solution.

B **18.** $48 = x + 9$ 39 or 57?   **19.** $a - 7 = 75$ 82 or 68?

20. $124 = 4b$ 31 or 120?   **21.** $\frac{1}{2}c = 80$ 40 or 160?

22. $2f + 1 = 5$ 2 or 3?   **23.** $7 = 3g - 2$ 3 or 5?

24. $5(h + 2) = 30$ 4 or 8?   **25.** $2k + 8 = 12$ 1 or 2?

26. $5(q - 3) = 0$ 3 or 12?   **27.** $2r + 9 = 19$ 3 or 5?

Solve the equation.

28. $24 + a = 32$ **29.** $46 = 2b$ **30.** $c + 17 = 100$ **31.** $d - 30 = 66$

32. $e + 13 = 51$ **33.** $5f = 65$ **34.** $130 = 13g$ **35.** $\frac{h}{3} = 10$

36. $j + 25 = 52$ **37.** $16 - k = 9$ **38.** $64 = m - 14$ **39.** $r + 7 = 92$

These equations have fractions or decimals. Solve.

40. $\frac{1}{5}x = 20$ **41.** $y - 0.1 = 2.1$ **42.** $v + \frac{1}{2} = 39\frac{1}{2}$ **43.** $s + 1.3 = 2.9$

44. $4\frac{1}{4} = q - \frac{1}{2}$ **45.** $1 = 2p$ **46.** $2r = 1.6$ **47.** $\frac{2}{3}w = 2$

Write another equation that has the same solution.

C **48.** $a + 8 = 10$ **49.** $12 = \frac{1}{2}e$ **50.** $7c = 35$ **51.** $8 - d = 7$

5-2 *Using Subtraction*

Julie wants to weigh her puppy. When she holds the puppy and steps on the scale, the scale shows 54 lb. Julie knows that she weighs 48 lb. How much does the puppy weigh?

$$\text{Puppy's weight} + \text{Julie's weight} = 54 \text{ lb}$$
$$p \quad + \quad 48 \quad = 54$$

To solve an equation like $p + 48 = 54$, you can picture a two-pan balance scale.

If you remove 48 lb from each side, the scales remain in balance. Likewise, if you subtract 48 from each side of the equation, the equation remains true.

$$p + 48 - 48 = 54 - 48$$
$$p + 0 = 6$$
$$p = 6$$

Julie's puppy weighs 6 lb.

Example 1   Solve $a + 4 = 27$.

Solution
$$a + 4 = 27$$
$$a + 4 - 4 = 27 - 4 \longleftarrow \quad \textit{To "undo" addition of 4,}$$
$$a + 0 = 23 \qquad\qquad \textit{you subtract 4.}$$
$$a = 23$$

Example 2   Solve $7 + b = 37$. Check the solution.

Solution
$$7 + b = 37$$
$$7 + b - 7 = 37 - 7$$
$$b = 30$$

Check:
$$7 + b = 37$$

$7 + 30$	37
37	37
$37 = 37$	$\checkmark$

So $b = 30$.

Your Turn
- Solve $c + 9 = 49$. Check the solution.
- Solve $6 + d = 58$. Check the solution.
- Solve $18 = 9 + e$. Check the solution.

Class Exercises

Tell what number must be subtracted from each side of the equation to solve for x.

1. $x + 3 = 5$ **2.** $x + 7 = 19$ **3.** $x + 6 = 21$ **4.** $9 + x = 25$

5. $6 + x = 37$ **6.** $54 = x + 13$ **7.** $8 + x = 10$ **8.** $30 = 29 + x$

Exercises

Solve the equation. Check the solution.

A **1.** $a + 2 = 11$ **2.** $b + 5 = 9$ **3.** $7 + c = 12$ **4.** $d + 8 = 15$

5. $13 = 9 + e$ **6.** $f + 6 = 20$ **7.** $9 + g = 15$ **8.** $21 = h + 10$

9. $4 + j = 6$ **10.** $k + 3 = 9$ **11.** $6 + m = 7$ **12.** $4 + p = 10$

13. $5 = 1 + q$ **14.** $10 + r = 23$ **15.** $9 + s = 20$ **16.** $17 = 9 + v$

B **17.** $96 + a = 99$ **18.** $63 = 17 + b$ **19.** $c + 12 = 51$ **20.** $d + 7 = 104$

21. $e + 17 = 33$ **22.** $29 + f = 72$ **23.** $37 = 19 + g$ **24.** $8 + h = 126$

25. $j + 9 = 206$ **26.** $113 = 29 + k$ **27.** $852 = 9 + m$ **28.** $37 + n = 94$

29. $p + 17 = 36$ **30.** $12 + q = 41$ **31.** $104 + r = 117$ **32.** $300 = 52 + s$

These equations have fractions and decimals. Solve and check.

33. $8.8 = b + 3.3$ **34.** $3.1 + u = 9.3$ **35.** $9.9 + v = 10.0$

36. $7\frac{2}{3} = w + 3\frac{1}{3}$ **37.** $x + 4\frac{1}{2} = 9\frac{3}{4}$ **38.** $0.81 = 0.06 + y$

39. $c + \frac{1}{2} = 3\frac{1}{2}$ **40.** $a + 8.8 = 16.16$ **41.** $9.75 + a = 12.99$

C **42.** In the equation $x + a = b$, b is greater than a. State a rule you can use to solve the equation for x.

5-3 *Using Division*

A pair of running shoes weighs 800 g. How much does each shoe weigh?

Of course you can solve this problem by dividing 800 by 2. We can also show how to solve it by writing an equation with a variable.

Let s represent the weight of one shoe. Then $2s$ represents the weight of the pair. So $2s$ equals 800.

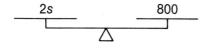

To solve the equation $2s = 800$, you can divide each side of the equation by 2. Then s equals 400. Each shoe weighs 400 g.

$$2s = 800$$
$$\frac{2s}{2} = \frac{800}{2}$$
$$s = 400$$

Some equations can be solved by dividing each side by the same non-zero number. Division by a non-zero number "undoes" multiplication by that number.

Example 1   Solve $24 = 4y$.

Solution
$$24 = 4y$$
$$\frac{24}{4} = \frac{4y}{4} \longleftarrow \quad \textit{To "undo" multiplication by 4, divide by 4.}$$
$$6 = 1y$$
$$6 = y$$

Example 2   Solve $2a = 9$. Check the solution.

Solution

$$2a = 9$$

$$\frac{2a}{2} = \frac{9}{2}$$

$$a = 4\frac{1}{2}$$

Check:

$$2a = 9$$

$2 \cdot 4\frac{1}{2}$	9
9	9

$$9 = 9 \quad \checkmark$$

Your Turn
- Solve $5b = 30$. Check the solution.
- Solve $27 = 3c$. Check the solution.
- Solve $9d = 152$. Check the solution.

How can you tell which operation to use to solve an equation? Remember to use the opposite operation.

Example 3   Solve $8d = 40$ and $e + 3 = 12$.

Solution   You must "undo" the operation performed on the variable.

$$8d = 40$$

To "undo" multiplication by 8, divide by 8.

$$\frac{8d}{8} = \frac{40}{8}$$

$$d = 5$$

$$e + 3 = 12$$

To "undo" addition of 3, subtract 3.

$$e + 3 - 3 = 12 - 3$$

$$e = 9$$

Your Turn
- To solve $5p = 20$, would you subtract 5 from each side, or would you divide each side by 5? Solve the equation.
- To solve $s + 10 = 16$ would you subtract 10 from each side, or would you divide each side by 10? Solve the equation.

Class Exercises

You are to solve for x. By what number would you divide each side of the equation?

1. $5x = 35$ **2.** $24 = 3x$ **3.** $7x = 14$ **4.** $4x = 48$

5. $27 = 9x$ **6.** $80 = 8x$ **7.** $13x = 13$ **8.** $30 = 2x$

Exercises

Use division to solve. Check the solution.

A 1. $2b = 6$ **2.** $18 = 9c$ **3.** $8d = 48$ **4.** $5e = 15$

  5. $72 = 8f$ **6.** $3g = 12$ **7.** $2h = 8$ **8.** $5j = 35$

  9. $3k = 21$ **10.** $20 = 4m$ **11.** $10n = 65$ **12.** $8p = 12$

Solve. Check the solution.

B 13. $9a = 270$ **14.** $186 = 6b$ **15.** $15c = 75$ **16.** $11d = 913$

  17. $244 = 4e$ **18.** $93 = 31f$ **19.** $14g = 420$ **20.** $9h = 972$

  21. $25j = 175$ **22.** $640 = 16k$ **23.** $960 = 30m$ **24.** $40n = 1600$

  25. $65p = 130$ **26.** $75q = 600$ **27.** $360 = 18r$ **28.** $19s = 76$

These equations have fractions or decimals. Solve and check.

29. $6a = 7.2$ **30.** $4.5 = 15w$ **31.** $0.6m = 120$ **32.** $1.2q = 36$

33. $1.8 = 0.2x$ **34.** $0.01y = 2$ **35.** $1.4d = 28$ **36.** $15 = 0.5r$

37. $\frac{1}{2}n = 8$ **38.** $\frac{4}{5} = 2e$ **39.** $\frac{1}{3}q = 4$ **40.** $18 = \frac{1}{4}s$

Decide which operation to use. Solve and check.

41. $10u = 400$ **42.** $v + 16 = 25$ **43.** $16 + s = 29$ **44.** $1100 = 20t$

45. $20x = 160$ **46.** $12 + y = 48$ **47.** $19 = j + 19$ **48.** $12k = 84$

Solve and check.

C 49. $\frac{2}{3}p = 165$ **50.** $\frac{3}{4}q = 180$ **51.** $\frac{1}{2}x = 4\frac{1}{2}$

  52. $3p + p = 16$ **53.** $4d - 3 + 2d = 5$ **54.** $9d - 3d = 30 + d$

5-4 *Using Addition*

To solve an equation like $x + 3 = 8$, we subtracted 3 from each side of the equation. Subtracting 3 is the opposite of adding 3.

$$\underline{x+3-3} \qquad \underline{8-3}$$

$$x + 3 = 8$$
$$x + 3 - 3 = 8 - 3$$
$$x + 0 = 5$$
$$x = 5$$

In the equation $y - 7 = 9$, the number 7 is subtracted from y. To "undo" the subtraction of 7, we add 7 to each side of the equation.

$$\underline{y-7+7} \qquad \underline{9+7}$$

$$y - 7 = 9$$
$$y - 7 + 7 = 9 + 7$$
$$y + 0 = 16$$
$$y = 16$$

Example Solve $a - 8 = 37$. Check the solution.

Solution
$$a - 8 = 37$$
$$a - 8 + 8 = 37 + 8$$
$$a = 45$$

Check: $a - 8 = 37$

$45 - 8$	37
37	37

$37 = 37$ ✓

Your Turn • Solve $a - 4 = 22$. Check the solution.
• Solve $29 = c - 7$. Check the solution.

Class Exercises

You are to solve for x. Tell what number you would add to each side of the equation.

1. $x - 4 = 6$ **2.** $12 = x - 10$ **3.** $2 = x - 9$ **4.** $x - 8 = 72$

5. $14 = x - 19$ **6.** $100 = x - 32$ **7.** $x - 3 = 5$ **8.** $x - 1 = 8$

Exercises

You are to solve the equation for *n*. What number must you add to each side of the equation?

A **1.** $n - 6 = 7$ **2.** $n - 3 = 7$ **3.** $12 = n - 5$ **4.** $n - 10 = 7$

  5. $6 = n - 14$ **6.** $n - 6 = 14$ **7.** $n - 18 = 2$ **8.** $3 = n - 5$

B **9–16.** Solve the equations in Exercises 1–8.

Solve and check.

17. $a - 7 = 39$ **18.** $84 = b - 19$ **19.** $c - 201 = 10$ **20.** $d - 13 = 156$

21. $13 = e - 47$ **22.** $117 = f - 39$ **23.** $g - 32 = 68$ **24.** $79 = h - 57$

25. $112 = j - 14$ **26.** $k - 76 = 52$ **27.** $m - 61 = 80$ **28.** $p - 17 = 118$

These equations have fractions or decimals. Solve and check.

29. $\frac{2}{3} = a - \frac{1}{3}$ **30.** $b - \frac{1}{2} = 1\frac{1}{2}$ **31.** $5 = c - \frac{1}{4}$ **32.** $d - \frac{1}{10} = \frac{3}{10}$

33. $e - 1.2 = 3.8$ **34.** $f - 1 = 3.6$ **35.** $4.5 = g - 0.5$ **36.** $0.01 = h - 1.10$

Solve and check.

37. $4p = 64$ **38.** $5 + q = 82$ **39.** $71 = r - 3$ **40.** $17h = 51$

41. $8 + v = 102$ **42.** $91 = w - 16$ **43.** $x - 21 = 52$ **44.** $a - 9 = 73$

45. $b - 3.9 = 7.1$ **46.** $c + 8.5 = 10$ **47.** $1.2d = 0.84$ **48.** $84 = 0.7h$

Problems

Write an expression for the number described.

  1. 5 times a number *n*   **2.** 3 less than the number *p*

  3. 5 more than the number *x*   **4.** a third of the number *q*

Solve.

  5. Pat delivered 16 papers more than Mike. Pat delivered 71 papers. In the equation $m + 16 = 71$, *m* represents the number of papers that Mike delivered. Solve the equation for *m*.

5-5 *Using Multiplication*

Nurses must record the pulse rate of patients often. Instead of counting the number of heartbeats in a minute, often they count the number of beats in 15 seconds, which is $\frac{1}{4}$ of a minute. Suppose a patient's heart beats 18 times in $\frac{1}{4}$ of a minute. How many times does it beat in 1 minute?

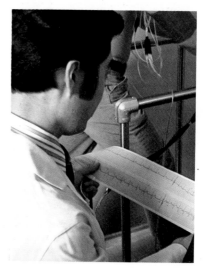

Let h = number of heartbeats in 1 minute.

Then $\frac{h}{4}$ = number of heartbeats in $\frac{1}{4}$ minute.

To solve an equation like $\frac{h}{4} = 18$, we can multiply each side of the equation by 4.

$$\frac{h}{4} \cdot 4 = 18 \cdot 4$$

$$h = 72$$

The patient's heart beats 72 times per minute.

Example 1 Solve $\frac{y}{3} = 9$.

Solution $\quad \dfrac{y}{3} = 9$

$\qquad 3 \cdot \dfrac{y}{3} = 9 \cdot 3 \longleftarrow$ *To "undo" division by 3,*

$\qquad\qquad y = 27 \qquad\qquad$ *multiply by 3.*

To "undo" multiplication by a fraction, you can multiply by the reciprocal of the fraction.

Example 2 Solve $\frac{2}{3}x = 16$. Check the solution.

Solution

$$\frac{2}{3}x = 16$$

$$\frac{3}{2} \cdot \frac{2}{3}x = 16 \cdot \frac{3}{2} \longleftarrow$$

$$x = 24$$

To "undo" multiplication by $\frac{2}{3}$, multiply by its reciprocal, $\frac{3}{2}$.

Check: $\frac{2}{3}x = 16$

$$\begin{array}{c|c} \frac{2}{3} \cdot 24 & 16 \\ \hline 16 & 16 \\ 16 = 16 & \checkmark \end{array}$$

Your Turn
- Solve $\frac{w}{5} = 7$. Check the solution.
- Solve $\frac{3}{4}v = 15$. Check the solution.
- Solve $10 = \frac{2}{3}n$. Check the solution.

Class Exercises

You are to solve each equation for *y*. Tell by what number you should multiply each side of the equation.

1. $\frac{y}{4} = 9$ 2. $\frac{y}{8} = 6$ 3. $3 = \frac{y}{5}$ 4. $\frac{y}{10} = 4$

5. $\frac{1}{2}y = 7$ 6. $10 = \frac{1}{3}y$ 7. $\frac{2}{3}y = 8$ 8. $\frac{3}{4}y = 12$

Exercises

You are to solve for *n*. By what number should you multiply each side of the equation?

A **1.** $\dfrac{n}{2} = 4$ **2.** $\dfrac{n}{3} = 5$ **3.** $6 = \dfrac{n}{4}$ **4.** $\dfrac{n}{5} = 7$

5. $\dfrac{1}{2}n = 5$ **6.** $6 = \dfrac{1}{3}n$ **7.** $\dfrac{1}{4}n = 7$ **8.** $\dfrac{1}{5}n = 8$

B **9–16.** Solve the equations in Exercises 1–8.

Solve each equation. Check the solution.

17. $\dfrac{x}{4} = 6$ **18.** $\dfrac{2}{5}y = 20$ **19.** $60 = \dfrac{a}{5}$ **20.** $7 = \dfrac{b}{4}$

21. $\dfrac{5}{8}c = 10$ **22.** $\dfrac{3}{5}d = 6$ **23.** $80 = \dfrac{2}{3}e$ **24.** $16 = \dfrac{f}{5}$

25. $\dfrac{2}{3}g = 24$ **26.** $48 = \dfrac{h}{5}$ **27.** $\dfrac{k}{4} = 19$ **28.** $\dfrac{1}{5}m = 32$

These equations have fractions or decimals. Solve and check.

29. $\dfrac{3}{5}d = 10.5$ **30.** $\dfrac{e}{4} = 1.2$ **31.** $3.7 = \dfrac{p}{2}$ **32.** $0.02 = \dfrac{2}{3}q$

33. $\dfrac{5}{6}g = 2.5$ **34.** $\dfrac{1}{2}h = \dfrac{5}{8}$ **35.** $1.7 = \dfrac{1}{3}r$ **36.** $\dfrac{5}{9} = \dfrac{1}{3}s$

Decide which operation to use. Solve and check.

37. $\dfrac{z}{5} = 12$ **38.** $y - 3 = 61$ **39.** $19 = \dfrac{1}{2}x$ **40.** $105 = 21d$

41. $w + 19 = 81$ **42.** $82 = v + 21$ **43.** $\dfrac{2}{3}u = 120$ **44.** $76 = e - 15$

45. $100 = a - 59$ **46.** $36 + b = 172$ **47.** $15c = 180$ **48.** $19q = 76$

Solve and check.

Example $\dfrac{250}{n} = 5$ ⟹ $\dfrac{250}{n} \cdot n = 5 \cdot n$ ⟹ $\dfrac{250}{5} = \dfrac{5 \cdot n}{5}$ ⟹ $50 = n$

C **49.** $\dfrac{18}{a} = 6$ **50.** $\dfrac{32}{b} = 8$ **51.** $\dfrac{75}{c} = 25$ **52.** $\dfrac{96}{d} = 2$

Using the Computer

This program will solve equations of the form $Ax + B = C$ and $Ax - B = C$. RUN it to solve $3x + 5 = 17$ and $5x - 3 = 27$. Now try some examples of your own.

```
10   REM: SOLVING SIMPLE EQUATIONS
20   PRINT "TO SOLVE AN EQUATION"
30   PRINT "    AX @ B = C"
40   PRINT "WHERE @ CAN BE + OR -:"
50   PRINT "WHAT SIGN DO YOU WISH FOR @"
60   PRINT "(TYPE + OR -)";
70   INPUT S$
80   IF S$="+" THEN 110
90   IF S$="-" THEN 110
100  GOTO 60
110  PRINT "WHAT IS A";
120  INPUT A
130  PRINT "WHAT IS B";
140  INPUT B
150  PRINT "WHAT IS C";
160  INPUT C
170  IF S$="-" THEN 200
180  PRINT "X =";(C-B)/A
190  GOTO 210
200  PRINT "X =";(C+B)/A
210  END
```

Self Test 1

Answer *Yes* or *No*.

1. Is 26 a solution of $x + 27 = 53$?   **5-1**

2. Is 36 a solution of $73 = 2y + 1$?

Solve the equation.

3. $a + 13 = 71$ 4. $32 = b + 7$ 5. $51 + c = 69$ **5-2**

6. $7d = 91$ 7. $400 = 8e$ 8. $3f = 693$ **5-3**

9. $g - 17 = 74$ 10. $102 = h - 21$ 11. $j - 43 = 12$ **5-4**

12. $\dfrac{k}{5} = 105$ 13. $\dfrac{1}{3}m = 13$ 14. $\dfrac{2}{5}p = 16$ **5-5**

5-6 *Using Two Steps*

Solving some equations may require more than one operation.

Example 1   Solve $2a + 6 = 14$. Check the solution.

Solution   First, subtract 6 from each side of the equation. That leaves only one term with the variable on the left of the "=" sign.

$$2a + 6 - 6 = 14 - 6$$
$$2a + 0 = 8$$
$$2a = 8$$

$$\underline{2a+6-6} \qquad \underline{14-6}$$

Then solve $2a = 8$ by dividing each side of the equation by 2.

$$2a = 8$$
$$\frac{2a}{2} = \frac{8}{2}$$
$$a = 4$$

$$\frac{2a}{2} \qquad \frac{8}{2}$$

Check:
$$2a + 6 = 14$$

$2 \cdot 4 + 6$	14
$8 + 6$	14
14	14
$14 = 14$	✓

Example 2   Solve $3b - 8 = 13$.

Solution
$$3b - 8 = 13$$
$$3b - 8 + 8 = 13 + 8 \quad \longleftarrow \quad \textit{Add 8 to each side.}$$
$$3b = 21$$
$$\frac{3b}{3} = \frac{21}{3} \quad \longleftarrow \quad \textit{Divide each side by 3.}$$
$$b = 7$$

149

Example 3 Solve $\frac{1}{2}c - 7 = 2$.

Solution

$$\frac{1}{2}c - 7 = 2$$

$$\frac{1}{2}c - 7 + 7 = 2 + 7 \quad \longleftarrow \quad \textit{Add 7 to each side.}$$

$$\frac{1}{2}c = 9$$

$$2 \cdot \frac{1}{2}c = 2 \cdot 9 \quad \longleftarrow \quad \textit{Multiply each side by 2.}$$

$$c = 18$$

Your Turn
- Solve $6 + 4d = 30$. Check the solution.
- Solve $3x - 2 = 13$. Check the solution.
- Solve $\frac{1}{3}e - 4 = 9$. Check the solution.

Some equations require you to use the distributive property to solve. Recall that the property states the following: $a(b + c) = ab + ac$.

Example 4 Solve $2(x + 3) = 18$.

Solution

$$2(x + 3) = 18 \quad \longleftarrow \quad \textit{Recall the distributive property.}$$
$$2x + 6 = 18$$
$$2x + 6 - 6 = 18 - 6 \quad \longleftarrow \quad \textit{Subtract 6 from each side.}$$
$$2x = 12$$
$$\frac{2x}{2} = \frac{12}{2} \quad \longleftarrow \quad \textit{Divide each side by 2.}$$
$$x = 6$$

Your Turn
- Solve $3(x - 1) = 21$. Check the solution.
- Solve $30 = 5(x + 2)$. Check the solution.

Class Exercises

Tell what steps you would take to solve the equation.

Example $5f + 9 = 54$

Solution   Subtract 9 from each side. Then divide each side by 5.

1. $3 + 2g = 11$ **2.** $4h - 3 = 21$ **3.** $\frac{1}{2}j - 3 = 4$ **4.** $2k - 7 = 19$

5. $\frac{1}{3}m - 4 = 9$ **6.** $7n - 13 = 22$ **7.** $8p + 9 = 65$ **8.** $3 + \frac{1}{4}q = 9$

9. $3q - 6 = 12$ **10.** $5r - 1 = 29$ **11.** $\frac{1}{3}s + 4 = 13$ **12.** $\frac{1}{4}t - 3 = 2$

Exercises

Follow the steps to solve the equation. Remember, the same operation must be performed on each side of the equation.

A   **1.** $2a + 4 = 16$ **2.** $3b - 5 = 28$ **3.** $\frac{x}{2} + 10 = 13$ **4.** $6c - 2 = 22$

Subtract 4.   Add 5.   Subtract 10.   Add 2.
Divide by 2.   Divide by 3.   Multiply by 2.   Divide by 6.

5. $5 + \frac{3}{4}d = 20$ **6.** $\frac{4}{5}e - 2 = 30$ **7.** $8f - 9 = 31$ **8.** $\frac{2}{3}g + 8 = 22$

Subtract 5.   Add 2.   Add 9.   Subtract 8.

Multiply by $\frac{4}{3}$.   Multiply by $\frac{5}{4}$.   Divide by 8.   Multiply by $\frac{3}{2}$.

Solve. Check the solution.

B   **9.** $\frac{1}{2}u - 9 = 9$ **10.** $4v + 3 = 15$ **11.** $7w - 9 = 19$ **12.** $9 + \frac{1}{2}x = 19$

13. $\frac{1}{6}y - 2 = 1$ **14.** $9 + 4n = 25$ **15.** $2a + 5 = 15$ **16.** $3b - 8 = 19$

17. $5 + 4c = 25$ **18.** $\frac{1}{4}d + 9 = 25$ **19.** $1 + 3e = 10$ **20.** $5f - 9 = 51$

21. $\frac{1}{4}g - 3 = 1$ **22.** $3h - 2 = 1$ **23.** $4j - 8 = 0$ **24.** $\frac{1}{3}k - 2 = 0$

25. $6 + 5m = 51$ **26.** $6n + 3 = 27$ **27.** $9p - 9 = 0$ **28.** $7q - 7 = 7$

29. $\frac{1}{2}r - 1 = 1$ **30.** $6s + 8 = 26$ **31.** $9 + 2n = 57$ **32.** $9 + 9u = 99$

33. $4 = \frac{1}{4}v - 4$ **34.** $1 = 3w + 1$ **35.** $2 = 5x + 2$ **36.** $24 = 2y + 22$

Solve the equation. Check the solution. Remember to use the distributive property.

37. $2(a - 2) = 6$ **38.** $5(b + 1) = 45$ **39.** $12 = 6(c - 7)$

40. $30 = 3(d + 8)$ **41.** $4(e + 3) = 32$ **42.** $10 = 2(f - 4)$

43. $42 = 7(g - 1)$ **44.** $9(h + 5) = 72$ **45.** $8(k + 5) = 96$

46. $3(x - 1) + 4 = 13$ **47.** $2(y + 1) - 3 = 9$ **48.** $5(z - 5) + 2 = 7$

49. $12 + 2(3p - 4) = 14$ **50.** $52 = 4(2q + 5) + 8$ **51.** $3(2r - 6) + 6 = 11$

Is the given number a solution? Write *Yes* or *No*.

Example $5x + 6 > 21$; 2   Check: $5x + 6 > 21$

$5(2) + 6$	21
$10 + 5$	21
16	21

No. 16 is not greater than 21, so 2 is not a solution.

52. $2a - 5 > 7$; 4   **53.** $3 + 5n > 24$; 5   **54.** $3x - 10 < 35$; 5

55. $2y + 3y < 18$; 4   **56.** $6(x + 2) < 20$; 5   **57.** $3(d + 25) < 100$; 2

Challenge

Alice has accepted a special 15-day assignment with ABC Corporation. She may choose to be paid according to one of two plans.

If Alice chooses Plan A, she will be paid $25,000 when the assignment is completed. If Alice chooses Plan B, she will be paid $1 on the first day. Each day thereafter she will be paid an amount equal to 2 times the amount she received the day before. Which plan should Alice choose?

5-7 *Equations with More Steps*

Some equations have the variable on both sides. To solve, you need to get the variable alone on one side of the equation.

Example 1   Solve $3a + 6 = 2a + 14$.

Solution   First, let's subtract $2a$ from each side of the equation. That will result in an equation having the variable on one side only.

$$3a - 2a + 6 = 2a - 2a + 14$$
$$a + 6 = 14$$

The variable is now on one side. Now, subtract 6 from each side.

$$a + 6 - 6 = 14 - 6$$
$$a = 8$$

In the example above, we first subtracted $2a$ from each side. Then we subtracted 6. You could solve the same equation by subtracting 6 from each side first and then subtracting $2a$ from each side. You would get the same answer.

Example 2   Solve $9b + 4 = 6b + 22$.

Solution

$$9b + 4 = 6b + 22$$
$$9b - 6b + 4 = 6b - 6b + 22 \longleftarrow \text{ \textit{Subtract 6b from each side.}}$$
$$3b + 4 = 22 \longleftarrow \text{ \textit{The variable is on one side.}}$$
$$3b + 4 - 4 = 22 - 4 \longleftarrow \text{ \textit{Subtract 4 from each side.}}$$
$$3b = 18$$
$$\frac{3b}{3} = \frac{18}{3} \longleftarrow \text{ \textit{Divide each side by 3.}}$$
$$b = 6$$

Your Turn
- Solve $2d = d + 3$. Check the solution.
- Solve $5f - 3 = 2f + 15$. Check the solution.

Class Exercises

Tell what steps you would take to solve the equation.

1. $5h = 12 - h$ **2.** $4k = 2k + 4$ **3.** $5m = 3m + 6$ **4.** $5x = 8 - 3x$

5. $8n = 2n + 6$ **6.** $5p = 4p + 4$ **7.** $2r = 5 - 3r$ **8.** $7y = 3y + 12$

Exercises

Follow the steps to solve the equation. Remember to perform the same operation on each side of the equation.

A
1. $5e = 3e + 8$
Subtract $3e$.
Divide by 2.

2. $4f = 10 - f$
Add f.
Divide by 5.

3. $3g = 12 - 3g$
Add $3g$.
Divide by 6.

4. $h = 9 - 2h$
Add $2h$.
Divide by 3.

Solve each equation. Check the solution.

B   **5.** $2a = a + 11$ **6.** $3b = 10 - b$ **7.** $4c = 2c + 8$ **8.** $8n = 6n + 4$

9. $3d = 2d + 9$ **10.** $5e = 2e + 12$ **11.** $7f = 18 - 3f$ **12.** $9g = 5g + 4$

13. $7h = 5h + 8$ **14.** $4j = 18 - 2j$ **15.** $6k = 9 - 3k$ **16.** $7m = 5m + 6$

17. $3t - 12 = 2t + 4$ **18.** $8u + 1 = 7u + 1$ **19.** $12v + 5 = 11v + 7$

20. $4w - 7 = 3w + 1$ **21.** $6x + 4 = 5x + 10$ **22.** $2y - 2 = y + 8$

23. $5z + 4 = 4z + 11$ **24.** $7s - 3 = 6s + 2$ **25.** $32t - 8 = 31t + 17$

26. $7g + 2 = 2g + 12$ **27.** $9h + 1 = 4h + 11$ **28.** $5k - 1 = 2k + 14$

29. $12m - 9 = 5m + 5$ **30.** $14n - 8 = 2n + 16$ **31.** $5p + 9 = p + 25$

32. $14q - 4 = 10q + 20$ **33.** $8r + 5 = 6r + 7$ **34.** $8s - 1 = 6s + 15$

35. $4t + 4 = 2t + 6$ **36.** $7u - 3 = 3u + 5$ **37.** $10v + 8 = 3v + 15$

38. $5w + 8 = 3w + 26$ **39.** $18 + 3x = x + 40$ **40.** $7 + 4y = 16 - 5y$

C   **41.** $5(a + 2) = 30 + a$ **42.** $b + 2(b + 1) = 7 - 2b$ **43.** $3(d - 4) + d = 2d$

5-8 *Writing Algebraic Expressions*

Algebra is a powerful tool. To develop skill in using algebra you will need more practice in writing expressions. The first step in writing an expression is to decide what the variable will represent.

Example 1 The Panthers won 7 more games than the Cougars. Write an algebraic expression to show the number of games won by the Panthers.

Solution Let c = number of games won by Cougars.
Panthers won 7 games more than Cougars.

$$7 \qquad + \qquad c$$

So, $7 + c$ = number of games won by Panthers.

Example 2 Susan is twice as tall as her brother John.
Paul is 5 in. taller than Susan.
Write an expression for the height of each person.

Solution First, decide what the variable will represent.
Let b = John's height in inches.
Then $2b$ = Susan's height in inches. ⟵ *2 times John's height*
And $2b + 5$ = Paul's height in inches. ⟵ *5 in. taller than Susan*

Your Turn Complete.
● Barney ran three times as far as Sal.
Ann ran 1 km farther than Barney.
Let s = number of kilometers that Sal ran.
Then __?__ = number of kilometers that Barney ran.
And __?__ = number of kilometers that Ann ran.

155

Using a pattern may help you write some expressions.

Example 3 Write expressions for three consecutive whole numbers.

Solution Think of examples of three consecutive whole numbers. Then look for a pattern.

First	Second	Third
3	4	5
17	18	19
216	217	218

Pattern: *Add 1.*   *Add 1 again.*

Let x = first of three whole numbers. Use the pattern.

First	Second	Third
x	$x + 1$	$(x + 1) + 1$ or $x + 2$

Add 1.   *Add 1 again.*

Your Turn
- Let $2x$ = an even number.
  Write an expression for the next even number.
- Let n be a number.
  Write an expression for twice the number.
  Write an expression for 1 more than twice the number.

Class Exercises

Lou is 5 years older than Sam. Complete the chart.

	If Sam's age is . . . ,	then Lou's age is . . .
1.	2	2 + 5, or __?__
2.	13	13 + __?__ or __?__
3.	a	__?__ + 5
4.	$2b$	__?__ + __?__

Complete the expression.

5. a. 1 less than 3 is 3 − _?_ ,
   or _?_
   b. 1 less than r is $r -$ _?_
   c. 1 less than $2r$ is _?_ − 1
   d. 1 less than 3 times n is
   _?_ − _?_

6. Let $2x$ be an even number.
   a. the next odd number is _?_
   b. the next even number is _?_
   c. 3 more than $2x$ is $2x +$ _?_
   d. 6 less than $2x$ is _?_ − _?_

Exercises

Choose the expression at the right that represents the number described.

A
1. 5 less than a number x
2. a number x decreased by 3
3. 5 times the number x
4. 3 more than x
5. the number x increased by 2
6. the number x doubled
7. the number x divided by 3
8. one-half of x
9. 3 more than 2 times x
10. 3 less than 4 times x

a. $x + 3$
b. $2x + 3$
c. $x + 2$
d. $x - 5$
e. $2x$
f. $4x - 3$
g. $\dfrac{x}{3}$
h. $x - 3$
j. $\dfrac{1}{2}x$
k. $5x$

Complete by writing an algebraic expression for each phrase.

B 11. a. 1 more than a number n is _?_
   b. 1 more than a number $2x$ is _?_
   c. 1 more than 3 times the number p is _?_
   d. 1 more than twice a number r is _?_
12. a. the sum of 2 and n is _?_
   b. the sum of 2 and $3y$ is _?_
   c. the sum of x and $3a$ is _?_
   d. the sum of x and 4 times h is _?_

157

13. **a.** 10 less than 50 is __?__
    b. 10 less than m is __?__
    c. 10 less than $2x$ is __?__
14. Let $2x$ be an even number.
    a. The next consecutive number is __?__
    b. The next consecutive even number is __?__
15. Paula is 3 in. taller than Sue.
    Let $s = $ Sue's height in inches.
    a. Paula's height $= $ __?__ in.
    b. Jim is 1 in. shorter than Paula. Jim's height $= $ __?__ in.
16. Carl is 6 years older than Jerry.
    Let $j = $ Jerry's age.
    a. Carl's age $= $ __?__
    b. Betty is 2 years older than Carl. Betty's age $= $ __?__
17. Sara ran half as far as Al.
    Let $a = $ number of km that Al ran.
    a. Sara ran __?__ km.
    b. Leo ran 3 km farther than Sara. Leo ran __?__ km.
18. Ned had twice as many strikes as Jo.
    Let $j = $ number of strikes by Jo.
    a. Number of strikes by Ned $= $ __?__
    b. Kim had 2 more strikes than Ned. Number of strikes by
    Kim $= $ __?__
19. Ken has twice as much money as Cora.
    Let $c = $ amount of money that Cora has.
    a. Amount of money that Ken has $= $ __?__.
    b. Max has $5 more than Ken. Amount of money that Max
    has $= $ __?__.

**To write an expression for these problems you will have to supply some
information. For example, 1 week has 7 days. Complete.**

C **20.** Let $c = $ number of weeks Paul will be at camp.
    Then __?__ $= $ number of days Paul will be at camp.
21. Let $d = $ number of dimes in a parking meter.
    Then __?__ $= $ the value of the dimes in dollars.
22. Let $w = $ number of words you can type in 1 hour.
    Then __?__ $= $ number of words you can type in 1 min.
23. Let $h = $ height of a rectangle in millimeters.
    Then __?__ $= $ height of the rectangle in centimeters.

5-9 *Writing Equations*

To write an equation you need to write two equal expressions. One of the expressions may be just a number.

Example 1
The Rams won 1 game more than the Trojans. If the Rams won 6 games, how many games did the Trojans win?

Solution
Let t = games won by Trojans.
Then $t + 1$ = games won by Rams.

Games won by Rams = Games won by Rams

$$t + 1 = 6 \longleftarrow \text{\textit{Write two equal expressions.}}$$
$$t + 1 - 1 = 6 - 1 \longleftarrow \text{\textit{Solve the equation.}}$$
$$t = 5$$

The Trojans won 5 games.

Example 2
An apple has about 70 calories. That is 2 times the number of calories in a peach. How many calories are in a peach?

Solution
Let p = calories in a peach.
Then $2p$ = calories in an apple.

Calories in an apple = Calories in an apple

$$2p = 70 \longleftarrow \text{\textit{Write two equal expressions.}}$$
$$\frac{2p}{2} = \frac{70}{2} \longleftarrow \text{\textit{Solve the equation.}}$$
$$p = 35$$

A peach has 35 calories.

Your Turn   • Josie is 2 in. shorter than Ken.
Josie is 58 in. tall.
How tall is Ken?

Class Exercises

Complete the steps to solve the problem.

Problem: Linda's bowling score is 50 points more than Al's.
Linda's score is 83.
What is Al's score?

1. Let a = Al's score. Then, Linda's score = $a +$ __?__.
2. The problem says that Linda's score = __?__.
3. State two equal expressions for Linda's bowling score.
4. An equation for the problem is $a +$ __?__ = __?__.

Problem: Sal has half as much money as Karen.
Sal has $16.
How much money does Karen have?

5. Let k = number of dollars that Karen has. Then, Sal has __?__ $\cdot k$ dollars.
6. The problem states that Sal has __?__ dollars.
7. State two equal expressions for the number of dollars that Sal has.
8. An equation for the problem is __?__ $\cdot k =$ __?__.

Exercises

Choose the equation that matches the problem. Write a, b, c, or d. Then solve the problem.

a. $m + 8 = 24$ **b.** $m - 8 = 24$ **c.** $8m = 24$ **d.** $\dfrac{m}{8} = 24$

A **1.** Vera is 8 years older than Marco.
Vera is 24 years old.
How old is Marco?
(Let m = Marco's age.)
  2. Kim has $8 less than Maria.
Kim has $24.
How much money does Maria have?
(Let m = money Maria has.)
  3. Mrs. Molloy divided a roll of twine evenly among 8 students in Home Ec class. Each student received a piece of twine 24 in. long. How many inches of twine did Mrs. Molloy have originally? (Let m = inches of twine that Mrs. Molloy had.)

4. The annual rainfall in Warwick last year was 24 in. That was 8 times the amount that fell in March. How many inches of rain fell in Warwick during March? (Let m = inches of rainfall in March.)

Write an equation and solve the problem.

B **5.** One number is 13 more than another. The greater number is 35. What is the lesser number? (Let s = lesser number.)

6. Mario's age is $\frac{1}{2}$ of his sister's. Mario is 8 years old. How old is his sister? (Let s = Mario's sister's age.)

7. Susan is 3 in. shorter than Tom. Susan's height is 62 in. How tall is Tom? (Let t = Tom's height in inches.)

8. One number is twice another. Their sum is 24. What are the two numbers? (Let n = the first number.)

9. A baseball team played 27 games. They won 5 games more than they lost. How many games did they win?

10. There are 28 students in Home Room 107. There were 3 students absent on Tuesday. How many students were present?

11. Suso sold three times as many tickets to the school play as Len. Suso sold 18 tickets to the school play. How many tickets did Len sell?

12. Ana swam 10 yd farther than Ivan. Ana swam 42 yd. How far did Ivan swim?

13. There are 952 students in Valley Central School. There are 38 more girls than boys. How many boys are there? How many girls are there?

5-10 Problem Solving

Architects and engineers solve many equations to make certain that designs for bridges and buildings will be safe. In this chapter we use simple equations to solve problems. These equations are much easier than many used by engineers. However, the methods you learn in this chapter are the bases of methods used to solve more difficult equations.

When you use an equation to solve a problem it is often helpful to let the variable represent the number you want to know. Using the variable, write two equal expressions. Then write an equation and solve it. Let's start with a simple example.

Example 1 Suppose so far this year you have saved $648. Your savings goal is $700. How much more must you save to meet your goal?

Solution Let x = amount of money you still need to save.

Write two equal expressions. Then write an equation and solve it.
Savings goal = $700
Savings goal = $648 + x ⟵ *amount already saved +*
amount you must save

$$\text{Savings goal} = \text{Savings goal}$$
$$\$648 + x = \$700$$
$$x + 648 = 700 \quad ⟵ \quad \textit{Commutative Property:}$$
$$x + 648 - 648 = 700 - 648 \quad \quad \textit{648 + x = x + 648}$$
$$x = 52$$

You must still save $52 to meet your goal.

Your Turn   Use an equation to solve the problem.
   • Marjorie's business is growing. She now employs 29
     people. In two months she plans to employ 57 people.
     How many more people does she need to hire?

Class Exercises

Complete the steps to write an equation for the problem.

1. There are 4 times as many carpenters as electricians at a construction site. If there are 52 carpenters, how many electricians are
   there? Let e = number of electricians. Then __?__ = number of carpenters.

2. Craig and Jane Conroy together earned $1200 last week. If Jane
   earned $110 more than Craig earned, how much did each earn last
   week? Let c = amount Craig earned. Then __?__ = amount Jane
   earned.

Exercises

Write an equation, solve it, then answer the question.

A 1. In Sue Crogen's office 14 people go to lunch at 12 noon. This leaves 29
   people in the office. How many people work in the office?

2. The sum of two numbers is 456. One of the numbers is 286. What is the
   other number?

3. Omega Manufacturers recently doubled the number of computers produced
   each day. They now make 54 computers each day. How many computers
   did Omega Manufacturers make each day before they increased production?

4. Herb traveled 279 miles on the first day of a 548-mile trip. How many
   more miles must he travel on the second day to reach his destination?

5. The product of two numbers is 486. One of the factors is 6. What is the
   other factor?

6. Madeline wants to delete 10 programs from a computer memory. This will
   leave 437 programs in memory. How many programs are in memory now?

B 7. The sum of two numbers is 687. One number is 2 times the other. What are the two numbers?

8. An auto dealer has 468 cars. He has 3 times as many four-door cars as two-door cars. How many 2-door cars does he have? How many 4-door cars does he have?

9. Jonas has a paper route. He has 20 customers more than Janice. Together they have 78 customers. How many customers does Jonas have? How many does Janice have?

10. Milton and Molton are neighboring towns. Molton has 18,112 more residents than Milton. Their combined populations is 478,658. What is the population of each town?

11. A pet store sold half as many pets in December as it sold during January through November. If 1272 pets were sold in all that year, how many pets were sold in December? What was the average number of pets sold in one month that year?

12. When you double a number and add 16, you get 86. What is the number?

Self Test 2

Solve the equation.

1. $2a - 3 = 27$ 2. $31 = 5b + 6$ 3. $97 = 9c + 16$ **5-6**

4. $3d - 4 = 2d + 7$ 5. $e + 4 = 8 - e$ 6. $5f - 1 = 2f + 8$ **5-7**

Fran is 5 years older than Jim. Let j = Jim's age. Complete by writing an algebraic expression.

7. Fran's age = __?__   **5-8**

8. Dean is 3 years younger than Fran. Dean's age = __?__

Write an equation. Solve the problem.

9. Marlene has twice as much money as Nick.   **5-9**
   Marlene has $16.
   How much money does Nick have?
   (Let n = amount Nick has.)   **5-10**

10. Choose a number. Add seven. Multiply the sum by 3. If the total is 54, what is the original number?

Algebra Practice

Write an equation. Solve.

1. The product of 308 and 280 is k.

2. The quotient of 156,204 and 36 is d.

3. 56.07 minus 9.806 is equal to w.

4. The numbers 12.83, 1.509, and 304.89 add to a sum of m.

5. 52.635 divided by 8.7 is equal to f.

6. 0.52 times 0.16 is equal to s.

7. The difference between $7\frac{1}{2}$ and $3\frac{5}{6}$ is some number.

8. How much is $2\frac{6}{7}$ times $2\frac{4}{5}$?

9. What is the sum of $3\frac{1}{2}$, $1\frac{1}{8}$, and $4\frac{3}{4}$?

10. $2\frac{4}{7}$ divided by $\frac{5}{3}$ is some number.

Use the distributive property to factor each expression.

11. $8uv^2 + 2(uv)^2 + u^3v$ 12. $6c^3 + 3c^2d^2 + 15c$ 13. $7qr + (qr)^2 + qr^2$

Write the LCM of each set of expressions.

14. m^3 and $5m$ 15. $3g$, $2h^2$, and 4   16. $6a^2$, $9ab$, and $2b^2$

Solve the equation. Check the solution.

17. $m + 5 = 12$ 18. $14 = r + 3\frac{1}{2}$ 19. $16.3 + w = 40$

20. $7l = 56$ 21. $4.3k = 7.74$ 22. $7\frac{1}{2} = 3\frac{1}{8}s$

23. $4.1 = g - 2.6$ 24. $a - 13 = 20$ 25. $t - \frac{1}{3} = \frac{1}{2}$

26. $\frac{e}{3} = 2.5$ 27. $4 = \frac{1}{6}v$ 28. $\frac{3}{8}l = \frac{15}{16}$

29. $4b - 5 = 23$ 30. $\frac{2}{3}d + 5 = 11$ 31. $54 = 6(j - 2)$

165

Problem Solving on the Job

Office Cashier

The duties of a cashier may vary somewhat from one company to another. In general, cashiers are responsible for handling money and for keeping accurate records of receipts and expenditures. Although computers are used to help with the recordkeeping, a cashier is still needed to handle the money and to deal with customers.

Mason Jones is a cashier at Franck Enterprises. He maintains a petty cash fund of $100, a fund used to pay for small, unforeseen expenses. The items in Exercises 1–3 were paid out of the petty cash fund one week.

```
OFFICE CASHIER

DESCRIPTION: Maintain petty-cash
fund; take money for goods and
services; keep records
QUALIFICATIONS: High school; good
recordkeeping skills; maturity and
good judgment
JOB OUTLOOK: Good, especially if
individual also has typing and
bookkeeping skills
```

What is the total of each kind of expense for the week?

1. Delivery Expenses: $12.83, $20.00, $4.60

2. Office Expenses: $1.50, $2.98, $4.98, $1.50

3. Miscellaneous Expenses: $7.50, $8.50, $21.80

4. When the petty cash fund has $25 or less, Mason must get more money to bring the fund up to $100. After paying for the items in Exercises 1–3, how much more money is needed to bring the fund up to $100?

A cashier in a store may accept payments from customers on charge accounts and layaway items. When merchandise is returned, it is often the cashier who refunds the money or credits the customer's account. The amount that a customer owes on an account is the balance due.

How much will the customer owe after the payment or credit shown?

5. Balance due: $68.37
   Payment: $25.00

6. Balance due: $107.82
   Credit: $58.63

7. Balance due: $306.34
   Credit: $19.98
   Payment: $45.00

8. A customer makes a down payment of $20 on a coat that costs $150 and puts the coat on layaway. There is a sales tax of $7.50 on the purchase. After the customer makes 3 payments of $15, what is the balance due?

Cashiers also count and wrap money and prepare bank deposit slips. The First Bank provides Costello Industries with the following types of wrappers.

BILLS

Bill	$1	$5	$10	$20
Number in Wrapper	50	50	50	50

COINS

Coin	Penny	Nickel	Dime	Quarter
Number in Roll	50	40	50	40

Sue Schwartz is the cashier at A-Z, Inc. At the end of each day she counts and wraps the money and sees to it that the money is deposited in the bank. The following cash and checks are to be deposited.

Bill	Wrapped	Unwrapped
$1	7 pkg	32
$5	2 pkg	16
$10	3 pkg	43
$20	1 pkg	3

Coin	Wrapped	Unwrapped
Pennies	4 rolls	37
Nickels	3 rolls	19
Dimes	5 rolls	8
Quarters	3 rolls	29

Checks
250.00
36.83
179.47
230.94

9. What is the total value of the bills to be deposited?
10. What is the total value of the coins to be deposited?
11. What is the total value of the checks to be deposited?
12. What is the total value of the money to be deposited?

Enrichment

Venn Diagrams

Sometimes it is useful to illustrate certain statements with diagrams like the ones shown below. Such diagrams are called **Venn Diagrams.**

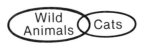      

Some cats are wild.   All whales are mammals.   No lion is a bird.

A single diagram can be used to picture several statements.

Some humans are vegetarians.
All giraffes are vegetarians.
No human is a giraffe.

Draw Venn Diagrams to illustrate the statements.

1. All kangaroos are marsupials.
2. No hares are carnivores.
3. Some tortoises weigh over 400 lb.
4. Penguins are found only in the southern hemisphere.
5. Frogs lay eggs. Birds lay eggs. Turtles lay eggs.
6. Some birds sing. Some whales sing.
7. Snakes and lizards have scales. Some fish have scales.
8. Some cats, like the lion, are social animals. Some cats, like the leopard, are solitary animals.
9. Mammals and birds are warm-blooded animals. No reptile is warm-blooded.
10. Some mammals are nocturnal animals (active at night). Leopards are nocturnal mammals. No bat is not a nocturnal mammal.

Chapter Review

Which number is a solution of the equation? Answer *a*, *b*, or *c*.

1. $171 + a = 376$
   a. 205   **b.** 547   **c.** 195

2. $247 = 19b$
   a. 228   **b.** 13   **c.** 4693

Solve the equation.

3. $c + 7 = 26$

4. $16 + d = 70$

5. $43 = e + 39$

6. $4f = 64$

7. $78 = 3g$

8. $15h = 225$

9. $j - 12 = 82$

10. $95 = k - 25$

11. $m - 48 = 102$

12. $\dfrac{n}{3} = 14$

13. $\dfrac{1}{2}p = 21$

14. $\dfrac{2}{5}q = 98$

15. $2r - 3 = 21$

16. $3s + 7 = 34$

17. $5u - 9 = 71$

18. $9v - 4 = 8v + 7$

19. $6w + 1 = 3w + 10$

20. $7x - 5 = 5x + 17$

Match the expression at the right with the number described.

21. a number divided by 3

22. a number increased by 3

23. 3 less than a number

24. 5 times a number

25. twice a number

26. 3 more than 5 times a number

27. twice a number decreased by 3

a. $n + 3$

b. $5n$

c. $\dfrac{n}{3}$

d. $2n$

e. $n - 3$

f. $2n - 3$

g. $3 + 5n$

Write an equation. Solve the problem.

28. Paulo is buying a calculator at half price.
   He pays $6 for the calculator.
   What is the regular price of the calculator?
   (Let n = regular price.)

29. Suppose you choose a number and multiply it by 6. Then
   you add 9 to the product. If the total is 345, what is
   the number?

Chapter Test

Which number is a solution of the equation?

1. $7a = 364$ 52 or 92?

2. $16 + b = 93$ 77 or 109?

5-1

Solve the equation.

3. $c + 8 = 17$

4. $6 + h = 21$

5. $57 = 19 + r$

5-2

6. $5e = 45$

7. $132 = 6p$

8. $9s = 540$

5-3

9. $d - 4 = 11$

10. $49 = n - 3$

11. $x - 12 = 39$

5-4

12. $\frac{1}{2}f = 7$

13. $\frac{g}{3} = 4$

14. $\frac{3}{5}y = 42$

5-5

15. $8a + 4 = 28$

16. $6x - 1 = 17$

17. $4v + 5 = 53$

5-6

18. $9w - 1 = 8w + 7$

19. $2b - 1 = 13 - b$

20. $5c + 5 = 2c + 23$

5-7

Complete by writing an algebraic expression.

21. a. 5 less than 12 = __?__

5-8

    b. 5 less than n = __?__

    c. 5 less than $3x$ = __?__

    d. 5 less than twice the number p = __?__

22. Carl worked 3 hours more than Janet last week.
Let j = number of hours Janet worked.

    a. Number of hours Carl worked = __?__

    b. Max worked 2 hours less than Carl worked last week.
Number of hours Max worked = __?__

Write an equation. Solve the problem.

23. Bart earned $15 more than Norma.
Bart earned $44.
How much did Norma earn?
(Let n = amount that Norma earned.)

5-9

24. Carol sold 6 more tickets to the school play than Bob.
Carol sold 13 tickets.
How many tickets did Bob sell?

25. A number is doubled, then the result is increased by 17. The final
result is 41. What is the original number?

5-10

Cumulative Review (Chapters 1–5)

Add, subtract, multiply, or divide.

1. $361 + 8.2 + 19$
2. $7491 \times 0 \times 84$
3. $17 + 35 \cdot 4$
4. $(4.2 + 5) \div 4$
5. $8002 - 36.21$
6. 100×483.21
7. $3 \cdot 3 \cdot 3 \cdot 3$
8. $10 \cdot 10 \cdot 10 \cdot 10 \cdot 10$
9. $36 \div 4 + 7$
10. $\dfrac{1}{2} \cdot \dfrac{1}{8} \cdot \dfrac{2}{3}$
11. $\dfrac{3}{4} \cdot \dfrac{2}{5} \cdot \dfrac{5}{6}$
12. $\dfrac{2}{3} + \dfrac{1}{4} - \dfrac{1}{4}$
13. $32 + 4 \cdot 9$
14. $(3645)(0)(1)$
15. $(6.2)(8) - 49.6$
16. $\dfrac{1}{8} \times 8$
17. $2004 \div 1.2$
18. $\left(1\dfrac{1}{2}\right) \div \left(1\dfrac{1}{2}\right) + 1$

Factor.

19. $\dfrac{1}{2}a + \dfrac{1}{2}b + \dfrac{1}{2}c$
20. $2a + 4n - 6x$
21. $5y^2 - 10y$
22. $2j^2 - 3j$
23. $5n^2 - 25$
24. $12p^2 - 6p - 2$

Solve.

25. A first number is 47 greater than a second number. Let $n =$ the first number. Then __?__ = the second number.
26. A florist told Rocky that his plant would grow from 10 to 12 cm during the summer if conditions were right. The plant grew from 14.7 cm to 15.9 cm. Was the growth within the limits the florist gave?

Write an equation. Solve the problem.

27. Sue worked 8 hours at $9.50 per hour, then at time and one half for 4 hours. How much did she earn in all?
28. I'm thinking of a number. If I multiply it by 6 and then add 7, the result is 31. What is my number?
29. There are 56 members of a volunteer group. Twenty-four are women. How many are men?
30. Mike's little sister has half as much money as Mike. Together they have $63. How much does each have?

6
Formulas

You often see skid marks on the road when a motorist brakes to come to a quick stop. Based on the condition of the road surface, you can use a formula to estimate the rate of speed before braking.

6-1 Perimeter

Some airports are enclosed by cyclone fencing. The diagram below is a drawing of one such airport. To find the total length of fencing around the airport, you add the lengths of the sides. If your addition is correct, you get 9487 ft.

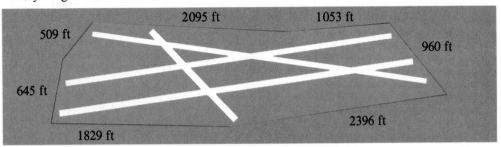

The distance around a figure is called its **perimeter.** Some figures have special formulas for their perimeters. To use any of these formulas, simply replace the variables in the formula with the numbers you know.

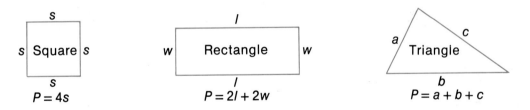

$$P = 4s \qquad\qquad P = 2l + 2w \qquad\qquad P = a + b + c$$

Example 1 A rectangle is 25 ft long and 15 ft wide. What is its perimeter?

Solution
$P = 2l + 2w$ ⟵ _Use the formula for rectangle._
$\quad = 2 \times 25 + 2 \times 15$ ⟵ _Replace the variables with the numbers._

So, the perimeter is 80 ft.

Your Turn • A rectangle is 24 ft long and 18 ft wide. What is its perimeter?

173

For some figures there are no general perimeter formulas. To find the perimeter of such figures, you simply add the lengths of the sides as we did when we found the distance around the airport. In some cases you might even need to find the lengths of one or more sides before you do the addition.

Example 2   Find the perimeter of the figure shown below.

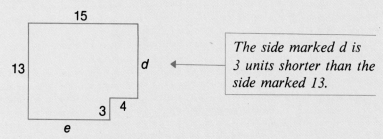

The side marked *d* is 3 units shorter than the side marked 13.

Solution   The distance marked *d* must be 13 − 3, or 10.
The distance marked *e* must be 15 − 4, or 11.
$P = 13 + 15 + 10 + 4 + 3 + 11 = 56$

Your Turn • Find the perimeter of the figure at the right.

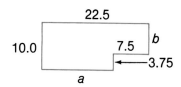

Class Exercises

Complete the formula for the perimeter of the figure.

1. rectangle
$P = 2l + \underline{\ ?\ }$

2. square
$P = \underline{\ ?\ }\, s$

3. triangle
$P = a + \underline{\ ?\ } + \underline{\ ?\ }$

Complete.

4. side *a* is 15 in. long
side *b* is 4 in. shorter than side *a*
side *b* is __?__ in. long

5. side *x* is 20 in. long
side *y* is 7 in. shorter than side *x*
side *y* is __?__ in. long

Exercises

Use a formula to find the perimeter of the figure.

A **1.**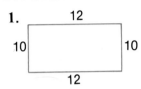

2.

3.

$$P = 2l + 2w \qquad\qquad P = a + b + c \qquad\qquad P = 4s$$

Use a formula to find the perimeter of the figure.

B **4.**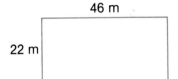

5.

6.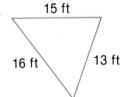

7.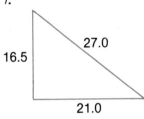

8.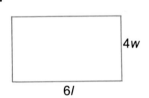

9.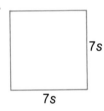

Find the perimeter of the figure.

10.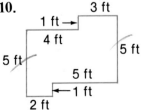

11.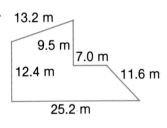

12.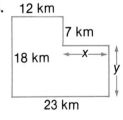

13. In the triangle shown, $a = 15$, $b = 18$, and $P = 49$. What is the length of c?

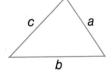

6-2 Circumference

The curved section in the drawing of the window is one half of a circle or a semicircle. The segment that passes through the center of the circle at point O and divides the circle in half is a **diameter** (d). The distance around the circle, called the **circumference** (C), has a special relationship to the diameter. A formula that relates the diameter to the circumference is $C \div d = \pi$. Complete each division in the table below to find an approximation for the Greek letter π (pi).

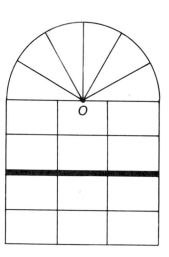

Round the quotient to the nearest hundredth			
Item	C	d	Quotient
bicycle tire	88 in.	28 in.	$88 \div 28 = 3.14$
flower bed	22 yd	7 yd	$22 \div 7 = ?$
manhole cover	125.6 in.	40 in.	$125.6 \div 40 = ?$

Since $C \div d$ results in the same number for all circles, you can use the formula $C \div d = \pi$ to find the circumference of any circle. We usually write $\pi \approx 3.14$. The symbol $\approx$ means *is approximately equal to*.

Example 1 A circular swimming pool has a diameter of 76 m. What is its circumference?

Solution
$$C \div d = \pi$$
$$C \div d \times d = \pi \times d \quad \longleftarrow \quad \textit{Rewrite the formula to get}$$
$$C = \pi d \qquad\qquad\qquad \textit{C on the left of the = sign.}$$
$$\approx 3.14 \times 76$$
$$\approx 238.64$$

Remember from the drawing of the window that the diameter touches the circle twice. A **radius** (*r*) is drawn from the center and touches the circle once. The diameter of a circle is twice as long as the radius. The formula that relates the radius to the circumference is $C = 2\pi r$.

Example 2   The radius of a circle is 32.5 cm. What is its circumference?

Solution
$$C = 2\pi r$$
$$\approx 2 \times 3.14 \times 32.5 \longleftarrow \quad \textit{Replace } \pi \textit{ with 3.14.}$$
$$\textit{Replace } r \textit{ with 32.5.}$$
$$\approx 3.14 \times 65$$
$$\approx 204.1$$

Sometimes the fraction $\frac{22}{7}$ is used as an approximation for π. Change $\frac{22}{7}$ to a decimal to see that $\frac{22}{7} \approx 3.14$. Here is an example of when it is more convenient to use $\pi \approx \frac{22}{7}$.

Example 3   The diameter of a circle is $3\frac{1}{2}$ in. What is its circumference?

Solution
$$C = \pi d$$
$$= \pi \times 3\frac{1}{2} \longleftarrow \quad \textit{Replace } d \textit{ with } 3\frac{1}{2}$$
$$\approx \frac{22}{7} \times \frac{7}{2} \longleftarrow \quad \textit{Replace } \pi \textit{ with } \frac{22}{7}.$$
$$\textit{Write } 3\frac{1}{2} \textit{ as the fraction } \frac{7}{2}.$$
$$\approx \frac{\overset{11}{\cancel{22}}}{\underset{1}{\cancel{7}}} \times \frac{\overset{1}{\cancel{7}}}{\underset{1}{\cancel{2}}}$$
$$\approx 11$$

Your Turn   What is the circumference of the circle? Use $\pi \approx \frac{22}{7}$, or $\pi \approx 3.14$.

- $d = 9$ yd   • $r = 5.5$ in.   • $d = 21$ m

Class Exercises

Complete.

1. $r = 8$
$C = 2 \times \pi \times \underline{\ ?\ }$

2. $r = 12$
$C = \underline{\ ?\ } \times \pi \times 12$

3. $r = 7.4$
$C = \underline{\ ?\ } \times \pi \times \underline{\ ?\ }$

4. $d = 20$
$C = \pi \times \underline{\ ?\ }$

5. $d = 16.70$
$\underline{\ ?\ } = \pi \times \underline{\ ?\ }$

6. $d = 0.98$
$C = \underline{\ ?\ } \times \underline{\ ?\ }$

Exercises

Use $\pi \approx 3.14$ to find the circumference.

A **1.**

10

2.

7

3.

2.5

Find the circumference of the circle described. Use $\pi \approx 3.14$ for 4–9.
Use $\pi \approx \frac{22}{7}$ for 10–15.

B **4.** diameter = 20 in.

5. diameter = 76 m

6. diameter = 40 in.

7. radius = 15 m

8. radius = 1.6 mi

9. radius = 3.12 km

10. diameter = 28

11. diameter = 42

12. diameter = 49

13. radius = $\frac{14}{33}$

14. radius = $\frac{21}{22}$

15. radius = $\frac{21}{44}$

Problems

1. The curved section of the figure is a semicircle. Which figure has the greater perimeter: the shaded figure or the unshaded one?

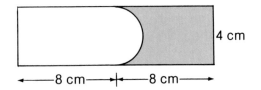

4 cm

←——8 cm——→|←——8 cm——→

2. The front wheel of a certain tricycle has a radius of 14 in. Each back wheel has a radius of 6 in. How much larger is the circumference of the front wheel than the circumference of a back wheel?

6-3 Angle Measurement

You are probably very familiar with the type of angle marked (⌐) in the square, the rectangle, and the right triangle shown below. The measure of this angle is 90°, and it is referred to as a **right angle.**

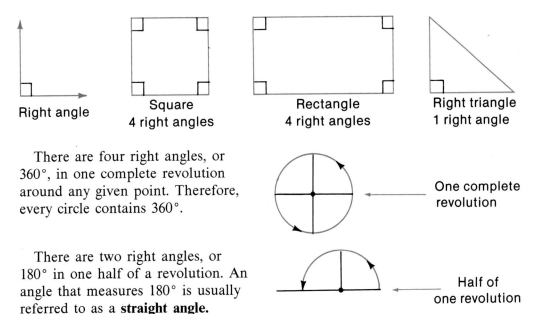

Right angle

Square
4 right angles

Rectangle
4 right angles

Right triangle
1 right angle

There are four right angles, or 360°, in one complete revolution around any given point. Therefore, every circle contains 360°.

One complete revolution

There are two right angles, or 180° in one half of a revolution. An angle that measures 180° is usually referred to as a **straight angle.**

Half of one revolution

Angles are classified by their measures. Study the table below to see how some angles are classified.

Angle	Name	Measure
	Acute	Less than 90°
	Right	90°
	Obtuse	More than 90° but less than 180°
	Straight	180°

We generally use three letters to name an angle, but an angle is sometimes named using only the letter at the **vertex.** The symbol for angle is ∠.

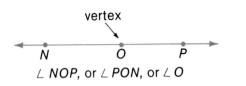

∠ NOP, or ∠ PON, or ∠ O

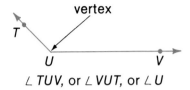

∠ TUV, or ∠ VUT, or ∠ U

Example 1   Use the diagram below to name an angle as indicated.
- right angle
- acute angle
- straight angle
- obtuse angle

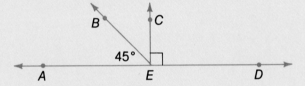

Solution   right angle: ∠ CED
straight angle: ∠ AED
acute angle: ∠ AEB
obtuse angle: ∠ BED
Notice that ∠ CEA is also a right angle and ∠ BEC is also an acute angle.

The diagram on the right shows two lines that intersect to form right angles. Two lines that form right angles are said to be **perpendicular.** The symbol for perpendicular is ⊥. We write $\overleftrightarrow{LM}$ to mean line LM.

Now look at these two lines. They are the same distance apart and will never intersect. Two lines in a plane that do not intersect are **parallel.** We write $\overleftrightarrow{AB} \| \overleftrightarrow{CD}$ to mean $\overleftrightarrow{AB}$ is parallel to $\overleftrightarrow{CD}$.

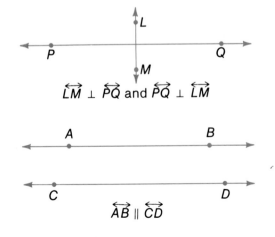

$\overleftrightarrow{LM} \perp \overleftrightarrow{PQ}$ and $\overleftrightarrow{PQ} \perp \overleftrightarrow{LM}$

$\overleftrightarrow{AB} \| \overleftrightarrow{CD}$

Example 2   In the diagram shown below, name two parallel lines and two perpendicular lines.

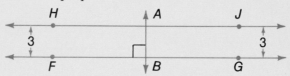

Solution   parallel lines: $\overleftrightarrow{HJ} \| \overleftrightarrow{FG}$
perpendicular lines: $\overleftrightarrow{AB} \perp \overleftrightarrow{FG}$

Your Turn   Use the diagram in Example 2 to complete the statement.

• $GF \| \underline{\quad?\quad}$ • $GF \perp \underline{\quad?\quad}$

Class Exercises

Match the statement with the correct letter.

1. The angle measure is less than 90°.

2. The angle measure is 90°.

3. The angle measure is 180°.

4. *B* is at the vertex of the angle.

5. *C* is at the vertex of the angle.

6. *A* is at the vertex of the angle.

A. right angle
B. straight angle
C. acute angle
D. ∠CAB
E. ∠CBA
F. ∠ACB

Exercises

**Write *yes* or *no* to answer the question.
Use the diagram shown at right.**

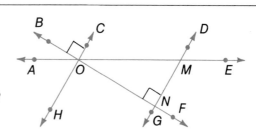

A **1.** Is $\overleftrightarrow{BF} \perp \overleftrightarrow{CH}$?   **2.** Is $\overleftrightarrow{HC} \| \overleftrightarrow{GD}$?

3. Is $\overleftrightarrow{AE} \| \overleftrightarrow{DG}$?   **4.** Is $\overleftrightarrow{GD} \perp \overleftrightarrow{BF}$?

5. Is $\overleftrightarrow{AE} \perp \overleftrightarrow{HC}$?   **6.** Is $\overleftrightarrow{CH} \| \overleftrightarrow{EA}$?

The measure of an angle is given below. Tell whether it is an acute, obtuse, right, or straight angle.

B **7.** 60° **8.** 25° **9.** 180° **10.** 92° **11.** 100°

12. 90° **13.** 105° **14.** 37° **15.** 172° **16.** 87°

 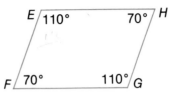 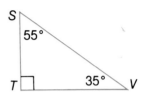

Use the figures above for Exercises 17–22.

17. Name two sides in the triangle that are perpendicular.

18. Name two pairs of parallel sides in the rectangle.

Find the sum of the angle measures.

19. $\angle STV + \angle TVS + \angle VST$

20. $\angle EFG + \angle FGH + \angle GHE$

21. $\angle ABC + \angle CDA + \angle DAB$

22. $\angle HEF + \angle TSV + \angle ADC$

In the diagram shown at right $\angle MPN = 50°$.

C **23.** What is the measure of $\angle MPR + \angle RPQ$?

24. Are $\overleftrightarrow{MQ}$ and $\overleftrightarrow{NR}$ parallel? Explain.

25. Are $\overleftrightarrow{MQ}$ and $\overleftrightarrow{NR}$ perpendicular? Explain.

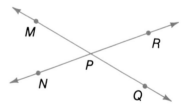

Self Test 1

Find the perimeter of the figure described.

1. square
$s = 11$ ft

2. rectangle
$l = 18$ in.
$w = 16$ in.

3. triangle
sides = 16.7 cm, 18.0 cm, 19.1 cm

6-1

Find the circumference of the circle described. Use $\pi \approx 3.14$.

4. diameter = 21 yd **5.** radius = 3.2 m **6.** radius = 1.1 mi **6-2**

Write whether the angle is acute, obtuse, right, or straight.

7. 90° **8.** 101° **9.** 180° **10.** 13° **11.** 3° **6-3**

6-4 *Area of Polygons*

When you measure the surface inside a closed figure, you are really finding the area (A) of the figure. Area is measured in square units. Some familiar square units used for measuring small areas are square inch (in.²) and square centimeter (cm²). Square miles (mi²) and square kilometers (km²) are generally used to measure large areas.

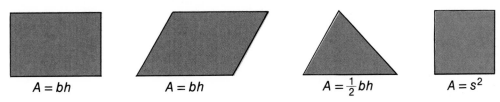

$A = bh$ $A = bh$ $A = \frac{1}{2}bh$ $A = s^2$

The area formulas for some common polygons are given above. We use b to represent the length of the base and h to represent the height. Any side of a parallelogram can be its base. Its height is the perpendicular distance between the base and the opposite side.

You can use the area formula $A = bh$ to find the area of a square, but the formula that is most often used is $A = s^2$. The variable s is used to represent the length of any side.

Example 1 A parallelogram is $4\frac{1}{2}$ in. high and its base is $3\frac{1}{2}$ in. What is its area?

Solution $A = bh$ ⟵ *Use the area formula for a parallelogram.*

$= 3\frac{1}{2} \times 4\frac{1}{2}$ ⟵ *Replace the variables with the numbers.*

$= \frac{7}{2} \times \frac{9}{2}$ ⟵ *Write the mixed numbers as fractions.*

$= \frac{63}{4} = 15\frac{3}{4}$ ⟵ *Multiply the fractions, then simplify the answer.*

The area is $15\frac{3}{4}$ in.²

Your Turn • A rectangle is $2\frac{1}{2}$ in. high and its base is 4 in. What is
its area?

Sometimes you may want to find the area of only a part of a figure.
Example 2 below shows one way to find a partial area.

Example 2 Find the area of the shaded
portion of the figure.

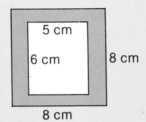

Solution Find the area of the square.

$$A = s^2$$
$$= 8^2$$
$$= 64 \longleftarrow \textit{Remember that } 8^2 \textit{ means } 8 \times 8.$$

Find the area of the rectangle.

$$A = bh = 5 \times 6$$
$$= 30$$

Find the difference in area.

$$(64 - 30) \text{ cm}^2 \longleftarrow$$
$$= 34 \text{ cm}^2$$

The area of the square is 64 cm².
The area of the rectangle is 30 cm².

Your Turn • Find the area of the shaded
portion of the figure.

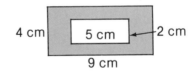

Class Exercises

Use a formula to find the area of the figure.

1.

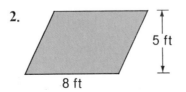

3.5 m

7.0 m

2.

5 ft

8 ft

3.

4 in.

4 in.

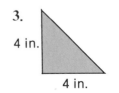

Exercises

Find the area of the figure described below.

A **1.** triangle
$b = 7$ ft
$h = 5$ ft

2. parallelogram
$b = 14$ cm
$h = 11$ cm

3. square
$s = 10.5$ m

B **4.** rectangle
$b = 6\frac{1}{2}$ ft
$h = 5\frac{1}{2}$ ft

5. triangle
$h = 8\frac{2}{5}$ in.
$b = 10\frac{1}{4}$ in.

6. parallelogram
$h = 12.9$ cm
$b = 16.2$ cm

Find the area of the shaded section.

7.
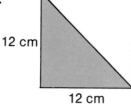
12 cm
12 cm

8.
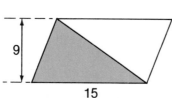
9
15

9.
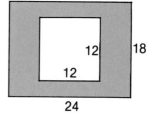
12   18
12
24

Problems

Solve.

1. 7 yd times some length is 63 yd².
2. $\frac{1}{2}b$ times 14 is 112.
3. 400 yd² divided by h is 20 yd.
4. A tennis court is 26 yd long and has an area of 312 yd². What is its width?
5. Table tennis is played on a table that has an area of 5 yd² and is 5 ft wide. What is the length of the table?
6. The area of the tennis court is how many times larger than the area of the table tennis table?

185

6-5 *Area of Circles*

You know that the area for many polygons is related to the height and base of the individual figure. The area formula of a circle is not usually described in terms of height and base, but the diagrams below suggest that we can use a parallelogram to show how a formula for finding the area of a circle is developed.

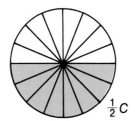
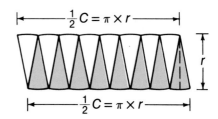

Notice that the base of the parallelogram is really the same as one half the circumference of the circle. The height of the parallelogram is just about the same as the radius of the circle. So, the area formula for the circle can be stated as $A = \frac{1}{2}C \cdot r$. You know that $C = 2\pi r$. So you can replace C with $2\pi r$ and rewrite the area formula.

$$A = \frac{1}{2} \cdot 2\pi r \cdot r$$

$$= \pi r^2 \longleftarrow \quad \frac{1}{\cancel{2}} \cdot \cancel{2}\pi r \cdot r = \pi r^2$$

Example 1   The radius of a circle is 8 cm. What is its area? Use $\pi \approx 3.14$.

Solution $A = \pi r^2 \longleftarrow$ *Use the area formula for a circle.*
$\approx 3.14 \times 8^2$
≈ 200.96
The area is 200.96 cm².

Your Turn   • The radius of a circle is 6 cm. What is its area? Use $\pi \approx 3.14$.

Sometimes you may have to find the area of two different figures, then subtract the smaller area from the larger area.

Example 2   Find the area of the shaded portion of the figure shown. Use $\pi \approx 3.14$.
Larger radius = 15 ft.
Smaller radius = 6 ft.

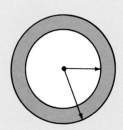

Solution   Find the area of the large circle.
$A \approx 3.14 \times 15^2$ ⟵——— *The radius of the large circle is 15 ft.*
$\approx 3.14 \times 225$
≈ 706.5
Find the area of the small circle.
$A \approx 3.14 \times 6^2$ ⟵——— *The radius of the small circle is 6 ft.*
$\approx 3.14 \times 36$
≈ 113.04
Find the difference.
$(706.5 - 113.04) = 593.46 \text{ ft}^2$
The area of the shaded portion is 593.46 ft².

Your Turn   Find the area of the shaded portion of the figure shown above. Use $\pi \approx 3.14$.
• larger radius = 6   • larger radius = 11
  smaller radius = 5   smaller radius = 7

Class Exercises

Write the correct number to complete. Use $\pi \approx 3.14$.

1. $r = 6$
$A \approx \underline{\ ?\ } \times 6^2$

2. $r = 8$
$A \approx 3.14 \times \underline{\ ?\ }$

3. $r = 10.2$
$A \approx 3.14 \times \underline{\ ?\ }$

4. $r = 7.5$
$A \approx \underline{\ ?\ } \times \underline{\ ?\ }$

5. $d = 14$
$A \approx \underline{\ ?\ } \times 7^2$

6. $d = 18$
$A \approx \underline{\ ?\ } \times \underline{\ ?\ }$

Exercises

Find the area of the circle. Use $\pi \approx 3.14$.

A **1.**

2.

3.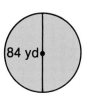

4.

Find the area of the circle described. Use $\pi \approx 3.14$.

B **5.** $r = 3.5$ in. **6.** $r = 6.1$ ft **7.** $r = 9$ yd

8. $d = 17$ m **9.** $d = 20$ mi **10.** $d = 24$ cm

Which figure has the greater area? What is the difference in area?

11.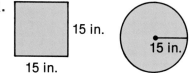

12.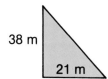

13.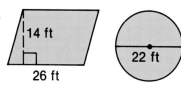

14.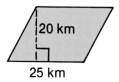

Find the area of the shaded part. Use $\pi \approx 3.14$.

15.

16.

17.

18.

Find the area of the figure. Each curve is a semicircle. Use $\pi \approx 3.14$.

C **19.**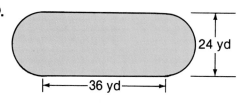

20.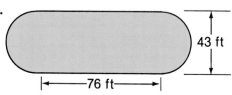

6-6 *Problem Solving*

The formulas you have studied so far are useful for solving problems that are related to perimeter and area. There are other useful formulas that you can use to solve a variety of problems.

When you use formulas, it is important that you know which number or quantity each variable represents. For example, a weight applied at one end of a lever can move a much heavier weight at the other end. To use the lever formula correctly you must be able to tell which variable represents the heavier weight and which represents the lighter weight.

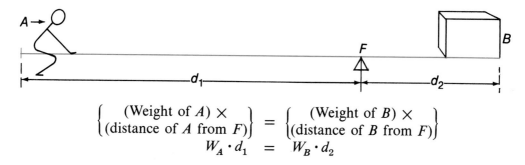

$$\left\{ \begin{array}{c} \text{(Weight of } A) \times \\ \text{(distance of } A \text{ from } F) \end{array} \right\} = \left\{ \begin{array}{c} \text{(Weight of } B) \times \\ \text{(distance of } B \text{ from } F) \end{array} \right\}$$
$$W_A \cdot d_1 = W_B \cdot d_2$$

Example   In the drawing shown above, B represents a 160-lb box and is 4 ft from F. The person, A, is 10 ft from F. Find the weight of the person that will move the box.

Solution   Use the lever formula.
$$W_A \cdot d_1 = W_B \cdot d_2$$
$$W_A \cdot 10 = 160 \cdot 4 \quad \longleftarrow \quad \textit{Divide both sides of the}$$
$$ \textit{equation by 10.}$$
$$10\, W_A = 640$$
$$W_A = 64$$
The weight of the person to move the box is 64 lb.

Your Turn   • In the drawing above, the box weighs 120 lb. F is 3 ft from the box and 8 ft from the person. Find the weight of the person that will move the box.

Class Exercises

Copy the table and use the formula to
complete. Round to the nearest thousandth.

$$\text{batting average} = \frac{\text{number of hits}}{\text{times at bat}}$$

$$A = \frac{h}{t}$$

	h	t	$h \div t$
1.	40	85	?
2.	60	210	?
3.	24	85	?
4.	32	94	?

Exercises

Use the formula $W = \dfrac{11(h - 40)}{2}$ to solve.

A **1.** $h = 64$ in.
   $W = \underline{\ ?\ }$

2. $h = 67$ in.
   $W = \underline{\ ?\ }$

3. $h = 59$ in.
   $W = \underline{\ ?\ }$

4. $h = 72$ in.

   $W = \underline{\ ?\ }$

5. $h = 49\frac{1}{2}$ in.

   $W = \underline{\ ?\ }$

6. $h = 60\frac{1}{3}$ in.

   $W = \underline{\ ?\ }$

More Formulas

Skid marks: $L = s^2 \div 18.6$ ⟵ L = length of skid marks (ft)
                               s = speed (mi/h)

Air speed: $a - h = g$ ⟵ a = airspeed (mi/h)
                        h = head wind speed
                        g = ground speed

Electricity: $I = W \div V$ ⟵ I = current (amps)
                          W = power (W)
                          V = force (volts)

Choose the correct formula and use it to solve problems 7–10.

B **7.** Two children are balanced on
   the seesaw (lever) as shown.
   How far is the heavier child
   from point F?

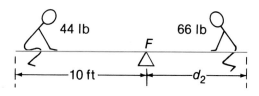

44 lb   66 lb F 10 ft d_2

8. Marsha was driving at a speed of 40 mi/h when a football rolled into the street. She applied her brakes immediately. How long were the skid marks made by the tires?

9. A pilot reports an air speed of 625 mi/h. If the plane is flying into a head wind of 30 mi/h, what is the ground speed?

10. A 20-amp fuse will burn out if more than 20 amps of current try to flow through the fuse. Suppose a hair styler using 1000 W of current, a humidifier using 177 W, and an electric iron using 1008 W are all connected to the same 110-volt line. If all these appliances are turned on at the same time, will the 20-amp fuse burn out?

Self Test 2

Find the area of the figure described.

1. triangle
   $b = 8$ ft
   $h = 4$ ft

2. rectangle
   $b = 13.6$ in.
   $h = 11.1$ in.

3. parallelogram
   $b = 11.8$ yd
   $h = 10.2$ yd

6-4

Find the area of the circle described. Use $\pi \approx 3.14$.

4. $r = 2.5$ ft

5. $d = 22$ yd

6. $r = 8.8$ cm

6-5

Use the formula $a - h = g$ to solve. (a = airspeed, h = head wind speed, and g = ground speed)

7. An airplane is flying into a 25 mi/h wind. Its air speed is 650 mi/h. What is the ground speed?

6-6

Self Test answers and Extra Practice are at the back of the book.

6-7 *Volume of Prisms and Pyramids*

A **prism** is a solid figure with two identical bases. The volume (V) of a prism is the amount of space it contains. Some standard units of volume are cubic centimeter (cm^3), cubic foot (ft^3), and cubic yard (yd^3). Study the prisms shown below.

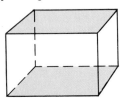

Rectangular Prism

Triangular Prism

height ← *The height is the distance between the bases.*

The shaded portions of each prism are the bases. Notice that the shape of the base determines the name of the prism. The base area of a prism is the area of one base. The formula for the volume of a prism is $V = Bh$. ← *B means base area; h means height.*

Example 1 Find the volume of the prism shown at the right.

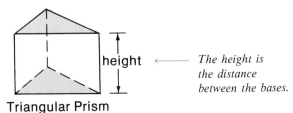

5 m

4 m   9 m

Solution
$$B = 9 \times 4 = 36 \text{ m}^2$$
$$V = Bh$$
$$= 36 \cdot 5$$
$$= 180 \text{ m}^3 \leftarrow \textit{Notice that we express the answer in cubic meters.}$$

The figures below are pyramids. Like the prisms you studied above, the base of a pyramid determines its name. The formula for the volume of a pyramid is $V = \frac{1}{3}Bh$. ← *B means base area; h means height.*

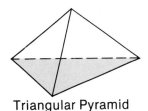

Triangular Pyramid

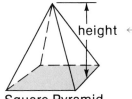

height ← *The height is the distance to the base.*

Square Pyramid

Example 2 The base area of the pyramid shown is 24 ft² and the height is 17 ft. What is its volume?

Solution $V = \dfrac{1}{3}Bh$

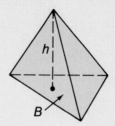

$= \dfrac{1}{3} \cdot \overset{8}{24} \cdot 17$

$= 8 \cdot 17 = 136$

The volume is 136 ft³.

Your Turn • The base area of a pyramid is 18 cm² and the height is 10 cm. What is its volume?

Class Exercises

Complete.

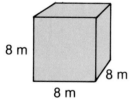

8 m

8 m

8 m

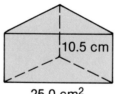

10.5 cm

25.0 cm²

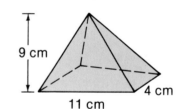

9 cm

11 cm

4 cm

1. $V = (\underline{\ ?\ } \cdot 8)\ m^3$

2. $V = (\underline{\ ?\ } \cdot \underline{\ ?\ })\ cm^3$

3. $V = \left(\dfrac{1}{3} \cdot \underline{\ ?\ } \cdot \underline{\ ?\ }\right) cm^3$

Exercises

Find the volume of the prism.

A
1. $h = 7$ in.
   $B = 22$ in.²

2. $h = 12$ cm
   $B = 19$ cm²

3. $h = 11$ yd
   $B = 30$ yd²

4. $h = 18$ cm
   $B = 49$ cm²

5. $h = 8$ ft
   $B = 25$ ft²

6. $h = 10$ yd
   $B = 26.75$ yd²

Find the volume of the figure.

B 7. square prism (cube)

$B = 18.5$ yd by 18.5 yd

$h = 18.5$ yd

8. triangular prism

$B = \left(\dfrac{1}{2} \cdot 17 \cdot 15\right)$ cm^2

$h = 14$ cm

9. triangular pyramid

$B = \left(\dfrac{1}{2} \cdot 21\right)$ m^2

$h = 8.5$ m

10. rectangular pyramid

$B = 19$ ft $\times$ 13.2 ft

$h = 20$ in.

11.

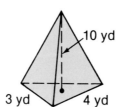

39 in.

7 in.   28 in.

12.

10 yd

3 yd   4 yd

13.

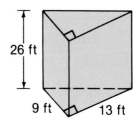

26 ft

9 ft   13 ft

Problems

Solve.

1. B times 8 yd is 128 yd^3.
2. B times 12 m is 216 m^3.
3. 49 m^2 times h is 294 m^3.
4. $\frac{1}{3}B$ times 9 ft is 54 ft^3.
5. 27 times $\frac{1}{3}$ times h is 99.
6. The original plan for the Great Pyramid of Cheops, near Cairo, Egypt, includes these measurements: $B = 765$ ft^2, $h = 320$ ft. Use the measurements to find the volume.
7. A plan for the Pyramid at Meydum includes these measurements: $B = 476$ ft^2, $h = 298$ ft. Find the volume.

6-8 *Volume of Cylinders and Cones*

Formulas for the volume of a cylinder and a cone are similar to the formulas for the volume of a prism and a pyramid.

The height is the distance between the bases. $\longrightarrow$ h

$B = \pi r^2$

Cylinder
$V = Bh$

$\longleftarrow$ *The height is the distance to the base.*

Cone
$V = \frac{1}{3}Bh$

Notice that the base of both a cylinder and a cone are circular. You will use what you know about the area of a circle to find the base area (B) of these two figures.

Example 1   The radius of a cylinder is 10 in. and the height is 12 in. What is the volume? Use $\pi \approx 3.14$.

Solution   Step 1: Solve for B.   Step 2: Solve for V.

$$B = \pi r^2 \qquad\qquad V = Bh$$
$$\approx 3.14 \cdot 10^2 \qquad\qquad \approx 314 \cdot 12$$
$$\approx 314 \text{ in.}^2 \qquad\qquad \approx 3768 \text{ in.}^3$$

Example 2   The base area and the height of a cone are the same as the base area and height of the cylinder in Example 1. What is the volume of the cone?

Solution   Step 1: $B = \pi r^2$ Step 2: $V = \dfrac{1}{3} \cdot 314 \cdot 12$

$$\approx 3.14 \cdot 10^2$$
$$\approx 314 \text{ in.}^2 \qquad\qquad\qquad = 1256 \text{ in.}^3$$

Your Turn   A cylinder and a cone have the same base area and the same height. The radius is 8 in. and the height is 20 in.
- Find the volume of the cylinder.
- Find the volume of the cone.

Class Exercises

Complete the formula by writing the correct number from the diagram on the right.

$$B = \pi r^2$$

8 in.
20 in.

$$V = Bh \qquad V = \frac{1}{3} Bh$$

1. $B = \underline{\ ?\ } \cdot 8^2$

2. $V = 3.14 \cdot 8^2 \cdot \underline{\ ?\ }$

3. $B = 3.14 \cdot \underline{\ ?\ }$

4. $V = \frac{1}{3} \cdot \underline{\ ?\ } \cdot 9^2 \cdot \underline{\ ?\ }$

14 cm

9 cm

5. $\underline{\ ?\ } = (3.14 \cdot 81)\ \text{cm}^2$

6. $V = \frac{1}{3} \cdot 3.14 \cdot 81 \cdot \underline{\ ?\ }$

Exercises

Find the volume of the figure. Use $\pi \approx 3.14$

A **1.**

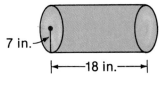

7 in.

|←——18 in.——→|

2.

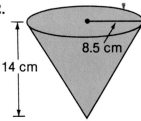

8.5 cm

14 cm

3.

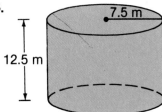

7.5 m

12.5 m

B **4.**

3 cm

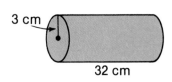

32 cm

5.

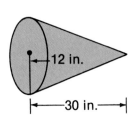

12 in.

|←——30 in.——→|

6.

15 yd

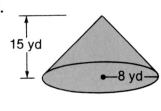

8 yd

7.

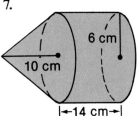

6 cm

10 cm

|←14 cm→|

8.

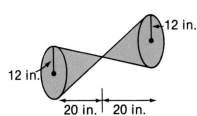

12 in.

12 in.

20 in. 20 in.

9. |←4 ft→|←6.5 ft→|

12 in.

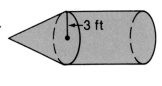

3 ft

10. The cylinder and the cone have equal radii and equal heights. The volume of the cylinder is 90 ft³. What is the volume of the cone?

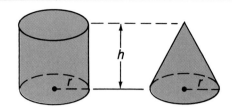

C 11. Assume that the top part of the bottle is a cone and the bottom part is a cylinder. What is the volume?

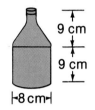

12. Find the total volume of the cup and the funnel.

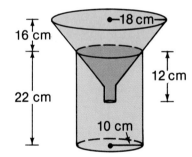

Using the Computer

Computer programs can be written to evaluate formulas. This program will find the volume of a cylinder when the height and the radius of the base are given. RUN this program for R = 3 and H = 6. Now try some examples of your own.

```
 5  REM: VOLUME OF A CYLINDER
10  PRINT "WHAT IS THE RADIUS OF THE BASE";
20  INPUT R
30  PRINT "WHAT IS THE HEIGHT";
40  INPUT H
50  LET P=3.14159
60  PRINT "VOLUME =";P*R*R*H
70  END
```

6-9 Surface Area

If you have to paint the entire outside surface of the birdhouse shown at right, you will probably paint one section at a time until all nine sections are painted. The entire outside surface of a figure is referred to as the **surface area** (S).

Like the birdhouse shown at right, many of the objects you see every day are made up of flat surfaces called **faces.**

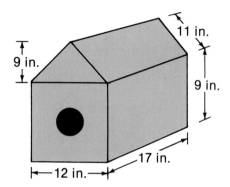

Example 1 Find the surface area of the birdhouse shown above.

Solution

Face	Area	How many?	Total
roof side (rectangle)	$11 \times 17 = 187$ in.^{2}	2	374 in.^{2}
roof end (triangle)	$\frac{1}{2} \times 12 \times 9 = 54$ in.^{2}	2	108 in.^{2}
front (rectangle)	$12 \times 9 = 108$ in.^{2}	1	108 in.^{2}
back (rectangle)	$12 \times 9 = 108$ in.^{2}	1	108 in.^{2}
side (rectangle)	$17 \times 9 = 153$ in.^{2}	2	306 in.^{2}
bottom (rectangle)	$17 \times 12 = 204$ in.^{2}	1	204 in.^{2}
		Surface area $= 1208$ in.^{2}	

The surface area is a little less than 1208 in.^{2} since the area of the entrance must be subtracted.

Your Turn • Find the surface area of the figure at right.

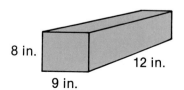

8 in. 12 in. 9 in.

The surface area for the soup can at right includes the area of the circular bases plus the area of the curved surface. Study the diagram to see that the curved surface can be flattened out to form a rectangle. Can you see that the base of the rectangle will have the same measure as the circumference of a circular base?

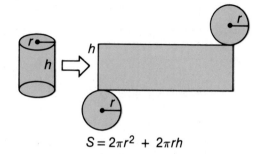

$$S = 2\pi r^2 + 2\pi rh$$

Example 2   Find the surface area of the object shown. Use $\pi \approx \frac{22}{7}$.

Solution
$$S = 2\pi r^2 + 2\,\pi rh$$

$$\approx \left(2 \cdot \frac{22}{7} \cdot 7^2\right) + \left(2 \cdot \frac{22}{7} \cdot 7 \cdot 30\right)$$

$$\approx \left(\frac{2}{1} \cdot \frac{22}{7} \cdot \frac{\overset{7}{\cancel{49}}}{1}\right) + \left(\frac{2}{1} \cdot \frac{22}{7} \cdot \frac{\overset{1}{\cancel{7}}}{1} \cdot \frac{30}{1}\right) \approx 1628 \text{ cm}^2$$

7 cm —

30 cm

Your Turn   • Find the surface area of a cylinder with a radius of $3\frac{1}{2}$ in. and a height of 8 in. Use $\pi \approx \frac{22}{7}$.

Class Exercises

Determine the number of faces in the figure.

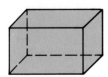   

1. __?__ rectangular faces
   __?__ triangular faces

2. __?__ triangular faces
   __?__ rectangular faces

3. __?__ circular faces
   __?__ rectangular faces

Exercises

Find the surface area. Use $\pi \approx 3.14$.

A **1.** 8 8 8   **2.** 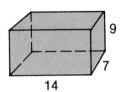 9 7 14   **3.** 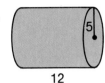 5 12

The height of each triangular face is 8 ft. Find the surface area.

B **4.** 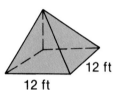 12 ft 12 ft   **5.** 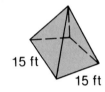 15 ft 15 ft   **6.** 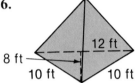 12 ft 8 ft 10 ft 10 ft

What is the least amount of paper needed to wrap the object? Use $\pi \approx 3.14$.

7. 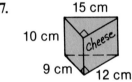 15 cm 10 cm 9 cm 12 cm cheese   **8.**  18 cm 5 cm GOLDEN CORN MEAL   **9.** 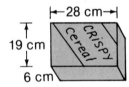 28 cm 19 cm 6 cm CRISPY Cereal

Self Test 3

Find the volume of the figure described. Use $\pi \approx 3.14$.

1. triangular prism
   $B = 80 \text{ cm}^2$, $h = 15 \text{ cm}$

2. rectangular prism   **6-7**
   $B = 45 \text{ mm}^2$, $h = 8 \text{ mm}$

3. cylinder
   $r = 6 \text{ in.}$, $h = 20 \text{ in.}$

4. cone   **6-8**
   $r = 7 \text{ cm}$, $h = 12 \text{ cm}$

Find the surface area of the figure. Use $\pi \approx 3.14$.

5. 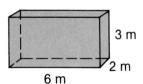 3 m 2 m 6 m   6. 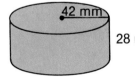 42 mm 28 mm   **6-9**

Algebra Practice

Write an equation. Solve.

1. The product of 355 and 982 is r.

2. The sum of 16.74, 39.1, and 26.492 is m.

3. 0.0861 divided by 0.07 is equal to x.

4. The difference between $5\frac{1}{2}$ and $2\frac{4}{9}$ is d.

5. $\frac{1}{8}$ of $4\frac{4}{5}$ is equal to q.

Use the distributive property to factor each expression.

6. $s^3t + (st)^2 + 5st^2$ 7. $6bc + 2b^2 + 4b^2c$ 8. $3gh^2 + g^2h + gh^3$

Solve the equation. Check the solution.

9. $a + 2\frac{3}{4} = 3\frac{1}{8}$ 10. $42 = 12t$ 11. $k - \frac{1}{8} = \frac{3}{4}$

12. $\frac{2}{5} = \frac{5}{6}d$ 13. $3(r - 2) = 12$ 14. $6z - 14 = z + 1$

Write an equation. Solve.

15. The perimeter of a rectangle 4.8 m long and 3.6 m wide is P.

16. The perimeter of a triangle with sides 5.3 ft, 4.4 ft, and 7.5 ft is P.

17. The circumference of a circle with a radius of 14 yd is C.

18. The area of a triangle with a base of 6.2 km and a height of 2.6 km is A.

19. The area of a circle with a radius of 5 m is A.

20. The area of a circle with a diameter of 28 ft is A.

21. The volume of a pyramid with a base area of 10 m² and a height of 6 m is V.

22. The volume of a cylinder with a radius of 4 yd and a height of 5 yd is V.

23. The surface area of a box 5 cm wide, 3 cm high, and 10 cm long is S.

24. The surface area of a cylinder with a radius of 5 ft and a height of 20 ft is S.

Problem Solving for the Consumer

The Do-It-Yourselfer

Have you ever wondered how much money people can save by doing at-home repairs themselves rather than hiring someone to do them? Doing the job yourself does cost money, though. Let's take a look at some of the hidden costs in do-it-yourself work.

Project: Renovate Dining Room
    Dimensions—13 ft long, 10 ft wide, 8 ft high

Problem: Peeling paint, cracking plaster in spots
Solution: Scrape existing paint, patch plaster, sand down rough spots, and paint

 First you need to figure how much paint is needed and how much it will cost.

1. How many square feet of ceiling space is there?
2. How many square feet of wall space is there? (Don't worry about doors and windows. Just pretend there are none.)
3. One quart of paint covers about 100 ft² on the first coat. How many quarts of ceiling paint will be needed for the first coat? How many quarts of wall paint will be needed?
4. Ceiling paint costs $14 per gallon and $3.75 per quart. What is the least amount of money the paint will cost?
5. Wall paint costs $15.50 per gallon and $4.50 per quart. What is the least amount of money the wall paint will cost?

6. Copy and complete the chart. Find the total costs of other materials to do the job.

A unit is one package, one can, etc. →		Price per Unit	Number of Units	Total
	scraper	$2.98	1	?
	plaster	$4.50	1	?
	sandpaper	$1.05 (pkg.)	2	?
	paint brush	$5.89	1	?
	paint roller covers	$1.69	2	?
	roller, roller pan	$6 (set)	1	?
Needed at the → *sanding stage*	goggles	$2.95	1	?
	protective masks	$1.99	1	?
				?

7. A professional painter will do the job for about $100 (including all materials). What is the difference in price when the job is done by the do-it-yourselfer instead of the professional? Does the do-it-yourselfer save money?

8. Suppose the do-it-yourselfer buys all the materials and pays an apprentice $5.50 an hour to do the work. If the apprentice completes the work in $6\frac{1}{2}$ h, what is the total cost of renovating the room?

9. If a second room needs to be done at another time, the do-it-yourselfer can use the following materials again: scraper, paint brush, roller and pan, goggles, protective masks. How much of a savings is this off the cost of supplies for the second room?

Enrichment

Functions

A function is sometimes referred to as a rule that takes a given value as an input and produces an output. For example, the rate of a certain long distance telephone call is 50¢ for the first minute and 34¢ for each additional minute. You can make up a rule or function for computing the costs of telephone calls. If you use n to represent the length of the phone call in minutes, then you can write a function as follows.

$$n \rightarrow 0.50 + (0.34)(n - 1)$$

Use the function to check the outputs in the table at right. For each input (n) there is one output (cost).

Input (n)	Output (cost)
5	$1.86
10	$3.56
15	$5.26

Use the given function to complete the table of outputs.

1. $n \rightarrow 4n + 1$
$1 \rightarrow 5$
$2 \rightarrow ?$
$3 \rightarrow ?$
$4 \rightarrow ?$
$5 \rightarrow ?$
$6 \rightarrow ?$
$7 \rightarrow ?$
$8 \rightarrow ?$

2. $n \rightarrow 5n - 3$
$1 \rightarrow 2$
$2 \rightarrow ?$
$3 \rightarrow ?$
$4 \rightarrow ?$
$5 \rightarrow ?$
$6 \rightarrow ?$
$7 \rightarrow ?$
$8 \rightarrow ?$

3. $n \rightarrow n^2 + 3$
$1 \rightarrow 4$
$2 \rightarrow ?$
$3 \rightarrow ?$
$4 \rightarrow ?$
$5 \rightarrow ?$
$6 \rightarrow ?$
$7 \rightarrow ?$
$8 \rightarrow ?$

4. $n \rightarrow 3n^2$
$1 \rightarrow 3$
$2 \rightarrow ?$
$3 \rightarrow ?$
$4 \rightarrow ?$
$5 \rightarrow ?$
$6 \rightarrow ?$
$7 \rightarrow ?$
$8 \rightarrow ?$

Complete the table of outputs, then write a function.

5.

Input	1	2	3	4	5	6	7	8
Output	3	8	15	?	?	?	?	?

Function: _____

6.

Input	1	2	3	4	5	6	7	8
Output	1	7	17	31	?	?	?	?

Function: _____

Chapter Review

Match.

1. perimeter of a rectangle
2. $A = \pi r^2$
3. perimeter of a square
4. $P = a + b + c$
5. perpendicular
6. $C = 2r \times \pi$
7. area of a parallelogram
8. $A = s^2$
9. an angle measuring less than 90°
10. obtuse angle
11. an angle that measures 90°
12. $V = \frac{1}{3}Bh$
13. parallel
14. volume of a prism
15. surface area of a prism

a. area of a circle
b. area of a square
c. lines that are the same distance apart
d. $P = 2l + 2w$
e. circumference of a circle
f. $V = Bh$
g. more than 90° but less than 180°
h. $P = 4s$
i. volume of a pyramid
j. $A = bh$
k. perimeter of a triangle
l. lines that form right angles
m. sum of the areas of all the faces
n. right angle
o. acute angle

Choose the correct answer. Write *a*, *b*, or *c*.

16. perimeter of a triangle with sides 4.5 ft, 4.4 ft, and 4.1 ft
    a. 11 ft   b. 12 ft   c. 13 ft

17. circumference of a circle with diameter = 9 mm. Use $\pi \approx 3.14$.
    a. 26.28 mm   b. 22.86 mm   c. 28.26 mm

18. an angle that measures 180°
    a. acute angle   b. straight angle   c. obtuse angle

19. the letter at the vertex of $\angle ABC$
    a. B b. C c. A

20. One circle has a radius of 25 in. Another circle has a radius of 12 in. Find the difference in the areas of the two circles.

Chapter Test

Find the perimeter of the figure described.

6-1

1. rectangle
   $l = 14.2$ ft, $w = 12.0$ ft

2. triangle
   sides $= 15.0$ m, 9.75 m, 11.3 m

What is the circumference of the circle? Use $\pi \approx 3.14$.

3. $d = 18$ in.   **4.** $r = 4$ yd   **5.** $d = 35$ ft   6-2

True or False? Write T or F.

6. An angle that measures $180°$ is a right angle.

7. An angle that measures $65°$ is an acute angle.

8. An angle that measures $108°$ is an obtuse angle.

What is the area of the polygon described?

9. rectangle
   $b = 23$ ft, $h = 16$ ft

10. triangle
   $b = 18$ cm, $h = 11$ cm

6-4

What is the area of the circle? Use $\pi \approx 3.14$.

11. $r = 15$ ft   **12.** $d = 118$ mi   **13.** $r = 8.3$ in.   6-5

Use the formula to solve: Power (kW) · Time (h) = Energy (kW · h).

14. In a certain experiment, 14 kW of power were used in 4 h. How much energy was used?

6-6

Find the volume of the figure described. Use $\pi \approx 3.14$.

15. rectangular prism
   $B = 15$ ft², $h = 4$ ft

16. triangular pyramid
   $B = 24$ in.², $h = 7\frac{1}{2}$ in.

6-7

17. cylinder
   $r = 6$ m, $h = 10$ m

18. cone
   $r = 3$ yd, $h = 13$ yd

6-8

What is the surface area of the figure shown? Use $\pi \approx 3.14$.

19.

20.

6-9

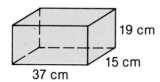

19 cm
15 cm
37 cm

 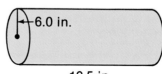

6.0 in.

12.5 in.

Cumulative Review *(Chapters 1–6)*

Solve the equation.

1. $y = 2704 - 96$

2. $0.4 \times 36 = x$

3. $6.2 \times 10 = n$

4. $\frac{1}{2} \times 84.2 = t$

5. $\frac{3}{4} \cdot 284 = s$

6. $\frac{4}{5} = \frac{x}{10}$

7. $(12)(4.2) - 6 = y$

8. $10 \cdot 10 \cdot 10 = z$

9. $A = (6.2)(5)$

10. $i = \$100 \cdot \frac{1}{8} \cdot 2$

11. $a = 4 + 7 \cdot 9$

12. $j = \frac{1}{3} \cdot 426$

13. $100 \times 0.46 = f$

14. $3.7 + 8 - 3.7 = w$

15. $1000 \times 8.92 = g$

16. $\frac{2}{5} = \frac{c}{100}$

17. $A = \frac{1}{2}(16)(7)$

18. $n = 5(2 + 5.6)$

19. $m = \frac{5}{8} \cdot 264$

20. $24 = 0.2n$

21. $0.25d = 1000$

Evaluate the expression.

22. ab; $a = 2, b = 7$

23. $x + y$; $x = 7, y = 9$

24. $4n - 1$; $n = 2.7$

25. n^2; $n = 4$

26. r^2; $r = 7$

27. $6p^2$; $p = 9$

28. $y^2 - 8$; $y = 8$

29. $2(x + 1)$; $x = 2.3$

30. $2xy$; $x = 5, y = 8$

Write an equation. Solve.

31. What number times 10 equals 4.6?

32. Some number is $\frac{3}{4}$ of 60. What is it?

33. $\frac{2}{3}$ of 63 is what number?

34. What number times 150 is 50?

35. $\frac{1}{4}$ of what number is 35?

Solve.

36. What is the sixth term of this sequence: 2, 5, 9, 14, 20, . . . ?

37. Tom's first computer had 16K memory. His next one had twice the memory. His third has twice the memory of the second. How much memory does the third one have? (One K equals 1000 units of memory.)

7
Ratio, Proportion, Percent

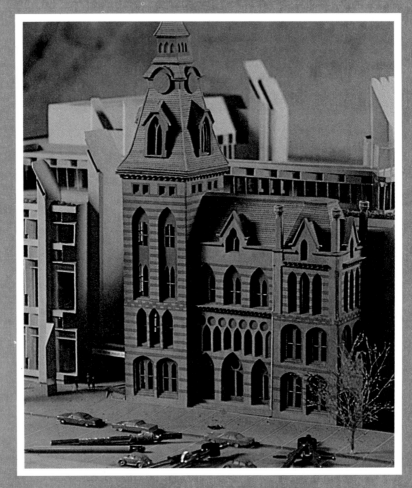

Scale models are used by architects to test their designs. The models make it possible to simulate various scenarios and test the overall appropriateness of the design.

7-1 Ratio

A theater owners group hired Pulse Polls, Inc. to conduct a survey in five cities. In each city 100 people were asked, "Would you attend the play It's a Short Time from September to April?" The results of the survey are shown in the table.

City	Brunswick	Eastway	Arbor Haven	Cuttiwell	Donnett
Favorable Responses	32	42	61	75	53

A **ratio** uses division to compare two numbers. The ratio of a to b can be written in three ways:

$$a \text{ to } b, \frac{a}{b}, \text{ or } a{:}b$$

You can write a ratio in lowest terms by dividing both terms by their GCF.

Example 1 Write a ratio in lowest terms to compare the number of favorable responses in Cuttiwell to all responses in that city.

Solution $\dfrac{75}{100} = \dfrac{75 \div 25}{100 \div 25}$ ⟵ *The GCF of 75 and 100 is 25.*

$= \dfrac{3}{4}$

Example 2 Write a ratio to compare the number of favorable responses in Brunswick to the favorable responses in Eastway.

Solution $\dfrac{\text{Brunswick}}{\text{Eastway}}$ ⇨ $\dfrac{32}{42} = \dfrac{32 \div 2}{42 \div 2} = \dfrac{16}{21}$

For every 16 favorable responses in Brunswick, there were 21 favorable responses in Eastway.

Your Turn
- Write a ratio as a fraction in lowest terms to compare 12 chickens to 4 eggs.
- Write a ratio as a fraction in lowest terms to compare 400 musicians to 250 instruments.
- A news report stated that 27 out of 30 people voted for Tom Mathews. Write this ratio in lowest terms.

Class Exercises

Write the ratio as a fraction in lowest terms.

1. 5 to 100   **2.** a to b **3.** 25 to 50

4. 35 to 80   **5.** 31 to 40   **6.** c to d

7. 65 to 15   **8.** 36 to 8   **9.** 40 to 120

10. $12d$ to $18b$ **11.** $7x$ to $56y$ **12.** $39t$ to $13n$

Exercises

Write the ratio as a fraction in lowest terms.

A
1. 35 patients to 10 children   **2.** 35 apartments to 5 vacant

3. 12 trucks to 3 pickups   **4.** 36 roads to 4 unpaved

5. 45 players to 5 injured   **6.** 16 puppets to 4 on strings

7. 27 departments to 9 sold out   **8.** 48 parking spaces to 8 empty

Write as a ratio in lowest terms using the ":" notation.

9. $\dfrac{2}{3}$ **10.** $\dfrac{4}{7}$ **11.** $\dfrac{19}{20}$ **12.** $\dfrac{m}{n}$ **13.** $\dfrac{47}{92}$

14. $\dfrac{95}{100}$ **15.** $\dfrac{x}{y}$ **16.** $\dfrac{95}{108}$ **17.** $\dfrac{46}{80}$ **18.** $\dfrac{p}{q}$

B **19.** Use the table to write a ratio in lowest terms to compare the responses in each category to the total responses.

Viewer Survey of Janet Effmar Show

Liked	Disliked	No Opinion
35	30	10

Use the table to complete the exercise.

20. Write a ratio in lowest terms to compare each response to the total responses for each candidate.

Voter Survey on Mayoral Candidates

Candidate	Like	Dislike	No Opinion
Carp	25	40	15
Nolan	34	34	12
Dorado	60	16	4
Owens	14	12	54
Thompson	15	45	20

Problems

Solve.

1. Of 120 people polled, 48 said they would not vote for Fenner for Congress. All others said they would.
  a. Write the following ratios:
     No Votes for Fenner : Total votes
     Yes Votes for Fenner : No Votes for Fenner.
  b. For each ratio in **a** write a fractional ratio in lowest terms. Express the ratio using the ":" notation.

2. In order to receive research funds, John Eston had to show that people favored his product by a ratio of at least 5:4 over the competition. (His product : competition = 5:4)
   In a survey of 4000 people, 2400 favored Eston's product while 1600 favored that of the competitor. Will Mr. Eston receive the research funds?

Sales of tennis rackets were studied by Delor Management Group. Since the totals were not available, they assigned letters to stand for the numbers, and constructed this table.

SALE OF TENNIS RACKETS

Tennis Racket	Super Serve	Smooth Hit	Volley
Number Sold	N	P	Q

3. a. Write a ratio comparing these sales, using the letters given.

Example $\dfrac{\text{Super Serve}}{\text{Smooth Hit}} = \dfrac{N}{P}$

  b. If $N = 4650$, $P = 8962$, and $Q = 2384$, write a ratio in lowest terms for each pair of tennis rackets.

211

7-2 *Proportions*

A national health agency asked 60 parents in a random sample if their children had had chickenpox. Of those asked, 50 said, "Yes." If 600 parents had been asked, about how many would you expect to say "Yes"? You can write equal ratios to solve the problem. An equation that shows equal ratios is called a **proportion.**

$$\frac{\text{Yes responses}}{\text{total responses}} = \frac{50}{60} \quad \Rightarrow \quad \frac{50}{60} = \frac{n}{600} \quad \Rightarrow \quad \frac{50 \times 10}{60 \times 10} \quad \Rightarrow \quad \frac{500}{600}$$

$$n = 500$$

You would expect 500 parents to say "Yes."

Example 1   A car model is being built with a scale of 3 cm to 20 cm. The length of the car is 400 cm. What is the length of the model?

Solution   Let n = length of the model in centimeters.
$$\frac{3}{20} = \frac{n}{400} \quad \Rightarrow \quad \frac{3}{20} = \frac{3 \times 20}{20 \times 20} = \frac{60}{400} \quad \Rightarrow \quad n = 60$$
The model is 60 cm long.

Example 2   The length of the model car's hood is 18 cm. How long is the hood of the actual car?

Solution   Let n = the length of the hood in centimeters.
$$\frac{3}{20} = \frac{18}{n} \quad \Rightarrow \quad \frac{3}{20} = \frac{3 \times 6}{20 \times 6} = \frac{18}{120} \quad \Rightarrow \quad n = 120$$
The hood is 120 cm long.

Your Turn   • A tire company sells 2 tires for $87. How much will 4 tires cost?

You can **cross multiply** to solve proportions that aren't easily solved by using equal ratios.

In Arithmetic

If $\dfrac{2}{5} = \dfrac{6}{15}$

Then $2 \times 15 = 5 \times 6$

In Algebra

If $\dfrac{a}{b} = \dfrac{c}{d}$ $(b \neq 0,\, d \neq 0)$

Then $ad = bc$

Example 3   Mr. Jones paid $9 for 4 kg of grass seed. How much would 10 kg cost?

Solution   Let $n =$ the cost of 10 kg of seed.

$$\dfrac{9}{4} = \dfrac{n}{10}$$

$9 \times 10 = 4n$ ⟵ *Cross multiply.*

$90 = 4n$

$\dfrac{90}{4} = \dfrac{4n}{4}$ ⟵ *Divide both sides by 4.*

$\$22.50 = n$

The cost of 10 kg of grass seed is $22.50.

Your Turn   • It took an equipment truck 3 hours to travel 200 km. How long will it take to travel 500 km?
  • A car maintains a speed of 80 km per hour. How far will it travel in 3.5 hours?

Class Exercises

Solve the proportion.

1. $\dfrac{3}{4} = \dfrac{n}{12}$

2. $\dfrac{2}{3} = \dfrac{6}{n}$

3. $\dfrac{n}{6} = \dfrac{21}{18}$

4. $\dfrac{5}{7} = \dfrac{n}{35}$

5. $\dfrac{1}{2} = \dfrac{n}{8}$

6. $\dfrac{5}{3} = \dfrac{20}{n}$

7. $\dfrac{3}{1} = \dfrac{n}{400}$

8. $\dfrac{n}{40} = \dfrac{3}{5}$

Exercises

Solve the proportion.

A **1.** $\dfrac{2}{14} = \dfrac{n}{21}$ **2.** $\dfrac{3}{15} = \dfrac{x}{35}$ **3.** $\dfrac{a}{6} = \dfrac{9}{54}$ **4.** $\dfrac{7}{y} = \dfrac{4}{9}$

5. $\dfrac{4}{18} = \dfrac{6}{n}$ **6.** $\dfrac{6}{n} = \dfrac{15}{35}$ **7.** $\dfrac{12}{27} = \dfrac{8}{y}$ **8.** $\dfrac{18}{21} = \dfrac{b}{28}$

Cars and trucks of various lengths are to have models made. The ratio will be $\dfrac{\text{model length}}{\text{vehicle length}} = \dfrac{1}{8}$. Complete the table for the length of each model.

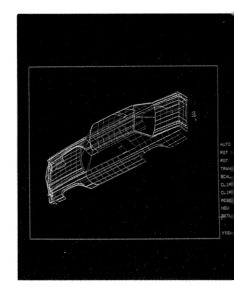

		Vehicle Name	Vehicle Length	Model Length
B	**9.**	Cat bird	360 cm	n
	10.	Forma	400 cm	n
	11.	Invit	560 cm	n
	12.	Torp	m	60 cm
	13.	Granda	m	90 cm
	14.	Macin	m	82 cm
	15.	Fash	433 cm	n
	16.	Sparka	528 cm	n
	17.	Dan Man	984 cm	n
	18.	Welton	288 cm	n

Solve.

C **19.** A square with sides of 5 cm has each side doubled. What effect does this have on the perimeter?

20. In a large room, the ratio of length to width is $3:1$.
   a. If the length is 18 m, how wide is the room?
   b. Find the perimeter and area of the room.
   c. The dimensions of the room are doubled. Does this affect the ratio of length to width?
   d. Find the perimeter of the room using the new dimensions. How does the "doubling" affect the perimeter?
   e. How does the "doubling" affect the area?

Problems

Solve. If necessary, round to the nearest tenth.

1. A truck driver can travel 200 miles in 3 days. How far will he travel in 12 days?
2. On a map of Fairview County, 12 km are represented by 1 cm. How many kilometers would be shown by 7 cm?
3. At Green's Department Store, shirts are on sale at 3 for $50. How much will 8 shirts cost?
4. In a sample of 75 light bulbs, a store owner found 9 that were broken. How many broken bulbs can she expect to find in a case of 200?
5. A baseball pitcher allowed 2 runs in 9 innings. About how many runs would he allow in 54 innings at the same rate?
6. A Metro car travels 450 km on 40 L of gas. How far will it travel on 110 L of gas?
7. A TV station broadcasts 5 out of 6 playoff games every year. This year there will be about 58 playoff games. About how many games will the station broadcast?
8. In Medfield, 3 out of 10 people attended the circus. The town has 900 residents. How many attended the circus? Estimate how many would have attended if the town had 1800 residents.

Using the Calculator

The diagram is drawn to a scale of $\dfrac{1 \text{ cm diagram length}}{48 \text{ cm tent length}}$. Measure the length on the diagram and calculate the corresponding length of the tent.

1. a **2.** b **3.** c **4.** d

If the scale were $\dfrac{1 \text{ cm diagram length}}{56 \text{ cm tent length}}$ what would the tent measurements be?

5. a **6.** b **7.** c **8.** d

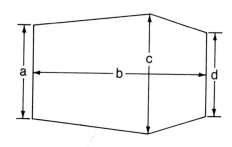

Floor plan of a
2-person tent

7-3 *Percents*

Two surveys were conducted to test voter opinion about a candidate for the state assembly. The candidate received 96 out of 400 votes in the first survey and 84 out of 300 votes in the second survey. In which survey did the candidate do better?

You can compare the ratios by writing a proportion with a denominator of 100.

First Survey

$$\frac{96}{400} = \frac{n}{100}$$

$400n = 96 \times 100$
$400n = 9600$
$n = 24$

In the first survey, the candidate received 24 out of 100 votes.

Second Survey

$$\frac{84}{300} = \frac{m}{100}$$

$300m = 84 \times 100$
$300m = 8400$
$m = 28$

In the second survey, the candidate received 28 out of 100 votes.

The candidate did slightly better in the second survey.

When you compare groups to 100 or find out how many out of 100, you're finding a **percent.** Percent means "per 100." The symbol % means percent.

Example 1   A candidate received 57 out of 300 votes. What percent is this?

Solution   Let $a =$ the number of votes out of 100.

$$\frac{57}{300} = \frac{a}{100}$$

$300a = 57 \times 100$ ⟵——— *Cross multiply to solve for* a.
$300a = 5700$
$a = 19$

The candidate received 19 out of every 100 votes.
The candidate received 19% of the vote by receiving 57 out of 300 votes.

Your Turn   Write the ratio as a percent.
- 7 out of 100
- 45 out of 300
- 56 votes out of 400

You can write percents as decimals as well as fractions.

Example 2   Write 44% as a fraction and as a decimal.

Solution $44\% = \dfrac{44}{100} = 0.44$

Your Turn   Write the percent as a fraction and as a decimal.
- 37%   • 9%   • 83%

Some fractions cannot easily be written with a denominator of 100. When this happens, use division to write the fraction as a decimal, then write the decimal as a percent.

Example 3   Write $\dfrac{9}{37}$ as a percent.

Solution $\dfrac{9}{37}$ $\Rightarrow$ $\begin{array}{r} 0.243\ldots \approx 0.24 \\ 37\overline{)9.000} \end{array}$ ← *Since the answer is a repeating decimal, it is convenient to round to the nearest hundredth.*

$0.24 = \dfrac{24}{100} = 24\%$ ← *Move the decimal point two places to the right.*

Class Exercises

Write as a percent. Round to the nearest hundredth.

1. 7 out of 100	**2.** 9 out of 100	**3.** 23 out of 100	**4.** 99 out of 100
5. 90 out of 225	**6.** 30 out of 75	**7.** 16 out of 85	**8.** 720 out of 1250

Exercises

Write as a percent.

A **1.** 0.07   **2.** 0.56   **3.** 0.99   **4.** 0.63   **5.** 0.12   **6.** 0.81

7. 0.03   **8.** 0.05   **9.** 0.15   **10.** 0.25   **11.** 0.90   **12.** 0.33

Write the percent as a fraction and a decimal.

13. 5%   **14.** 18%   **15.** 45%   **16.** 28%   **17.** 8%

18. 10%   **19.** 91%   **20.** 77%   **21.** 50%   **22.** 1%

Write as a percent. Round to the nearest hundredth if necessary.

B **23.** $\dfrac{1}{5}$ **24.** $\dfrac{5}{8}$ **25.** $\dfrac{9}{10}$ **26.** $\dfrac{3}{4}$ **27.** $\dfrac{3}{8}$

28. $\dfrac{5}{12}$ **29.** $\dfrac{3}{25}$ **30.** $\dfrac{39}{50}$ **31.** $\dfrac{1}{3}$ **32.** $\dfrac{5}{6}$

33. $\dfrac{1}{7}$ **34.** $\dfrac{4}{9}$ **35.** $\dfrac{15}{62}$ **36.** $\dfrac{18}{75}$ **37.** $\dfrac{68}{135}$

Write the percent.

38. What percent of 100 is 22?   **39.** What percent of 200 is 150?

40. What percent of 50 is 6?   **41.** What percent of 58 is 25?

42. What percent of 75 is 25?   **43.** What percent of 550 is 250?

Using the Computer

This program will find one number as a percent of another. RUN it
using P = 27, B = 135, and R = 20. Now try some other examples.

```
10  PRINT "TO FIND WHAT PERCENT P IS OF B:"
20  PRINT "WHAT IS P";
30  INPUT P
40  PRINT "WHAT IS B";
50  INPUT B
60  PRINT P;" IS";(P/B)*100;"% OF";B;"."
70  END
```

7-4 *Percent of a Number*

To approve of a new camera at least 40% of the people in a survey must respond favorably. If there are 300 people in the survey, how many favorable responses are needed?

You can write and solve an equation to answer the question "What number is 40% of 300?"

Let n = the number of favorable responses needed

Then n is 40% of 300 ⟵ *In percent problems, "of" means multiply.*
$\quad n = 0.40 \times 300$ ⟵ *Write the percent as a decimal and solve.*
$\quad n = 120$

At least 120 people must give favorable responses.

When you compute sales tax and interest you're finding a percent of a number.

Example 1   The Wilcox family bought camping gear that cost $439.98. If the sales tax rate is 5%, how much was the sales tax? What was the price including tax?

Solution   Let n = the amount of the sales tax
Then n = 5% of $439.98
$\quad n = 0.05 \times 439.98$
$\quad n = \$21.999$ ⟵ *Round to the nearest cent.*
$439.98 + \$22 = \$461.98$ ⟵ *Add the tax to the total cost.*
The sales tax was $22 and the price including tax was $461.98.

Your Turn   • Riverdale High chartered a bus for a class trip. The cost of the charter was $178.45 plus 4% sales tax. What was the amount of the tax?

You can use the following formula to find the amount of interest paid on a loan or deposit.

$$\text{interest } (i) = \text{principal } (p) \times \text{rate } (r) \times \text{time } (t)$$

The principal is the amount on which the interest is being computed, the rate is the rate of interest being paid, and the time is the length of time, in years, over which interest is being earned.

Example 2   Jeremy borrowed $400 for 6 months to buy a computer. The rate of interest is 15%. How much interest does he pay?

Solution $i = p \times r \times t$
$i = 400 \times 0.15 \times 0.5$ ⟵ *Use t = 0.5 since 6 months is 0.5 year.*
$i = 30$
Jeremy pays $30 interest.

Your Turn   Compute the interest
- Max has $375 in a savings account. How much interest will he earn in one year if the annual rate is 5%?
- Maria borrowed $950 to take a course in TV production. The annual interest rate is 14%. How much interest will she pay if she borrows the money for two years?

Class Exercises

Find the number. Round to the nearest whole number.

1. 10% of 200 is n

2. 32% of 25 is y

3. 9% of 800 is c

4. 42% of 50 is z

5. 63% of 1500 is p

6. 5% of 20 is d

7. 99% of 2500 is t

8. 1% of 2500 is w

9. 14% of 25 is m

Exercises

Write an equation. Solve. Round to the nearest whole number.

A
1. 32% of 900 is *r*   **2.** 14% of 800 is *s*   **3.** 90% of 20 is *z*

4. 12% of 25 is *q*   **5.** 3% of 4682 is *d*   **6.** 18% of 775 is *c*

7. 3% of 35,482 is *n*   **8.** 9% of 1186 is *y*   **9.** 20% of 689 is *b*

10. 89% of 1752 is *m*   **11.** 43% of 27 is *t*   **12.** 55% of 3002 is *z*

Allied Appliance is having a 20% off sale. What is the sale price for the item?

B
13. Toaster   —$39.98   **14.** Microwave oven   —$578

15. Refrigerator   —$637.50   **16.** Hair dryer   —$42.50

17. Food processor —$138.98   **18.** Stove   —$588.88

Find the total price for each item after a 20% price reduction and a 7% sales tax.

19. Electric wok   —$45.50   **20.** Foot massager   —$33.75

21. Electric shaver —$29.98   **22.** Electric pencil sharpener—$15.88

23. Vacuum cleaner—$158.59   **24.** Blender   —$95.89

Problems

Solve.

1. A team of 8 people was paid $4000 for work completed on a project. The payment was divided as follows: Malloy—30%; Hodge—25%; Hemmer—18%; Sullivan—13%; Miller—10%; Belini—8%; Cowens—6%; and Ling—3%. How much did each person receive? What is wrong with the total amount?

2. An author received $8000 in advances for the sale of a book. She paid 25% of the advance to an assistant. The assistant paid 50% of his share to a researcher. The researcher paid 50% of her share to a fact finding company. The company paid 50% of its share to a systems analyst who set up their system. How much did each person receive? How much did each person have left after paying out part of his or her share?

7-5 Problem Solving

Suppose you want to open a NOW account at a bank. In order to compare the features of each bank, it's helpful to make a chart to **organize the information.**

NOW ACCOUNT – SNOW BANK

The checking account that pays you 5.25% annual interest compounded daily from day of deposit to day of withdrawal.

- NO CHARGE for this service as long as you keep a balance of $500.
- If your balance falls below $500 at any time, there will be a $2.00 account maintenance fee plus a 15¢ charge per check for all checks written during that month's statement period.

River Bank   **NOW Accounts.**

There will be no monthly service charge if the minimum balance in your NOW Account is $500 or more during the statement period. Otherwise, the service charge will be $1.75 and an additional $.25 for each check.

Earn 5¼% compounded daily.

Example 1 Which bank offers the more economical plan if you write about 15 checks per month?

Solution Make a chart to organize the information.

Bank	Interest Rate	Service Charge	Minimum Balance	Check Charge
Snow	5.25%	$2.00	$500	$.15/check
River	5.25%	$1.75	$500	$.25/check

The interest rates and minimum balances are the same, so look at the other two columns. You need to compare the service charges and check charges.

Snow Bank

Checks		Check Charge		Fee
15	×	$.15	=	$2.25
		Service Charge		+2.00
				$4.25

River Bank

Checks		Check Charge		Fee
15	×	$.25	=	$3.75
		Service Charge		+1.75
				$5.50

Snow Bank is slightly more economical.

Your Turn • At the Diamond Bank, the service charge is $3.00 per month. The check charge is $.15 per check. If you write 20 checks per month, will you pay less at River Bank or Diamond Bank?

Class Exercises

Use the chart for Exercises 1 and 2.

Bank	Interest Rate	Service Charge	Min. Bal.	Check Charge
Wilde	6.09	$1.75	$300	$.20/check
Data	6.09	$2.15	$300	$.15/check
Roman	6.09	$1.60	$300	$.25/check
Justice	6.09	$2.10	$300	$.12/check

1. If you write about 30 checks per month, which bank's plan is more economical, Wilde's or Data's?
2. If you write about 45 checks per month, which bank's plan is more economical, Roman's or Justice's?

Exercises

Solve.

A 1. At a sporting goods store golf balls are sold in packages of 3. Each package costs $3.75. Make a chart to show the cost of from 1 to 10 packages.
2. Bus tokens cost 55¢ apiece. Make a chart to show the cost of from 1 to 10 tokens.
3. A math teacher gave a quiz containing 15 questions. She figured each student's grade by multiplying the number of correct answers by 6 and adding 10. Make a chart to show the possible grades.
B 4. At a photo shop reprints cost 65¢ each plus 3% sales tax. Make a chart to show the cost of from 1 to 10 reprints. How much will 18 reprints cost?

5. A billing clerk must calculate interest charges for charge customers before preparing their bills. The charges are as follows: $1\frac{1}{2}\%$ for the first $500, 1% for the amounts over $500. Make a chart to show the interest charge and total amount of bills for customers owing the following amounts: Mr. Wang—$186.50, Mrs. Toon—$468.25, Miss Saunders—$86.75, Ms. Kopek—$675.82, Mr. Laffer—$572.39, Miss Golden—$888.69.

Self Test 1

1. Use the table to write a ratio in lowest terms to compare the responses in each category to the total responses for each brand.   **7-1**

Customer Survey of Toothpaste Preference

Brand	Like	Dislike	No Opinion
Glisten	55	22	8
PearlyWhite	63	18	7
Shine'n'Smile	19	75	2

Solve the proportion.

2. $\dfrac{1}{7} = \dfrac{n}{28}$ 3. $\dfrac{6}{y} = \dfrac{3}{9}$ 4. $\dfrac{b}{22} = \dfrac{5}{10}$ 5. $\dfrac{5}{15} = \dfrac{7}{q}$ **7-2**

Write as a percent.

6. $\dfrac{4}{25}$ 7. $\dfrac{6}{24}$ 8. $\dfrac{15}{75}$ 9. $\dfrac{5}{20}$ **7-3**

A-1 Auto parts offers a 15% discount on all purchases. Find the discounted price for the item.

10. case of oil—$9.00   11. windshield wipers—$4.85   **7-4**

12. antenna—$12.50

13. At the Town Rink, it costs $.75 to rent ice skates and $1.85 for an afternoon's skating pass. Make a chart to show the cost of skating for from 1 to 10 days. If you can get a 12% discount for buying a 5-day pass that includes skate rental, how much will you save?   **7-5**

7-6 *More Percents*

The Alair Rug Company installed 200 m² of carpeting. In the process of trimming the rug, 1 m² was wasted. What percent of the rug was wasted?

Let w = the percent of the rug wasted

Then $w = \dfrac{1}{200}$

$ = 0.005 \longleftarrow$ *Move the*
$ = 0.5\%$ *decimal point two places to the right.*

We say that 0.5% or $\dfrac{1}{2}\%$ of the rug was wasted. This means that

5 tenths of 1% or $\dfrac{1}{2}$ of 1% was wasted. Decimal or fractional percents

mean an amount less than $\dfrac{1}{100}$ or 1%.

Example 1   What is $\dfrac{3}{4}\%$ of 582?

Solution   Let $n = \dfrac{3}{4}\%$ of 582

$\dfrac{3}{4}\% = 0.75\% = 0.0075 \longleftarrow$
$n = 0.0075 \times 582$
$n = 4.365$

> *Write the fraction as a decimal. Move the decimal point two places to the left. Add zeros as needed.*

Your Turn   • What is $\dfrac{1}{5}\%$ of 750?   • What is $4\dfrac{1}{2}\%$ of 88?

Be careful not to confuse fractions and decimals with fractional and decimal percents.

Example 2   Are $\frac{1}{3}\%$ of 600 and $\frac{1}{3}$ of 600 the same?

Solution   Let $n = \frac{1}{3}\%$ of 600   Let $m = \frac{1}{3}$ of 600.

$\frac{1}{3}\% \approx 0.33\% = 0.0033$ $\frac{1}{3} = 0.333$

$n = 0.0033 \times 600$ $m = 0.333 \times 600$
$n = 1.98 \approx 2$ $m = 199.8 \approx 200$

Since 2 is less than 200, $\frac{1}{3}\%$ of 600 is not the same as $\frac{1}{3}$ of 600.

Your Turn   • Sturn Corporation earned $8000 one week. Of the week's earnings, $\frac{3}{4}\%$ was spent on office supplies and $\frac{3}{4}$ was spent on salaries. How much was spent in each area?

Last week's sales at Walder Company totalled $6000. The sales manager would like to see this week's sales be 150% of last week's. This means that the sales manager would like to see sales increase by 50% or be $\frac{1}{2}$ again as large as last week's. Percents greater than 100 are equal to numbers greater than 1. Here's how to find the actual amount of this week's sales goal.

$150\% = \dfrac{150}{100} = 1.50$ ⟵ —— *Move the decimal point two places to the left.*

$n = 1.5 \times 6000$
$n = 9000$ This week's sales goal is $9000.

Your Turn   • What is 135% of 86?   • What is 200% of 75?

Class Exercises

Write as a decimal.

1. $\frac{1}{4}\%$ 2. 0.4% 3. $\frac{1}{10}\%$ 4. 0.7% 5. $\frac{1}{3}\%$

6. $4\frac{1}{2}\%$ 7. $12\frac{3}{5}\%$ 8. 9.5% 9. $15\frac{7}{8}\%$ 10. 75.75%

Exercises

Write as a percent.

A 1. $\frac{4}{50}$ 2. $\frac{6}{500}$ 3. $7\frac{1}{2}$ 4. $\frac{3}{750}$ 5. $5\frac{1}{8}$

6. 0.56 7. 0.003 8. 0.469 9. 3.62 10. 4.615

11. $\frac{15}{1200}$ 12. 0.0909 13. $\frac{12}{1500}$ 14. 15.2 15. $\frac{15}{3000}$

Find the number. Round to the nearest hundredth.

B 16. $\frac{1}{3}\%$ of 450 is y 17. $\frac{2}{5}\%$ of 500 is n 18. $\frac{9}{10}\%$ of 368 is r

19. 125% of 993 is b 20. 250% of 2166 is t 21. $3\frac{2}{3}\%$ of 1230 is z

Problems

Solve.

1. Compedium, Inc. sold 3598 electric mixers last year. This year they hope to have sales 133% of last years sales. What is the company's goal in units?

2. A six-month bank certificate pays $11\frac{1}{2}\%$ interest. How much interest will be earned on a $10,000 investment?

3. The population of Aldersville is increasing at the rate of $3\frac{1}{2}\%$ per year. If last year's population was 7816, what should this year's be?

7-7 *Finding the Base*

In a recent shipment, a local supermarket received 75 cases of soup. The soup represented only 25% of the cases in the shipment. How many cases were in the shipment?

To solve this problem you need to find the **base** or total number of cases that were to be in the shipment.

Let s = the base
Then 25% of s = 75
$0.25 s = 75$ ⟵ *Write the percent as a decimal and solve.*
$$s = \frac{75}{0.25}$$
$$s = 300$$

The base number of cases was 300.

Example An agent for a musical group earns a 12% commission based on the group's earnings. How much must the group earn for the agent to earn $36,000?

Solution Let g = group's earnings
Then $12\% \, g = 36{,}000$
$0.12 \, g = 36{,}000$
$$g = \frac{36{,}000}{0.12}$$
$$g = \$300{,}000$$
The group must earn $300,000 for the agent's commission to be $36,000.

Your Turn
- Season's attendance at Larchmont's football games was 2876. This is a 130% of last years attendance. What was last year's attendance?
- There are 386 books on loan from the East High library. This represents 9% of the books owned by the library. How many books does the library own?

Class Exercises

Solve. Round to the nearest whole number.

1. $60 = 0.10\,n$

2. $90 = 0.30\,b$

3. $45 = 0.05x$

4. $66 = 0.66\,a$

5. $44 = 0.15\,z$

6. $37 = 0.99\,y$

7. $570 = 0.25\,n$

8. $56 = 0.75\,b$

9. $999 = 0.12\,x$

Exercises

Write and solve an equation to find the base. Round to the nearest whole number.

A

1. 120 is 8% of what number?

2. 1200 is 10% of what number?

3. 1586 is 2% of what number?

4. 2100 is 7% of what number?

5. 108 is 9% of what number?

6. 725 is 8% of what number?

B

7. 6 is 0.5% of what number?

8. 14 is $\frac{3}{4}$% of what number?

9. 25 is 0.25% of what number?

10. 82 is $\frac{1}{2}$% of what number?

11. 65 is 125% of what number?

12. 130 is 200% of what number?

13. 88 is 33.5% of what number?

14. 235 is 5.5% of what number?

Problems

Solve.

1. On the average, families should expect to pay about 25% of income for housing costs. If the rent on an apartment is $575 per month, how much must a family earn to afford the rent?

2. Union Utilities requested a 17% increase in electricity rates. This will raise the Cooper Company's monthly bill to $380. How much are they now paying?

3. "With an 8% increase in production we can produce 400 more transmissions per month," said the production manager at Central Transmission Co. How many are they currently producing?

229

7-8 *Percent of Increase or Decrease*

Sue and Mary work for a law firm. Sue earns $400 per week and Mary earns $650 per week. Sue recently received a raise to $436 while Mary received a raise to $700. Mary received the higher dollar amount but you should compare the percent of increase to see who received a higher percent of salary.

To find the percent of increase, first subtract to find the amount of increase.

Sue: $436 - 400 = 36$

Mary: $700 - 650 = 50$

Next write an equation and solve for the percent of increase.

Sue: Let t = the % of increase

$400\, t = 36$

$$t = \frac{36}{400} = 0.09 = 9\%$$

Sue's raise was 9%.

Mary: Let m = the % of increase

$650\, m = 50$

$$m = \frac{50}{650} = 0.077 = 7.7\%$$

Mary's raise was 7.7%.

Sue's percent of increase was greater than Mary's.

Example 1   Jerry bought a table for $480 that originally cost $600. What was the percent of discount?

Solution $600 - 480 = 120$ ⟵ *Subtract to find the amount of discount.*
Let n = % of discount
$600\, n = 120$

$$n = \frac{120}{600} = 0.20 = 20\%$$

Jerry received a 20% decrease or discount.

Your Turn
- A news staff director must reduce his staff from 40 to 25. What is the percent of decrease?
- Joan scored 80 on her first test and 96 on her second test. What was the percent of increase?

Class Exercises

Complete to find the percent of change. Round to the nearest whole percent.

	Original Price	New Price	Amount of Change	Percent of Change
1.	$485	$425	?	?
2.	$ 65	$ 75	?	?
3.	$188	$206	?	?
4.	$238	$200	?	?
5.	$344	$328	?	?

Exercises

Find the percent of decrease. Round to the nearest whole percent.

A **1.** old price: $170
decrease: $45

2. old price: $45
decrease: $6

3. old price: $3591
decrease: $659

4. old price: $15.98
decrease: $2.75

5. old price $239.50
decrease: $23.95

6. old price: $87.99
decrease: $22.00

Find the percent of increase. Round to the nearest whole percent.

7. old price: $65
increase: $10

8. old price: $359
increase: $36

9. old price: $1500
increase: $300

10. old price: $69.85
increase: $3.50

11. old price: $78.35
increase: $11.75

12. old price: $1.29
increase: $1.94

Find the percent of change. Round to the nearest whole percent.

13. old price: $18.25
new price: $15.79

14. old price: $33.12
new price: $42.26

15. old price: $56.89
new price: $47.59

Find the percent of increase or decrease to the nearest tenth of a percent.

B **16.** from 675 to 580 **17.** from 33 to 77 **18.** from 186 to 200

19. from 300 to 178 **20.** from 5 to 65 **21.** from 78 to 56

22. from 87 to 35 **23.** from 4786 to 3955 **24.** from 1350 to 866

25. from 98 to 125 **26.** from 116 to 88 **27.** from 760 to 670

C **28.** A company employs 100 people. If the staff is decreased to 90, what is the percent of decrease? If the staff is increased from 90 to 100 what is the percent of increase? Explain.

Problems

Solve.

1. The value of a mining stock rose from $36 to $40 per share. What is the percent of increase?
2. The price of a stereo set was raised from $300 to $315. What is the percent of increase?
3. A salesperson drove 450 miles in October. In November she drove 360 miles. What is the percent of decrease?

Self Test 2

Find the number. Round to the nearest hundredth.

1. $\frac{1}{2}\%$ of 300 is x **2.** 150% of 450 is y **3.** $4\frac{3}{4}\%$ of 80 is b **7-6**

Write and solve an equation to find the base. Round to the nearest whole number.

4. 15 is 10% of what number?   **5.** 7 is 2% of what number?   **7-7**

Find the percent of change to the nearest tenth of a percent.

6. from 48 to 84   **7.** from 100 to 35   **8.** from 650 to 605   **7-8**

Algebra Practice

Write an equation. Solve.

1. 487 times 503 is equal to *s*.

2. 229,392 divided by 72 is equal to *n*.

3. The difference between 20 and 5.683 is *d*.

4. The quotient of 0.0567 and 0.09 is *g*.

5. The sum of $\frac{2}{15}$ and $\frac{29}{30}$ is *w*.

6. $11\frac{3}{4}$ times $8\frac{1}{3}$ is equal to *b*.

7. The perimeter of a rectangle $2\frac{5}{8}$ in. long and $1\frac{2}{8}$ in. wide is *p*.

8. The circumference of a circle with a diameter of 2.5 cm is *c*.

9. The area of a triangle with a base of $4\frac{1}{2}$ in. and a height of $2\frac{1}{4}$ in. is *a*.

10. The volume of a pyramid with base area of 12 m² and a height of 3 m is *v*.

11. The surface area of a cylinder with a radius of 2 ft and a height of 4 ft is *s*.

Write the LCM of each set of expressions.

12. a^2 and $3ab$

13. $2r$, 3, and s^2

14. $9h^3$, $6gh$, and g^2

Solve the equation. Check the solution.

15. $3\frac{1}{2}m = \frac{3}{4}$

16. $12 = 9a - 15$

17. $2s + 5 = 6s - 9$

Solve. Round to the nearest hundredth.

18. $\frac{3}{8} = \frac{b}{32}$

19. $\frac{4}{1} = \frac{y}{144}$

20. $\frac{12}{4} = \frac{m}{10}$

23. 10% of 40 is *c*.

24. 20% of 60 is *w*.

25. 4% of 40 is *s*.

26. $\frac{3}{4}$% of 1200 is *n*.

27. $33\frac{1}{3}$% of 600 is *h*.

28. 125% of 64 is *e*.

29. 9 is 12.5% of *d*.

30. 6 is $\frac{3}{5}$% of *f*.

31. 2.9 is 116% of *x*.

233

Problem Solving on the Job

Landscape Architect

LANDSCAPE ARCHITECT

DESCRIPTION: Plan the type and lo-
cation of plants and shrubs for
homes and buildings.
QUALIFICATIONS: 4–5 years of college
JOB OUTLOOK: Openings are expected
to grow faster than average
throughout the next decade.

A landscaper submitted the following estimate for planting and lawn care services to the National Boat Company.

Greentree Landscapers

To: *National Boat Company*   Date: 4/7

Item	Description	Price
Trees	4-5 ft diameter, 20 ft tall	$450/ per tree
Top-Soil + Seed	Grassy areas	$ 75/ per 1000 ft²
Maintenance	Weekly cutting, dressing beds	$ 40/ per acre (43,560 ft²)

Use the estimate to figure out some costs for National Boat Company.

1. The company expects to plant 30 large trees. How much will this cost?
2. The area to be seeded will cover 200,000 ft². How much will this cost?
3. The grassy area covers about 5 acres. How much will grass cutting cost per week? for a 26 week season?

4. What is the total cost of planting, seeding, and maintenance for one season?

5. The landscape company receives a 12% commission on the total bill for their design work. How much is the commission?

6. The National Boat Company is planning on planting an additional 23 trees and seeding an additional 18 acres of lawn at another site. What would the cost be, including maintenance, for one season?

Solve.

7. Each spring and fall the landscaper's crew fertilizes the grassy areas. They use 10 lb of fertilizer for every 1000 ft² of grass. If fertilizer costs $.20 per pound how much will it cost to fertilize 200,000 ft²?

8. Greentree Landscapers employs 20 workers to do their planting and maintenance work. Each worker is paid $7.58 per hour. Compute the total payroll if each worker works a 40 hour week.

9. Greentree Landscapers recently purchased two trucks for $6982 each. They traded in an old truck and received a trade-in allowance of $735. What was the final cost of the new trucks?

Greentree Landscapers does planting, seeding, and maintenance work for private homes as well as businesses. Complete the monthly bills below.

10.

Greentree Landscapers	
Mr. Thomas Hawkins 48 Leland Road Hackensack, MA 02167	
Bill for June	
Monthly maintenance	$75.00
Fertilizer	18.00
Tree Trimming	65.00
Total	?

11.

Greentree Landscapers	
Ms. Leona Chan 19 Waverly Rd. Myopia, MA 02212	
Bill for June	
Monthly maintenance	$60.00
Mulch	55.00
2 Azaleas @ $27.50 each	?
Total	?

Enrichment

Direct and Inverse Variation

The area of the rectangle shown is
$3 \times 8 = 24$ sq. ft. What happens to
the area if one of the dimensions is
doubled while the other remains the
same? What if the one dimension is
tripled?

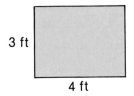

3 ft

4 ft

$A = 1 \times w$
$\quad = 3 \times 8$
$\quad = 24$

$A = 1 \times w$
$\quad = (2 \times 3) \times 8$
$\quad = 48$

$A = 1 \times w$
$\quad = (3 \times 3) \times 8$
$\quad = 72$

As one dimension varies by a factor of 2, the area varies by a factor
of 2. As one dimension varies by a factor of 3, the area varies by a
factor of 3. This is an example of **direct variation.**

Direct Variation

One factor remains unchanged as a second factor and the product
change by the same factor.

Inverse Variation

The product remains constant as one factor increases and the other
factor decreases by the same factor.

Complete.

1.

Length	×	Width	=	Area
1		3		?
5		3 ·		?
25		3		?
125		3		?

Length	×	Width	=	Area
1		36		36
2		18		?
3		12		?
4		9		?

3. Is Exercise 1 an example of direct or inverse variation? Exercise 2?

Chapter Review

True or false? Write *T* or *F*.

1. A ratio compares two numbers using division.

2. Percent means per one thousand.

3. An equation that shows equal ratios is called a percent.

4. In the proportion $\dfrac{3}{4} = \dfrac{9}{x}$, x is equal to 12.

5. When computing 150% of y, the product will be greater than y.

Match.

6. $\dfrac{150}{300} = y\%$

7. 0.3% of 3000 is y

8. 15 is 75% of y

9. $0.009 = y\%$

10. $\dfrac{6}{5} = y$

a. $y = 9$
b. $y = 120\%$
c. $y = 50$
d. $y = 20$
e. $y = 0.9$

Choose the best answer. Write *a*, *b*, or *c*.

11. Two tires cost $87, so 8 tires will cost __?__.
    a. $174   **b.** $870   **c.** $348

12. In a shipment of TVs, 4 out of 80 had defects. This is $x\%$.
    a. $x = 50$ **b.** $x = 5$ **c.** $x = 20$

13. In a recent fund raising campaign, 35% of the 820 people contacted made contributions. Which equation could be solved to find the number of contributors?
    a. $0.35x = 820$ **b.** $(0.35)(820) = x$ **c.** $0.35 = 820 + x$

14. Max's rent went from $585 per month to $731.25 per month. This is a __?__% increase.
    a. 20   **b.** 125   **c.** 25

Chapter Test

1. Use the table to write a ratio in lowest terms to compare the attendance each day to the total for the week.

Symphony Attendance

Day	Mon.	Tues.	Wed.	Thurs.
Attendance	725	430	520	320

2. Four out of 5 people attending a concert were adults. Use the table to find how many adults attended a concert each night.

Write as a percent.

3. $\dfrac{2}{5}$ 4. $\dfrac{8}{200}$ 5. $\dfrac{35}{75}$ 6. $\dfrac{100}{300}$ 7. $\dfrac{14}{28}$

Dudley saves 7% of his pay each month. How much does he have left from each check after deducting his savings?

8. March—$1588   9. April—$1176   10. May—$1701

11. At Pippin's Garage it costs $.75 to park for the first half-hour and $1.35 for each additional hour or fraction of an hour up to a maximum of $7.00. Make a chart to show the cost of parking for from 1 to 4 hours. How much would it cost to park for 1 hour 20 min?

Find the number. Round to the nearest hundredth.

12. $\dfrac{4}{5}$% of 75 is y 13. 135% of 68 is b 14. $2\dfrac{1}{2}$% of 55 is z

Write and solve an equation to find the base. Round to the nearest whole number.

15. 26 is 50% of what number?   16. 150 is 8% of what number?

Find the percent of increase or decrease to the nearest tenth of a percent.

17. from 155 to 100   18. from 17 to 25   19. from 12 to 5

Cumulative Review (Chapters 1–7)

Add, subtract, multiply, or divide.

1. $9\frac{2}{3} - 6$

2. $8 - 3\frac{3}{5}$

3. $14\frac{1}{8} - 6\frac{2}{3}$

4. $5.6 - 4 \div 2$

5. $(8.1)(9.6)(100)(0)$

6. 1000×0.83

7. $(100)(0.2)$

8. $62 - 4.837$

9. $1.06 \div 0.2$

10. $2.04 \div 2.4$

11. $\frac{1}{4} \cdot 36 \cdot \frac{5}{8}$

12. $\left(\frac{25}{4.7}\right)(4.7)$

13. $3 - 1 - 1 - 1$

14. $(10)(3.52)(10)$

15. $(10^3)(8.34)$

Find the area.

16.
8 cm, 8 cm

17.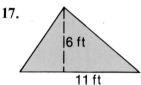
6 ft, 11 ft

18.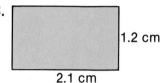
1.2 cm, 2.1 cm

19.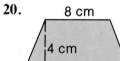
5.2 m, 5.4 m

20.
8 cm, 4 cm, 13 cm

21.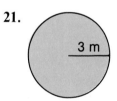
3 m

The formula $D = 16t^2$ relates the distance in feet (D) an object will fall with the time (t) in seconds it will fall. Solve the problem.

22. A broken tree branch fell directly from a tree to the ground. The branch took 1.1 seconds to hit the ground. How far did it fall?

23. How long will it take an object to fall 256 ft from the roof of a building?

Write an equation. Solve.

24. An empty truck weighs 8200 lb. Its load weighs 4100 lb less than the truck. How much do both the truck and the load weigh in all?

25. A tire for Charlie's van cost $42 two years ago. This year the price is $5 less than $1\frac{1}{2}$ times the old price. What's this year's price?

26. The regular price of a camera is $259. It's now on sale for 20% off the regular price. What's the savings?

8

Integers and Rational Numbers

Negative charges of electricity at the bottom of a thunderhead are attracted by positive charges in the ground causing the powerful burst of energy that is a lightning bolt.

8-1 Integers on the Number Line

The highest spot on Earth is the top of Mt. Everest, about 29,000 ft above sea level. The lowest spot is the bottom of the Mariana Trench in the Pacific Ocean, about 36,000 ft below sea level.

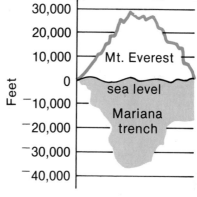

We can think of sea level as being 0 on a vertical number line. Numbers above 0 are called positive numbers. Numbers below 0 are called negative numbers.

Positive 29,000 is written as 29,000 or $^+29{,}000$. Negative 36,000 is written as $^-36{,}000$.

On a horizontal number line, we show negative numbers to the left of 0, and positive numbers to the right of 0.

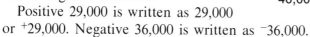

Numbers such as $^-2$ and 2 that are the same distance from 0 but on opposite sides of it are called **opposite** numbers, or **additive inverses.** The numbers 1, 2, 3, 4, ..., together with their opposites $^-1$, $^-2$, $^-3$, $^-4$, ..., and 0 are called the **integers.**

On the number line below, points A, B, and C are the **graphs** of $^-5$, $^-2$, and 3. The numbers $^-5$, $^-2$, and 3 are the **coordinates** of their graphs.

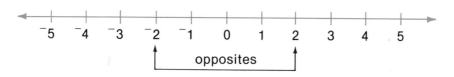

Of any two numbers on the number line, the one at the right is the greater. Thus, for example:

$$^-2 > ^-5 \qquad 3 > ^-2 \qquad 0 > ^-4$$

This also means that the number at the left is the lesser:

$$^-3 < ^-1 \qquad ^-2 < 1 \qquad 0 < 3$$

Example 1 Complete. Compare the numbers, 4 _?_ 0.

Solution   4 $>$ 0

Your Turn Complete. Compare the numbers, $^-$12 _?_ $^-$2.

Class Exercises

What is the opposite?

1. Up   **2.** East   **3.** Hot   **4.** Win   **5.** North

Exercises

What is the opposite?

A **1.** 8   **2.** $^-$9   **3.** $^-$23   **4.** 17   **5.** $^-$17

On the number line, what is the co-ordinate of the graph?

B **6.** Point A **7.** Point B

8. Point C **9.** Point D

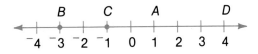

Draw a number line and on it graph the number.

10. Point A at 4   **11.** Point B at $^-$2   **12.** Point C at 0   **13.** Point D at 1

Write a number sentence using $>$ to show which number is greater.

14. 8 or $^-$9   **15.** 12 or 0   **16.** $^-$18 or 18   **17.** $^-$57 or 67

How many units do you travel, and in which direction do you go, to make the move on the number line?

C **18.** From 3 to 10   **19.** From 10 to 3   **20.** From $^-$5 to 2

21. From 2 to $^-$7   **22.** From 0 to $^-$4   **23.** From $^-$3 to $^-$12

Let n be negative and t be positive. Is the statement true or false?

24. $n > t$ **25.** $t > n$ **26.** $t < 0$ **27.** $0 < n$

8-2 *Adding Negative Integers*

Pete carried the ball twice during the first quarter of the game. He lost 4 yd the first time and 2 yd the second time. He lost 6 yd in all. We add two negative integers to find the total lost yardage.

$$^-4 + {}^-2 = {}^-6$$

By using a number line you can see that the sum of two negative integers is always negative.

Example 1   Use a number line to find the sum $^-3 + {}^-5$.

Solution   You start at $^-3$ on the number line and move 5 units to the left, stopping at $^-8$.

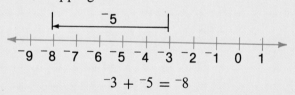

$$^-3 + {}^-5 = {}^-8$$

Your Turn   • Use a number line to find the sum $^-2 + {}^-7$.

You already know from past experience that the sum of two positive numbers is positive. When you add two negative numbers on a number line you always start to the left of 0 and then move farther to the left. Thus the sum of two negative numbers is always negative.

The sum of two positive numbers is positive.
The sum of two negative numbers is negative.

Class Exercises

What sum is shown on the number line?

1.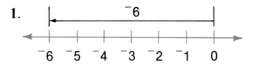

2.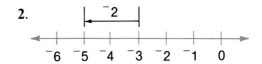

3.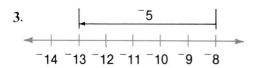

4.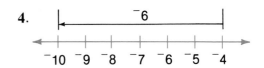

Exercises

Use a number line to find the sum.

A
1. $^-4 + ^-3$
2. $^-1 + ^-7$
3. $^-6 + ^-6$
4. $^-8 + ^-5$

5. $^-2 + ^-10$
6. $^-8 + ^-1$
7. $^-11 + ^-3$
8. $^-4 + ^-9$

Find the sum.

B
9. $^-6 + ^-7$
10. $^-8 + ^-2$
11. $^-5 + ^-9$
12. $^-10 + ^-6$

13. $^-2 + ^-18$
14. $^-26 + ^-5$
15. $^-4 + ^-30$
16. $^-11 + ^-7$

Solve the equation.

17. $m = ^-12 + ^-8$
18. $b = ^-4 + ^-15$
19. $^-18 + ^-17 = d$

20. $h = ^-8 + ^-25$
21. $f = ^-13 + ^-31$
22. $^-26 + ^-15 = t$

23. $^-9 + ^-42 = x$
24. $w = ^-27 + ^-32$
25. $n = ^-43 + ^-38$

26. $c = ^-56 + ^-17$
27. $^-31 + ^-42 = t$
28. $^-65 + ^-19 = m$

Combine like terms.

29. $^-5x + ^-2x$
30. $^-h + ^-8h$
31. $^-8t + ^-12t$
32. $^-7d + ^-16d$

33. $^-13g + ^-12g$
34. $^-26m + ^-7m$
35. $^-19y + ^-y$
36. $^-14p + ^-17p$

C
37. $^-6n + ^-7p + ^-5p + ^-3n$

38. $^-8k + ^-t + ^-9k + ^-14t$

39. $^-12a + ^-14x + ^-4x + ^-8a$

40. $^-3f + ^-12g + ^-21f + ^-18g$

8-3 *Adding Integers of Opposite Sign*

During the first week of August, Jim and Jean's Yard Care Service took in $75 for work done and paid out $84 for gasoline, fertilizer and repair of equipment. Since they paid out more than they took in they had a loss of $9. We can show this as ⁻$9.

Number lines are helpful when learning to add positive and negative numbers.

Example 1 Find the sum $5 + {}^-7$.

Solution Start at 5 on the number line. To add ⁻7, move 7 units to the left. You stop at ⁻2.

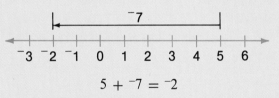

$$5 + {}^-7 = {}^-2$$

Example 2 Find the sum $^-5 + 9$.

Solution Start at ⁻5. To add 9, move 9 units to the right. You stop at 4.

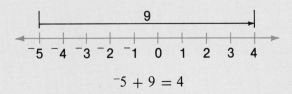

$$^-5 + 9 = 4$$

245

Example 3 Find the sum $6 + {}^-6$.

Solution Start at 6. To add $^-6$, move 6 units to the left. You
stop at 0.

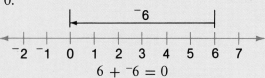

$$6 + {}^-6 = 0$$

The last example above displays the following property.

The sum of two opposite numbers is 0.
$$a + {}^-a = 0 \qquad {}^-a + a = 0$$

You can use the following rules to add a positive and negative integer without drawing a number line.

1. Think of both numbers as positive and subtract the smaller from the larger.
2. Notice the sign of the addend that is farther from 0. This will be the sign of the sum.
3. Write the number found in Step 1 with the sign found in Step 2. This is the sum.

Example 4 Solve the equation $n = 5 + {}^-8$.

Solution Step 1. If both numbers were positive they would be 5 and 8. Subtracting, $8 - 5 = 3$.
Step 2. $^-8$ is farther from 0 than 5. The sign of the sum is negative.
Step 3. The sum is $^-3$. So, $n = {}^-3$.

Your Turn • Solve the equation $y = {}^-6 + 2$.

Class Exercises

What sum is shown on the number line?

1.

2.

Exercises

Use a number line to find the sum.

A **1.** $8 + {}^-2$ **2.** $^-5 + 4$ **3.** $^-8 + 6$ **4.** $^-4 + 5$

5. $9 + {}^-9$ **6.** $12 + {}^-8$ **7.** $4 + {}^-7$ **8.** $^-6 + 9$

Find the sum.

B **9.** $^-2 + 3$ **10.** $^-6 + 1$ **11.** $2 + {}^-7$ **12.** $8 + {}^-10$

13. $^-12 + 12$ **14.** $13 + {}^-4$ **15.** $12 + {}^-10$ **16.** $^-9 + 18$

17. $^-11 + 7$ **18.** $3 + {}^-20$ **19.** $^-14 + 16$ **20.** $9 + {}^-21$

Solve the equation.

21. $x = 15 + {}^-4$ **22.** $b = {}^-6 + 13$ **23.** $14 + {}^-19 = t$

24. $d = {}^-2 + 18$ **25.** $g = 23 + {}^-8$ **26.** $^-28 + 16 = m$

27. $c = {}^-5 + 32$ **28.** $56 + {}^-18 = y$ **29.** $x = 84 + {}^-92$

30. $g = {}^-37 + 67$ **31.** $n = 41 + {}^-37$ **32.** $33 + {}^-53 = p$

Combine like terms.

33. $3a + {}^-2a$ **34.** $^-d + 5d$ **35.** $7y + {}^-8y$ **36.** $^-4n + 4n$

37. $^-8x + 12x$ **38.** $10g + {}^-16g$ **39.** $14m + {}^-8m$ **40.** $^-k + {}^-16k$

41. $^-13p + 12p$ **42.** $^-16y + {}^-4y$ **43.** $18n + 12n$ **44.** $7a + {}^-a$

C **45.** $^-3x + 2y + 4x + 5y + 6$ **46.** $5d + {}^-2h + {}^-6d + 5h + 7$

47. $9t + 12p + 10 + {}^-6p + {}^-11t$ **48.** $^-12 + {}^-8k + 7m + {}^-2k + {}^-9m$

49. $4g + {}^-15n + {}^-g + {}^-8 + 12n$ **50.** $^-6x + 13y + {}^-5x + {}^-13y + 4$

8-4 *Subtracting Integers*

The number line below can be read as either $4 - 6 = {}^-2$ or $4 + {}^-6 = {}^-2$.

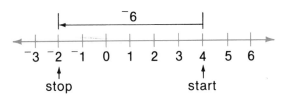

For any subtraction equation we can always write a related addition equation.

$$12 - 9 = 3 \qquad\qquad 12 + {}^-9 = 3$$
$$24 - 11 = 13 \qquad\quad 24 + {}^-11 = 13$$
$$18 - 12 = 6 \qquad\quad 18 + {}^-12 = 6$$

We use this fact to subtract integers.

To subtract a number, we add its opposite.
$$a - b = a + {}^-b$$

Example 1   Subtract $3 - {}^-5$.

Solution $3 - {}^-5 = 3 + 5 = 8$

Example 2   Subtract ${}^-5 - 2$.

Solution ${}^-5 - 2 = {}^-5 + {}^-2 = {}^-7$

Your Turn   Subtract
  • $2 - 5$ • $6 - {}^-8$ • ${}^-3 - 4$ • $9 - {}^-2$

Class Exercises

Write the opposite.

1. 5   **2.** $^-12$ **3.** 56   **4.** $^-75$ **5.** 101   **6.** $^-92$

Complete the equation.

7. $7 - 5 = 7 + \underline{\ ?\ }$ **8.** $8 - {}^-1 = 8 + \underline{\ ?\ }$ **9.** $^-5 - 3 = {}^-5 + \underline{\ ?\ }$

10. $^-12 - 14 = {}^-12 + \underline{\ ?\ }$ **11.** $^-7 - {}^-4 = {}^-7 + \underline{\ ?\ }$ **12.** $13 - 23 = 13 + \underline{\ ?\ }$

Exercises

Write a related addition equation.

A   **1.** $12 - 3 = 9$ **2.** $^-5 - 6 = {}^-11$ **3.** $4 - {}^-10 = 14$

     4. $^-18 - 20 = {}^-38$ **5.** $^-9 - 6 = {}^-15$ **6.** $14 - 23 = {}^-9$

     7. $17 - {}^-8 = 25$ **8.** $^-13 - {}^-8 = {}^-5$ **9.** $^-8 - {}^-19 = 11$

     10. $22 - 34 = {}^-12$ **11.** $16 - {}^-23 = 39$ **12.** $^-15 - 8 = {}^-23$

Subtract.

B   **13.** $8 - {}^-2$ **14.** $^-3 - {}^-2$ **15.** $12 - 22$ **16.** $7 - {}^-16$

     17. $^-9 - 18$ **18.** $4 - 21$ **19.** $^-13 - 5$ **20.** $^-25 - {}^-16$

     21. $38 - 38$ **22.** $^-15 - 12$ **23.** $21 - {}^-21$ **24.** $^-17 - 34$

Solve the equation.

     25. $t = {}^-4 - 8$ **26.** $n = 12 - 9$ **27.** $13 - {}^-9 = p$

     28. $x = {}^-18 - 12$ **29.** $b = {}^-21 - {}^-21$ **30.** $g = 27 - 32$

     31. $19 - {}^-28 = m$ **32.** $33 - {}^-16 = y$ **33.** $n = {}^-42 - 5$

Combine like terms.

     34. $8n - 2n$ **35.** $^-6d - 3d$ **36.** $4t - {}^-5t$ **37.** $12x - {}^-6x$

     38. $^-7k - {}^-4k$ **39.** $^-8h - 3h$ **40.** $10y - {}^-y$ **41.** $^-9p - {}^-11p$

C   **42.** $7t - 5n + {}^-3t - 4n$ **43.** $12x - {}^-3y - 2x + 4y$

     44. $^-9p + 12d - {}^-6d - 3p$ **45.** $15g + {}^-8n - {}^-4g + 5n$

8-5 Multiplying Integers

Monica lost 3 points on each of 5 rounds in a card game. Dale lost 3 yd on each of 5 running plays in a football game. In each case we could use addition to find the total loss.

$$^-3 + {}^-3 + {}^-3 + {}^-3 + {}^-3 = {}^-15$$

Since $^-3$ is used as an addend 5 times, we could also use multiplication.

$$5 \times {}^-3 = {}^-15$$

On the number line it looks like this.

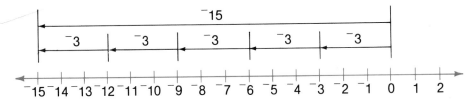

Example 1 Solve the equation $y = 6(^-2)$.

Solution $y = {}^-12$

Example 2 Solve the equation $12(^-4) = n$.

Solution $^-48 = n$

Your Turn Solve the equation.
- $y = 8(^-7)$
- $15(^-2) = p$

To multiply a negative number times a positive number we use the commutative property.

We know that ⟶ $^-6 \times 2 = 2 \times {}^-6$ ⟵ *the commutative property*
and that ⟶ $2 \times {}^-6 = {}^-12$
so ⟶ $^-6 \times 2 = {}^-12$

Example 3 Solve the equation $d = {}^-4(5)$.

Solution $d = {}^-20$

When you multiply a positive and a negative number, the product is negative.

In the pattern at the right, as the multiplier of $^-2$ decreases from 4 to 0, the product increases. This pattern does not stop when the multiplier becomes a negative number. You can see that the product becomes positive.

$$4 \times {}^-2 = {}^-8$$
$$3 \times {}^-2 = {}^-6 \quad \downarrow + 2$$
$$2 \times {}^-2 = {}^-4 \quad \downarrow + 2$$
$$1 \times {}^-2 = {}^-2 \quad \downarrow + 2$$
$$0 \times {}^-2 = \phantom{{}^-}0 \quad \downarrow + 2$$
$${}^-1 \times {}^-2 = \phantom{{}^-}2 \quad \downarrow + 2$$
$$-2 \times {}^-2 = \phantom{{}^-}4 \quad \downarrow + 2$$

When you multiply two negative numbers the product is positive.

Example 4 Solve the equation.
  a. $x = {}^-7({}^-10)$ **b.** $n = {}^-6({}^-8)$

Solution   **a.** $x = 70$ **b.** $n = 48$

Your Turn Solve the equation.
  • $p = {}^-9(3)$ • ${}^-8({}^-7) = g$

The following table summarizes the rules for signs when multiplying integers.

Factors	Product
Both positive	Positive
Both negative	Positive
One positive, one negative	Negative

251

Class Exercises

Is the product positive or negative?

1. $^-5(2)$ 2. $6 \cdot 10$ 3. $^-8(^-12)$ 4. $9(^-4)$

5. $^-15 \cdot ^-21$ 6. $30(^-7)$ 7. $28 \cdot 43$ 8. $^-32(3)$

Exercises

Multiply.

A
1. $8(^-11)$ 2. $^-9 \cdot 3$ 3. $^-12(4)$ 4. $^-5(^-3)$

5. $7 \cdot 6$ 6. $^-3 \cdot 4$ 7. $^-5(^-6)$ 8. $2(^-7)$

9. $8(^-4)$ 10. $^-6(^-2)$ 11. $3 \cdot 5$ 12. $6(^-3)$

13. $12(^-10x)$ 14. $^-2 \cdot 9d$ 15. $7 \cdot ^-4h$ 16. $2 \cdot 13p$

Solve the equation.

B
17. $h = ^-8(^-11)$ 18. $x = 3(^-6)$ 19. $t = 4(^-30)$

20. $^-2(8) = m$ 21. $p = ^-8 \cdot 6$ 22. $^-4(^-5) = d$

23. $f = ^-6(2)$ 24. $y = 6 \cdot 4$ 25. $2(^-15) = n$

26. $^-3(8) = g$ 27. $^-7(^-5) = m$ 28. $h = ^-8(5)$

Evaluate the expression using the given value for the variable.

29. $^-7y; \, y = 4$ 30. $8x; \, x = ^-3$ 31. $^-4d; \, d = ^-2$

32. $5m; \, m = ^-5$ 33. $9t; \, t = ^-2$ 34. $^-6g; \, g = 9$

35. $^-10p; \, p = ^-3$ 36. $8y; \, y = ^-6$ 37. $^-2k; \, k = 20$

Solve the equation.

38. $x = ^-1(3)(^-2)(4)$ 39. $h = 2(^-3)(2)(^-1)$ 40. $p = ^-3(7)(2)(^-2)(^-1)$

41. $n = 2(^-5)(3)(4)$ 42. $^-3(2)(^-1)(^-5) = m$ 43. $k = 4(^-2)(3)(^-5)$

Multiply.

C
44. $^-2(^-a + ^-3)$ 45. $(^-4x + 7)4$ 46. $^-4(2y - 6)$

47. $(5d - ^-4)3$ 48. $^-7(^-h + 6)$ 49. $8(^-3k + ^-4)$

50. $^-6(g + ^-8)$ 51. $4(^-9t - ^-7)$ 52. $(^-7n + 10)(^-5)$

8-6 *Dividing Integers*

We can use what we know about multiplying integers to help us learn how to divide integers.

Using whole numbers you can write a related division equation for any multiplication equation.

$$4 \times 3 = 12 \qquad 12 \div 3 = 4$$
$$7 \times 2 = 14 \qquad 14 \div 2 = 7$$
$$a \times b = c \qquad c \div b = a$$

The same pattern holds true for integers.

$$a \times b = c \qquad\qquad c \div b = a$$
$$^-5 \times 6 = ^-30 \qquad ^-30 \div 6 = ^-5 \longleftarrow \text{\textit{negative} ÷ \textit{positive} = \textit{negative}}$$
$$^-8 \times ^-4 = 32 \qquad 32 \div ^-4 = ^-8 \longleftarrow \text{\textit{positive} ÷ \textit{negative} = \textit{negative}}$$
$$3 \times ^-2 = ^-6 \qquad ^-6 \div ^-2 = 3 \longleftarrow \text{\textit{negative} ÷ \textit{negative} = \textit{positive}}$$

From the above we can write the following rules for the sign of the quotient when dividing integers.

Numbers being divided	Quotient
Both positive	Positive
Both negative	Positive
One positive and one negative	Negative

Example 1 Solve the equation $x = \dfrac{^-15}{5}$.

Solution $\dfrac{^-15}{5} = ^-15 \div 5$. We are dividing a negative integer by a positive integer, therefore the quotient is negative.

$$x = \frac{^-15}{5} = ^-15 \div 5 = ^-3$$

Your Turn • Solve the equation $y = \dfrac{24}{^-4}$.

Class Exercises

Tell whether the quotient will be positive or negative.

1. $8 \div {}^-2$ **2.** ${}^-10 \div 5$ **3.** $12 \div 6$ **4.** ${}^-18 \div 9$

5. $\dfrac{-24}{-6}$ **6.** $\dfrac{22}{-11}$ **7.** $\dfrac{-32}{8}$ **8.** $\dfrac{42}{21}$

Exercises

Divide.

A   **1.** $63 \div {}^-9$ **2.** $35 \div {}^-7$ **3.** ${}^-77 \div {}^-7$ **4.** ${}^-32 \div {}^-8$

   5. $\dfrac{-42}{6}$ **6.** ${}^-90 \div 10$ **7.** $\dfrac{49}{-7}$ **8.** ${}^-36 \div {}^-4$

   9. ${}^-39 \div 13$ **10.** $\dfrac{-84}{4}$ **11.** $121 \div {}^-11$ **12.** ${}^-75 \div {}^-5$

B   **13.** $6x \div 3$ **14.** ${}^-12m \div 4$ **15.** $10d \div {}^-5$ **16.** ${}^-8t \div {}^-8$

   17. $\dfrac{18m}{-9}$ **18.** $\dfrac{-21h}{3}$ **19.** $\dfrac{26k}{-2}$ **20.** $\dfrac{-72a}{9}$

   21. ${}^-50x \div 25$ **22.** $84n \div {}^-4$ **23.** ${}^-180y \div 9$ **24.** ${}^-69g \div 3$

   25. $\dfrac{45h}{5}$ **26.** $\dfrac{-120t}{-60}$ **27.** $\dfrac{-200n}{25}$ **28.** $\dfrac{124b}{-4}$

Solve the equation.

29. $y = {}^-64 \div 8$ **30.** ${}^-45 \div {}^-3 = n$ **31.** $a = 108 \div {}^-9$

32. $\dfrac{-144}{-12} = x$ **33.** $d = \dfrac{150}{-15}$ **34.** $\dfrac{-36}{-18} = g$

35. $88 \div {}^-11 = n$ **36.** $x = {}^-180 \div 9$ **37.** $p = 175 \div {}^-5$

38. $n = \dfrac{-84}{-4}$ **39.** $d = \dfrac{-210}{-70}$ **40.** $\dfrac{-160}{20} = y$

41. $110 \div 10 = m$ **42.** ${}^-140 \div 7 = y$ **43.** $a = {}^-96 \div {}^-3$

44. $x = \dfrac{320}{-8}$ **45.** $t = \dfrac{-360}{-18}$ **46.** $\dfrac{-284}{4} = b$

Solve the equation. Remember to perform the operations in the parentheses first.

C **47.** $d = (^-65 \div {}^-5) \times (^-3)$

48. $(95 \div 5) - (117 \div {}^-3) = x$

49. $n = \dfrac{48}{^-4} - \dfrac{^-60}{^-5}$

50. $y = \dfrac{36}{^-2} + (^-15 \div {}^-3)$

51. $\dfrac{^-60}{5} - (^-48 \div 4) = t$

52. $(^-72 \div 9) + \dfrac{^-35}{7} = b$

Problems

Decide which operation to use, then solve.

1. At noon the temperature was 3°C. Between noon and 10 P.M. it fell 6°, rose 8°, and then fell 12°. What was the temperature at 10 P.M.?
2. Jane had $126 in a checking account on March 1. For each of the next 8 weeks she withdrew $16. How did her account stand then?
3. Sue and Sam borrowed $4000 to open a sports equipment shop. Over the next 6 months they had fixed expenses of $1535 per month. During those months they took in the following amounts: $1289, $1562, $2874, $3925, and $5201. How much could Sue and Sam pay off on the $4000? What profit, if any, did they make?

Self Test 1

Write a number sentence using $>$ to show which number is greater.

1. 5 or $^-6$	**2.** $^-8$ or 3	**3.** $^-12$ or 20	**4.** 18 or $^-4$	**8-1**

Solve the equation.

5. $x = {}^-3 + {}^-8$	**6.** $n = {}^-12 + {}^-10$	**7.** $t = {}^-22 + {}^-7$	**8-2**
8. $d = 7 + {}^-4$	**9.** $k = {}^-15 + 9$	**10.** $y = 12 + {}^-20$	**8-3**
11. $m = 4 - 12$	**12.** $a = {}^-6 - 12$	**13.** $h = {}^-8 - {}^-5$	**8-4**
14. $p = 7(^-8)$	**15.** $t = {}^-10(4)$	**16.** $x = {}^-6(^-11)$	**8-5**
17. $n = 24 \div {}^-3$	**18.** $d = {}^-15 \div 3$	**19.** $x = {}^-36 \div {}^-9$	**8-6**

Self Test answers and Extra Practice are at the back of the book.

8-7 *Adding/Subtracting Rationals*

A share of Satellite Communications Company stock was selling for $60. Over the next 2 days it went up $2.50 and then down $3.38. We can show changes as positive and negative decimals and operate with them as we did with integers.

Example 1 By how much did the price of a share of the stock change? What was the final price?

Solution Let c = the amount by which the price changed.
$$c = 2.50 + {}^-3.38$$
$$c = {}^-0.88 \longleftarrow$$
The price dropped $0.88.

Let f = the final price
$$f = 60 + {}^-0.88$$
$$= 59.12$$
The final price was $59.12.

To add, follow the rules for adding integers.
$$\begin{array}{r} {}^-3.38 \\ + \ 2.50 \\ \hline {}^-0.88 \end{array}$$

A number represented by a fraction whose numerator is an integer and whose denominator is any integer other than 0 is called a **rational** number.

Example 2 Solve the equation $n = \dfrac{{}^-1}{2} + \dfrac{{}^-3}{4}$.

Solution $$n = \frac{{}^-1}{2} + \frac{{}^-3}{4} = \frac{{}^-2}{4} + \frac{{}^-3}{4}$$
$$= \frac{{}^-5}{4} = {}^-1\frac{1}{4}$$

Your Turn Solve the equation.

- $t = {}^-6.17 + {}^-2.52$
- ${}^{-3}\over 8$ $+$ ${}^{-1}\over 4$ $= n$

You can also subtract positive and negative decimals and fractions just as you do integers.

Example 3 Solve the equation $y = 8.62 - {}^-4.31$.

Solution
$y = 8.62 - {}^-4.31$
$= 8.62 + 4.31$ ◄ —— To subtract, you add the opposite.
$= 12.93$

Example 4 Solve the equation ${}^-6\frac{2}{3} - 4\frac{1}{6} = k$.

Solution
$${}^-6\frac{2}{3} + {}^-4\frac{1}{6} = k$$

$${}^-6\frac{4}{6} + {}^-4\frac{1}{6} = k$$

$${}^-10\frac{5}{6} = k$$

Your Turn • Solve the equation ${}^-5\frac{3}{8} - 4\frac{1}{4} = h$.

Class Exercises

The graph of which number is farther from 0 on the number line?

1. 8.3 or 10.2   2. 5.6 or ${}^-6.3$ 3. ${}^-4.5$ or ${}^-5.4$ 4. 12.7 or ${}^-14.5$

5. 9.61 or ${}^-9.62$ 6. 0.05 or ${}^-0.50$ 7. ${}^-18.59$ or ${}^-18.63$ 8. ${}^-23.92$ or 29.63

9. $8\frac{2}{3}$ or $8\frac{1}{3}$ 10. ${}^-5\frac{3}{8}$ or $5\frac{5}{8}$ 11. ${}^-6\frac{2}{5}$ or ${}^-6\frac{4}{5}$ 12. $12\frac{5}{6}$ or ${}^-12\frac{1}{6}$

13. $3\frac{1}{4}$ or ${}^-3\frac{1}{3}$ 14. ${}^-8\frac{2}{5}$ or ${}^-8\frac{3}{10}$ 15. $14\frac{5}{6}$ or ${}^-14\frac{7}{12}$ 16. ${}^-11\frac{3}{8}$ or $11\frac{5}{16}$

257

Exercises

Is the sum positive or negative?

A   **1.** $5.4 + {}^-6.2$ **2.** $8.7 + 0.3$ **3.** ${}^-6.1 + 7.4$ **4.** ${}^-5.9 + {}^-2.1$

    5. ${}^-8\frac{1}{3} + 6\frac{2}{3}$ **6.** $12\frac{1}{2} + {}^-8\frac{3}{4}$ **7.** $13\frac{1}{8} + 5\frac{11}{16}$ **8.** ${}^-7\frac{1}{5} + {}^-1\frac{7}{10}$

Solve the equation.

B   **9.** $t = 9.8 + {}^-9.2$ **10.** $y = {}^-4.7 + 8.5$ **11.** $7.5 + 0.8 = m$

   12. $x = {}^-8.8 - {}^-9.7$ **13.** $6.2 - {}^-6.1 = h$ **14.** $b = {}^-8.3 - 5.2$

   15. ${}^-5.01 + 6.52 = n$ **16.** $t = {}^-7.85 + {}^-12.31$ **17.** $y = 12.41 + {}^-12.41$

   18. $9.55 - {}^-10.71 = b$ **19.** $t = {}^-12.76 - {}^-19.18$ **20.** $h = {}^-13.44 - 23.44$

   21. $d = {}^-4\frac{1}{3} + 5\frac{2}{3}$ **22.** ${}^-7\frac{2}{5} + {}^-3\frac{3}{5} = f$ **23.** $x = 6\frac{3}{8} + {}^-7\frac{5}{8}$

   24. $n = 8\frac{3}{8} - {}^-9\frac{1}{8}$ **25.** $y = {}^-5\frac{1}{3} - 8\frac{2}{3}$ **26.** ${}^-7\frac{5}{6} - {}^-10\frac{1}{6} = n$

   27. $m = 6\frac{2}{3} + {}^-5\frac{1}{6}$ **28.** $x = {}^-12\frac{2}{5} + {}^-6\frac{7}{10}$ **29.** $t = {}^-15\frac{3}{8} + 5\frac{3}{4}$

   30. $12\frac{1}{2} - {}^-9\frac{3}{4} = t$ **31.** $h = {}^-15\frac{2}{3} - 10\frac{4}{9}$ **32.** $b = {}^-8\frac{1}{2} - {}^-16\frac{5}{8}$

C   **33.** $n = 0.8 - (3.6 - {}^-8.4)$ **34.** $t = {}^-12.3 + ({}^-7.4 + {}^-5.6)$

Using the Calculator

Here is one week's record of the checking account for the Music Club.

    Use a calculator to find the final balance. Has the club lost or gained money during the week?

Date	Deposits (Add)	Checks (Subtract)	Balance Forward
			$396.17
3/7		$95.16	
3/10	$23.71		
3/11		$189.05	
3/12		$74.19	
	Final Balance		

8-8 *Multiplying/Dividing Rationals*

A chemist was preparing 5 samples for an experiment. For each sample he used 2.7 mL of a starch solution that he took from a bottle containing 94.5 mL of the solution. How much less starch solution was in the bottle when he finished?

We can show the amount he took for each sample as a negative decimal and multiply.

$$5(^-2.7) = ^-13.5$$

The amount of starch solution in the bottle had changed by $^-13.5$ mL.

We use the same rules of signs when multiplying or dividing decimals and fractions as we do for integers.

Example 1   Solve the equation $d = ^-3.6 \div 0.4$.

Solution   Since a negative number is divided by a positive number, the quotient is negative.

$$d = ^-3.6 \div 0.4 = ^-9 \qquad 0.4\overline{)^-3.6}^{\,^-9}$$

Example 2   Solve the equation $n = \dfrac{^-3}{5}\left(\dfrac{^-1}{2}\right)$.

Solution   Since both factors are negative, the product is positive.

$$n = \frac{^-3}{5}\left(\frac{^-1}{2}\right) = \frac{3}{10}$$

259

Example 3   Solve the equation $g = \dfrac{3}{8} \div \dfrac{^-2}{3}$.

Solution   Since a positive number is being divided by a negative number, the quotient is negative.

$$g = \frac{3}{8} \div \frac{^-2}{3} = \frac{3}{8} \times \frac{^-3}{2} = \frac{^-9}{16}$$

Class Exercises

Is the product or quotient positive or negative?

1. $7.5(^-8.2)$ **2.** $^-12.3(^-6.4)$ **3.** $^-16.48(3.02)$ **4.** $12.59(10.61)$

5. $\dfrac{^-5}{8}\left(\dfrac{3}{4}\right)$ **6.** $\dfrac{1}{3}\left(\dfrac{3}{16}\right)$ **7.** $\dfrac{^-2}{5}\left(\dfrac{^-1}{2}\right)$ **8.** $\dfrac{4}{9}\left(\dfrac{^-2}{3}\right)$

9. $^-9.2 \div ^-6.3$ **10.** $^-6.7 \div 9.4$ **11.** $12.3 \div 3.1$ **12.** $15.6 \div ^-0.03$

13. $\dfrac{4}{5} \div \dfrac{^-1}{3}$ **14.** $\dfrac{^-2}{5} \div \dfrac{3}{8}$ **15.** $\dfrac{^-5}{6} \div \dfrac{^-3}{4}$ **16.** $\dfrac{5}{8} \div \dfrac{1}{2}$

Exercises

Multiply.

A   **1.** $2.3(^-4)$ **2.** $^-1.7(3)$ **3.** $^-5.6(^-3.1)$ **4.** $^-4.2(^-5.1)$

5. $\dfrac{^-2}{3}\left(\dfrac{3}{8}\right)$ **6.** $\dfrac{^-5}{6}\left(\dfrac{^-3}{4}\right)$ **7.** $\dfrac{^-1}{2}\left(\dfrac{4}{5}\right)$ **8.** $\dfrac{4}{9}\left(\dfrac{^-3}{10}\right)$

Divide.

9. $^-5.7 \div 3$ **10.** $8.4 \div ^-2.1$ **11.** $^-6.9 \div 1.5$ **12.** $^-9.1 \div ^-1.3$

13. $\dfrac{1}{2} \div \dfrac{^-5}{8}$ **14.** $\dfrac{^-4}{5} \div \dfrac{^-3}{10}$ **15.** $\dfrac{7}{8} \div \dfrac{3}{4}$ **16.** $\dfrac{^-2}{9} \div \dfrac{4}{3}$

Solve the equation.

B   **17.** $d = 16.2(3.1)$ **18.** $t = 5.8(^-7.6)$ **19.** $^-4.6(^-9.3) = g$

20. $n = 12.5(^-6.3)$ **21.** $p = ^-22.3(^-0.7)$ **22.** $m = ^-14.4(8.5)$

23. $g = \dfrac{3}{5}\left(\dfrac{^-5}{6}\right)$

24. $\dfrac{^-2}{9}\left(\dfrac{^-3}{8}\right) = d$

25. $\dfrac{3}{4}\left(\dfrac{^-1}{6}\right) = n$

26. $k = \dfrac{^-3}{16}\left(\dfrac{^-8}{9}\right)$

27. $h = \dfrac{^-7}{8}\left(\dfrac{3}{14}\right)$

28. $\dfrac{^-5}{16}\left(\dfrac{4}{5}\right) = p$

29. $t = 12.8 \div {^-3.2}$

30. $8.4 \div {^-1.4} = n$

31. $g = {^-11.2} \div 2.8$

32. $\dfrac{3}{8} \div \dfrac{^-1}{4} = p$

33. $n = \dfrac{^-5}{6} \div \dfrac{^-2}{3}$

34. $t = \dfrac{1}{2} \div \dfrac{^-7}{8}$

Simplify.

C **35.** $\dfrac{^-2a(^-3 + {^-1.5})}{3}$

36. $\dfrac{(2.4d - 0.8d)1.2}{^-4}$

37. $\dfrac{0.5n(4.2 - {^-1.8})}{^-0.2}$

Using the Computer

This program demonstrates multiplication and division of rational numbers. RUN it for $\dfrac{^-3}{4} \times \dfrac{4}{5}$ and then try some other exercises from this chapter.

```
  5  REM: TO MULTIPLY OR DIVIDE RATIONAL NUMBERS
 10  PRINT "FOR MULTIPLICATION, TYPE 1."
 20  PRINT "FOR DIVISION, TYPE 2."
 30  PRINT "WHICH";
 40  INPUT C
 50  PRINT "WHAT ARE NUMERATOR AND"
 60  PRINT "DENOMINATOR OF FIRST FRACTION";
 70  INPUT N1,D1
 80  PRINT "WHAT ARE NUMERATOR AND"
 90  PRINT "DENOMINATOR OF SECOND FRACTION";
100  INPUT N2,D2
110  IF C=1 THEN 160
120  PRINT N1;" /";D1;" /";N2;" /";D2;
130  PRINT " =";N1*D2;" /";D1*N2;
140  PRINT " OR";N1*D2/(D1*N2)
150  GOTO 190
160  PRINT N1;" /";D1;" X ";N2;" /";D2;
170  PRINT " =";N1*N2;" /";D1*D2;
180  PRINT " OR";N1*N2/(D1*D2)
190  END
```

8-9 Problem Solving

An elevator started at the 29th floor, went up 2 floors, down 3 floors, down 1 more floor, up 5 floors, and down 6 floors. At what floor did the elevator stop last?

A diagram is useful in solving a problem like this. It helps you to picture the situation more clearly.

$$2 + {}^-3 + {}^-1 + 5 + {}^-6 = {}^-3$$
$$29 + {}^-3 = 26$$

The last stop was on the 26th floor.

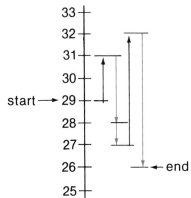

Following are other situations in which a diagram is helpful.

Example 1 You have invited 3 friends to dinner. In how many different ways can the four of you be seated at a round table if you occupy a particular position.

Solution Let's number your friends 1, 2, and 3. First draw diagrams showing friend 1 on your left.
Next show friend 2 on your left.
Finally show friend 3 on your left.
There are 6 arrangements in all.

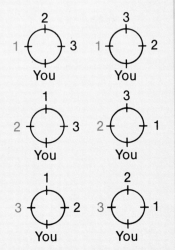

Example 2   The Computer Club entered a game contest. To determine the best 3 players to represent the club, all 8 members were to play each other. How many games would be played?

Solution   We can represent the members of the club by 8 dots, labeled *A* through *H*, placed in a circle. We show a game between two members by a line segment. Now our problem is to count the number of line segments connecting two dots.

There are 7 segments connecting *A* and the other dots, shown in black in the diagram. For *B* there are only 6 segments, the red ones, because *B* has already been connected to *A*. For *C* there will be only 5 segments because *C* is already connected to *A* and *B*. Now we can make a table to continue the count and find the total.

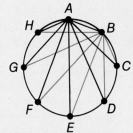

Dot	*A* *B* *C* *D* *E* *F* *G* *H*
Number of segments	$7 + 6 + 5 + 4 + 3 + 2 + 1 + 0 = 28$

A total of 28 games would be played.

Class Exercises

Draw a diagram for the problem.

1. A dolphin started at the surface of the ocean, dove 20 ft down, rose 12 ft, dove 16 ft down, rose 18 ft, and dove 33 ft down. How far below the surface was the dolphin then?

2. In how many ways can you, Al, Mimi, and Pat be seated at a round table if you always occupy the same seat and Mimi sits opposite you?

Problems

Draw a diagram to help you solve the problem.

1. There are 5 people at a meeting. If each person shakes the hand of everyone else, how many handshakes will there be?

2. A group of 6 friends play tennis together. They plan to have each person play all the others. How many games will be played?

3. In how many ways can 3 people be seated at a round table if each person is free to occupy any seat?

4. In how many ways can Elsie, John, Lucia, and Manuel be seated at a round table so that boys and girls alternate?

5. In how many ways can Jill, Lisa, Chris, and Pedro be seated at a round table if Jill and Chris want to sit next to each other?

6. Each Saturday a pair of members of the Nature Club leads a field trip. There are 8 members in the club. If all the different possible pairs are made, how many field trips can be lead?

7. A ball is dropped from a height of 16 ft. Each time it bounces it reaches half its previous height. How high is the 5th bounce? How far has the ball traveled at the top of that bounce?

8. In a 100-m race, Marie is given a headstart of 20 m over Rena. If Rena runs 30 m in the same time that Marie runs 20 m, who wins the race? If Marie can run 10 m in 2 s, how long does it take her to finish the race? How long does it take Rena?

9. A 50-ft rope is cut into 3 sections. The longest section is 5 times as long as the shortest section, and the remaining section is 3 times as long as the shortest section. How long is each section?

Copy and label the diagram at the right to help solve the problem.

10. Scott knows the following distances:

home to Steve's house	0.5 mi
home to the library	0.6 mi
home to the store	0.7 mi
library to the store	0.2 mi
library to Steve's house	0.4 mi
store to Steve's house	0.3 mi

What is the shortest round trip that Scott can make to go from home to the store, to the library, and to Steve's, not necessarily in that order, and then back home?

8-10 *Equations with Rationals*

Thirty shares of Vidword stock dropped $150 in value. How much did each share drop?

You can use an equation to solve this problem.

Let s = amount each share dropped.

$$\text{Then } 30s = {}^-150$$

$$\frac{30s}{30} = \frac{{}^-150}{30} \longleftarrow \textit{Divide each side by 30.}$$

$$s = {}^-5$$

Each share dropped $5 in value.

We use the same rules for solving equations involving both negative and positive numbers as we do for equations involving just positive numbers.

Example 1   Solve the equation $y + {}^-6 = 8$

Solution
$$y + {}^-6 = 8$$
$$y + {}^-6 - {}^-6 = 8 - {}^-6 \longleftarrow \textit{Subtract } {}^-6 \textit{ from each side.}$$
$$y + 0 = 8 + 6$$
$$y = 14$$

Example 2   Solve the equation $\frac{{}^-2}{3}t = 18$

Solution
$$\frac{{}^-2}{3}t = 18$$

$$\left(\frac{{}^-3}{2}\right)\left(\frac{{}^-2}{3}\right)t = \frac{{}^-3}{2}(18) \longleftarrow \textit{Multiply each side by } \frac{{}^-3}{2}.$$

$$t = {}^-27$$

Your Turn   Solve the equation.

• $3n = {}^-21$ • $k + {}^-6 = {}^-4$ • $\frac{1}{4}m = {}^-6.2$ • $\frac{a}{{}^-7} = 3.1$

Class Exercises

What number would you add or subtract to solve the equation?

1. $t - 6 = {}^-2$ **2.** $p - {}^-3 = 8$ **3.** $x - 4 = {}^-7$ **4.** $y - {}^-10 = 9$

5. $n + 12 = 13$ **6.** $k + {}^-4 = 15$ **7.** $m + {}^-9 = {}^-3$ **8.** $d + 11 = {}^-7$

What number would you multiply or divide by to solve the equation?

9. $\dfrac{h}{5} = {}^-20$ **10.** $\dfrac{{}^-n}{6} = 3$ **11.** $\dfrac{{}^-d}{4} = {}^-5$ **12.** $\dfrac{x}{{}^-3} = 8$

13. ${}^-6t = 4$ **14.** $8m = {}^-2$ **15.** ${}^-y = 7$ **16.** ${}^-5n = {}^-9$

Exercises

Solve the equation.

A   **1.** $x - 5 = {}^-6$ **2.** $m - {}^-3 = 9$ **3.** $t - {}^-7 = {}^-5$ **4.** $k - 12 = 13$

5. $d + 8 = {}^-4$ **6.** $y + {}^-5 = 7$ **7.** $a + {}^-9 = {}^-2$ **8.** $n + 11 = {}^-7$

9. $\dfrac{n}{{}^-2} = 6$ **10.** $\dfrac{t}{7} = {}^-4$ **11.** $\dfrac{h}{{}^-5} = {}^-3$ **12.** $\dfrac{b}{{}^-8} = 6$

13. ${}^-10p = 3$ **14.** ${}^-6d = {}^-5$ **15.** ${}^-k = {}^-7$ **16.** ${}^-t = {}^-12$

B   **17.** $p - 3.2 = 4.7$ **18.** $h - {}^-4.3 = 2.5$

19. $x - 7.1 = {}^-8.2$ **20.** $y - {}^-6.5 = {}^-1.4$

21. $h + {}^-5.4 = 7.2$ **22.** $y + 6.9 = {}^-3.8$

23. $m + {}^-7.4 = 2.6$ **24.** $n + {}^-6.3 = {}^-5.1$

25. $\dfrac{x}{4} = {}^-1.2$ **26.** $\dfrac{k}{{}^-5} = 1.3$ **27.** $\dfrac{d}{{}^-3} = {}^-2.2$ **28.** $\dfrac{h}{6} = {}^-4.2$

29. ${}^-5m = 12.5$ **30.** $7t = {}^-14.7$ **31.** ${}^-8g = {}^-7.2$ **32.** ${}^-6n = 8.4$

33. $h - \dfrac{3}{4} = \dfrac{{}^-1}{4}$ **34.** $y - \dfrac{{}^-1}{8} = \dfrac{3}{8}$ **35.** $t + \dfrac{2}{3} = \dfrac{{}^-1}{6}$ **36.** $x + \dfrac{{}^-5}{8} = \dfrac{1}{4}$

37. $\dfrac{1}{3}p = \dfrac{{}^-1}{2}$ **38.** $\dfrac{{}^-3}{4}m = \dfrac{1}{8}$ **39.** $\dfrac{{}^-2}{5}h = \dfrac{{}^-3}{5}$ **40.** $\dfrac{3}{8}t = \dfrac{{}^-5}{8}$

C   **41.** $2.4k + {}^-1.2 = {}^-3.6$ **42.** ${}^-1.5t - 7.4 = 1.6$ **43.** $6.2n - {}^-10.4 = {}^-8.2$

8-11 *Inequalities*

There is a greater number of tape cassettes in stock than stereos.

65 is greater than 32.

Stock Inventory

Tape Cassettes	Stereos
65	32

This is an inequality. An **inequality** is a statement that two quantities are unequal.

Inequalities	
In words	In symbols
65 is greater than 32.	$65 > 32$
n is less than $^-5$.	$n < ^-5$
y is equal to or greater than $^-17$.	$y \geq ^-17$
t is equal to or less than $5\frac{1}{2}$.	$t \leq 5\frac{1}{2}$

We use the same rules to solve inequalities as we do to solve equations *except* when we multiply or divide the inequality by a negative number.

Rules for Solving Inequalities

Add or subtract the same number.	$n - 3 > 10$ $n - 3 + 3 > 10 + 3$ $n > 13$	$d + ^-5 > 7$ $d + ^-5 - ^-5 > 7 - ^-5$ $d > 12$
Multiply or divide by the same positive number.	$\frac{x}{5} \geq ^-7$ $5\left(\frac{x}{5}\right) \geq 5(^-7)$ $x \geq ^-35$	$6t \leq 18$ $\frac{6t}{6} \leq \frac{18}{6}$ $t \leq 3$
Multiply or divide by the same *negative* number and *reverse* the inequality symbol.	$\frac{a}{^-3} > 6$ $^-3\left(\frac{a}{^-3}\right) < ^-3(6)$ $a < ^-18$	$^-4m < 8$ $\frac{^-4m}{^-4} > \frac{8}{^-4}$ $m > ^-2$

The last rule tells you to reverse the inequality symbol when you multiply or divide by a negative number. You can see why this is so by looking at the following.

$$8 > 6$$
$$^-2(8) \underline{\quad ? \quad} ^-2(6)$$
$$^-16 \underline{\quad ? \quad} ^-12$$
$$^-16 < ^-12$$

Example 1 Solve the inequality $y - 7 > ^-5$.

Solution
$$y - 7 > ^-5$$
$$y - 7 + 7 > ^-5 + 7$$
$$y > 2$$

Example 2 Solve the inequality $^-8d < 16$.

Solution
$$^-8d < 16$$
$$\frac{^-8d}{^-8} > \frac{16}{^-8}$$
$$d > ^-2$$

Your Turn Solve the inequality.
  • $t + 6 < 10$ • $x - 5 > 2$

We can graph inequalities on the number line.

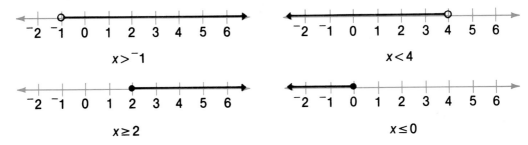

$x > ^-1$ $x < 4$

$x \geq 2$ $x \leq 0$

Notice that an open dot shows that a number *is not* included in a graph and a filled-in dot shows that a number *is* included in a graph.

Example 3   Write an inequality for the graph.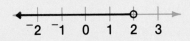

Solution $x < 2$

Your Turn   • Write an inequality for the graph.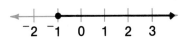

Class Exercises

Write the inequality in symbols.

1. 8 is greater than 3.
2. 6 is less than 9.
3. $^-7$ is less than $^-1$.

4. 2 is greater than $^-5$.
5. $^-3$ is greater than $^-6$.
6. $^-3$ is less than 6.

Is the given number a solution of the inequality? Write Yes or No.

7. $x > 6$; 12
8. $n > {}^-3$; 2
9. $m < 4$; 8
10. $d < 5$; 0

11. $t > {}^-5$; $^-6$
12. $p < 9$; $^-9$
13. $h > 0$; $^-7$
14. $p < 6$; 1

Exercises

What step would you take to solve the inequality?

A   **1.** $t - 7 > 4$
2. $x + {}^-5 < 10$
3. $6d > {}^-12$
4. $\dfrac{a}{^-6} > 2$

Would you reverse the inequality symbol when solving the inequality? Write Yes or No.

5. $x + 7 > 4$
6. $^-3y < 6$
7. $\dfrac{n}{2} \geq {}^-5$
8. $m + 3 \leq {}^-4$

9. $\dfrac{t}{^-8} > 3$
10. $h + 2 \leq 3$
11. $^-9d > {}^-3$
12. $b - 2 \geq 6$

269

Solve the inequality.

B 13. $a + 5 < 2$ 14. $d - 6 \geq 3$ 15. $p + {}^-4 > 2$ 16. $5t \leq 10$

17. $^-2b > 8$ 18. $h - {}^-7 < 8$ 19. $\dfrac{n}{9} < 3$ 20. $\dfrac{x}{^-4} > 5$

21. $n + 8 > {}^-2$ 22. $p - 4 < {}^-5$ 23. $^-5y < {}^-3$ 24. $6n \leq {}^-18$

25. $\dfrac{a}{4} \geq {}^-5$ 26. $t + {}^-3 > {}^-7$ 27. $x - 6 \leq {}^-1$ 28. $8n \geq 0$

29. $\dfrac{d}{^-5} > 6$ 30. $^-x < {}^-9$ 31. $h - {}^-3 \leq 7$ 32. $^-4n < {}^-2$

Write an inequality for the graph.

33. 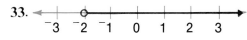   34.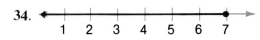

35.    36.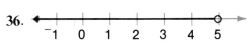

Solve the inequality.

C 37. $2x - 5 > 3$ 38. $7n + {}^-4 \geq 6$ 39. $^-3m + 4 > {}^-9$

Self Test 2

Solve the equation.

1. $b = 7.5 + {}^-7.1$ 2. $x = {}^-8.3 + 4.6$ 3. $n = {}^-2.4 - {}^-8.3$ **8-7**

4. $t = {}^-3(2.1)$ 5. $y = 1.2({}^-1.2)$ 6. $h = {}^-3.9 \div {}^-3$ **8-8**

Draw a diagram to help you solve the problem.

7. In how many ways can Lee, Mike, Nan, and Oliver be seated at a round table if Mike and Nan do not sit next to each other?   **8-9**

Solve the equation.

8. $m - 1.5 = 6.2$ 9. $d + {}^-3.4 = 4.2$ 10. $^-6t = 12.6$ **8-10**

Solve the inequality.

11. $y + 3 > 5$ 12. $x - {}^-9 < 12$ 13. $^-4b > 12$ **8-11**

Algebra Practice

Write an equation. Solve.

1. The product of 761 and 1002 is equal to r.
2. 156,204 divided by 36 is equal to p.
3. The sum of 364.8, 1.065, and 98.37 is d.
4. 9.6 times 0.875 is equal to b.
5. The difference between $17\frac{8}{9}$ and $5\frac{3}{8}$ is m.
6. $2\frac{11}{14}$ divided by $\frac{3}{7}$ is equal to y.
7. The perimeter of a triangle with sides $2\frac{1}{2}$, $3\frac{5}{8}$, and $5\frac{3}{4}$ is P.
8. The area of a parallelogram with a base of 12.8 cm and a height of 7.2 cm is A.
9. The volume of a cylinder with a radius of 2 m and a height of 4 m is V.

Use the distributive property to factor each expression.

10. $c^3d + cd^2 + 3c^2d$
11. $5r^2q + 5rq^2 + 10q^2$
12. $2b^2h + (bh)^2 + bh^2$

Solve.

13. $\frac{3}{8}j = \frac{1}{6}$
14. $7 = 6w + 3$
15. $2t + 2 = 5t - 1.6$

16. 52% of 70 is h.
17. $4\frac{3}{8}$% of 200 is y.
18. 6.12 is 7.2% of a.

19. $r = {}^-4 + {}^-5$
20. $d = {}^-29 + {}^-18$
21. $g = {}^-7 + {}^-4$

22. $j = {}^-6 - 3$
23. $m = 7 - {}^-5$
24. $c = {}^-10 - {}^-8$

25. $t = {}^-9({}^-3)$
26. $k = -2 \cdot 5$
27. $u = 7({}^-9)$

28. $f = 6 + ({}^-3 \times 3)$
29. $j = ({}^-9 \times 2) + 4$
30. $p = 8 - (6 \div 2) + {}^-2$

31. $v = {}^-8.63 - 4.68$
32. $q = \frac{{}^-1}{8} + \frac{{}^-3}{4}$
33. $z = 3\frac{1}{6} - {}^-2\frac{1}{2}$

34. $e = 88.56 \div {}^-2.7$
35. $g = {}^-1\frac{1}{6} \times {}^-1\frac{1}{3}$
36. $r = {}^-3\frac{1}{3} \div 2\frac{1}{3}$

37. $2x + {}^-3 > 5$
38. ${}^-4 - 2b < 6$
39. $\frac{m}{{}^-4} + 2 > {}^-3$

Problem Solving for the Consumer

Travel Rates

If air travel is ever in your plans, it pays to think smart and to compare prices on different airlines and special deals. Listed below are sample air fares from City A to City B. All the prices are for coach class. Notice that you can save quite a bit of money if you can take advantage of a special deal.

Airline	Regular Fare	Night Fare
XYZ	$269 one way	$236 one way
ABC	$295 one way	$236 one way
FUN	$280 one way	—

Airline	Special Fare
XYZ	$99 one way
FUN	$99 one way

← Limited seats. Fares only good through a specified date.

Airline	Advance Purchase Fare
XYZ	$189 one way
ABC	$189 one way
FUN	$189 one way

← Must be purchased at least 7 days before flight.

Children fly free on XYZ Airline.

Solve.

1. What's the least expensive rate available for a round-trip ticket?
2. Suppose the special fares on airlines XYZ and FUN are good through April 1, and you wish to fly on April 10. What is the least expensive fare? When must you purchase your ticket?
3. To travel first class costs 20% more than to travel in coach at the regular fare. What is the first class fare on ABC Airline?
4. A parent with two children wants to make a round trip and leave in two days. If there are no special fare seats left, what is the least expensive way to make the trip? How much will it cost?

Hotels, resorts, and airlines all want the consumer's business, and sometimes they offer good deals to get it. Rates can vary quite a bit, so it pays to examine these deals carefully. Here's an example of a tour package that includes all transportation and hotel space.

| Hotel | 8 Days, 7 Nights* | |
	Feb–April	May–June
Economy	$419	$399
Superior	$519	$499
Deluxe	$569	$489

* All rates per person, double occupancy. All rates plus 15% tax.

5. How much would the package deal cost per person, for an economy hotel in April, including the tax?
6. Suppose meals will average $35 per day, and the traveler wants to bring $200 extra for spending money. How much will the trip at a superior hotel in April cost?

Package rates are usually based upon the fact that people will arrive and leave on a Saturday or Sunday. If they don't, usually they'll have to pay individual rates. Here are some typical ones for the same trip described above.

Individual Rates, April*	
Round-trip air fare	$359
Superior hotel per night	$ 50
Limousine to hotel**	$ 6

* Prices include taxes.   **One way

7. How much would the trip cost if you paid the individual rates? Include $35 per day for meals and $200 for spending money.
8. By taking the package deal in April and staying at the superior hotel, how much money do you save?

273

Enrichment

Absolute Value

On a number line, ⁻3 and 3 are each 3 units in distance from 0. The distance of a number from 0 is called its **absolute value**. It is always a positive number or zero. The absolute value of both ⁻3 and 3 is 3.

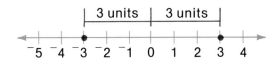

The absolute value of a number x is written $|x|$. We write $|{-}3| = 3$ and $|3| = 3$.

Here is a complete definition of absolute value.

$$|x| = x \text{ if } x \geq 0 \text{ and } |x| = {}^-x \text{ if } x < 0$$

Thus
$$|19| = 19 \text{ because } 19 \geq 0$$
$$|{-}23| = 23 \text{ because } {}^-23 < 0$$

What is the value of the expression?

1. ⁻|7| **2.** |⁻50| **3.** ⁻|⁻12| **4.** |18.3| **5.** $-\left|\dfrac{7}{10}\right|$

Which is greater?

6. |⁻7| or |5| **7.** ⁻|8| or |⁻8| **8.** |12| or ⁻|22| **9.** |⁻9| or |⁻15|

10. ⁻|⁻2| or |⁻2| **11.** |25| or |⁻50| **12.** |⁻18| or ⁻|16| **13.** ⁻|⁻23| or |⁻24|

Arrange in order from smallest to largest.

14. |⁻9|, ⁻|8|, ⁻|⁻7|, |6|, |⁻10| **15.** |5|, ⁻|4|, |⁻3|, |⁻4|, |0|

16. $\left|\dfrac{1}{2}\right|, \left|\dfrac{-3}{4}\right|, -\left|\dfrac{3}{8}\right|, \left|\dfrac{1}{4}\right|, \left|\dfrac{-1}{8}\right|$ **17.** |⁻4.3|, |5.6|, ⁻|3.8|, |7.1|, |⁻6.2|

Let $a = {}^-5$ and $b = 3$. Evaluate the expression.

18. |3a| **19.** |3a + 4b| **20.** |⁻3a + b| − |2b|

21. |⁻5b| **22.** |5a + 2b| **23.** |⁻9a + 3b| − |3a|

24. |a − b| **25.** |⁻4b + a| **26.** |3b + ⁻2a| − |6b|

When will these be true statements?

27. |a| = |b| **28.** |a + b| = |a| + |b| **29.** |a + b| = |a| − |b|

Chapter Review

Complete, using < or >, to make a true statement.

1. $^-12$ __?__ 3

2. $^-5$ __?__ 8

3. 14 __?__ $^-3$

4. 2 __?__ $^-20$

Choose the correct answer. Write a, b, or c.

5. $x = ^-7 + 5$
    a. $x = 2$
    b. $x = 12$
    c. $x = ^-2$

6. $n = 8 - ^-4$
    a. $n = 12$
    b. $n = 4$
    c. $n = ^-4$

7. $d = ^-6(^-3)$
    a. $d = ^-18$
    b. $d = 18$
    c. $d = 9$

8. $t = 16 \div ^-2$
    a. $t = ^-32$
    b. $t = ^-8$
    c. $t = 8$

Is the given number a solution of the equation? Write Yes or No.

9. $m + 8 = ^-6$; 2

10. $g - 5 = ^-7$; $^-12$

11. $k + ^-3 = 9$; 12

12. $t - ^-4 = 5$; 9

13. $h + ^-10 = ^-12$; 2

14. $b - 7 = 0$; 7

15. $\dfrac{n}{6} = ^-3$; $^-2$

16. $\dfrac{b}{^-4} = 2$; $^-8$

17. $\dfrac{a}{^-3} = ^-5$; $^-15$

18. $7x = ^-21$; 3

19. $^-4g = 20$; $^-5$

20. $^-3d = ^-33$; $^-11$

Solve the equation.

21. $n + 7 = ^-3$

22. $t - ^-12 = ^-8$

23. $m + ^-9 = 1$

24. $^-3h = 24$

25. $\dfrac{a}{7} = ^-5$

26. $5t = ^-40$

27. $d + ^-3.3 = 7.4$

28. $k - 7.1 = 5.5$

29. $b + 9.6 = ^-0.4$

30. $\dfrac{n}{11} = ^-1.21$

31. $^-4x = ^-4.8$

32. $\dfrac{m}{^-4.3} = 5$

Solve the inequality.

33. $y - 6 < ^-3$

34. $x + ^-12 > 6$

35. $g + ^-4 > ^-18$

36. $^-8t > ^-4$

37. $\dfrac{a}{6} < ^-9$

38. $7h < ^-12$

39. $\dfrac{x}{^-5} < 4$

40. $^-6d < ^-36$

41. $13y > ^-39$

Chapter Test

Write a number sentence using $>$ to show which number is greater.

8-1

1. $^-3$ or 2   2. 10 or $^-12$ 3. $^-4$ or $^-14$ 4. $^-30$ or 20

Solve the equation.

8-2

5. $t = ^-8 + ^-4$ 6. $b = ^-2 + ^-13$ 7. $h = ^-20 + ^-9$

8. $x = ^-7 + ^-12$ 9. $n = ^-9 + ^-21$ 10. $d = ^-23 + ^-14$

8-3

11. $x = ^-3 + 8$ 12. $n = 12 + ^-1$ 13. $y = 4 + ^-16$

14. $d = 15 + ^-18$ 15. $k = ^-32 + 4$ 16. $t = 35 + ^-16$

8-4

17. $m = 8 - ^-3$ 18. $g = ^-7 - 10$ 19. $d = ^-2 - ^-9$

20. $a = 2 - 13$ 21. $t = 15 - ^-8$ 22. $b = ^-4 - 2$

8-5

23. $b = 5(^-6)$ 24. $k = ^-8(^-3)$ 25. $g = ^-12(5)$

26. $x = ^-9(3)$ 27. $m = 12(^-6)$ 28. $y = ^-14(^-3)$

8-6

29. $h = ^-18 \div 2$ 30. $m = 24 \div ^-6$ 31. $t = ^-20 \div 5$

32. $y = 16 \div ^-8$ 33. $d = ^-32 \div 4$ 34. $b = ^-38 \div ^-2$

8-7

35. $d = 4.5 + ^-3.2$ 36. $p = ^-3.8 + ^-2.1$ 37. $n = ^-9.6 - 5.4$

38. $x = ^-6.3 - 7.4$ 39. $y = 8.2 - ^-4.3$ 40. $t = ^-5.1 + ^-3.4$

8-8

41. $x = 5(^-4.1)$ 42. $b = ^-7(^-3.4)$ 43. $y = ^-3.5 \div 7$

44. $n = ^-6.3 \div ^-9$ 45. $p = 2.4(^-4)$ 46. $a = 12.6 \div ^-3$

Draw a diagram to help you solve the problem.

8-9

47. A board 12 ft long is cut into 2 pieces, one piece 3 times longer than the other. How long is each piece?

Solve the equation.

8-10

48. $t + ^-5.6 = 3.3$ 49. $x - 4.5 = ^-2.5$ 50. $^-3n = 36$

51. $\frac{a}{4} = ^-1.2$ 52. $5n = ^-2.5$ 53. $\frac{m}{^-6} = 2.3$

Solve the inequality.

8-11

54. $n + ^-6 < 4$ 55. $t - 5 > ^-9$ 56. $^-6d < 30$

57. $\frac{x}{^-5} > 7$ 58. $b - ^-7 > 10$ 59. $3a < ^-12$

Cumulative Review (Chapters 1–8)

Add, subtract, multiply, or divide.

1. $6.2 - 0.465$
2. $4.25 \div 0.17$
3. 6.3×2.09
4. $16 - 5 \times 3.1$
5. $6(12 + 8)$
6. 0.25×100
7. $1\frac{1}{3} - \frac{4}{5}$
8. $\frac{5}{9} \times \frac{3}{8} \times 1\frac{1}{2}$
9. $(10^2)\left(1\frac{3}{5}\right)$
10. $(12)(10)(0.6)$
11. $26 + 3 \div 2$
12. $(10)(17.1)(10)$
13. $\frac{1}{5} \div \frac{1}{8}$
14. $\frac{1}{8} \cdot \frac{2}{3} \cdot 15$
15. $\left(2\frac{1}{2}\right)\left(\frac{1}{2}\right)$
16. $(4.6)^2$
17. $42 + 8.5$
18. $12 \div 4 + 4 \cdot 8$

Solve.

19. 76% of 50 is what number?
20. 50 out of 80 is what percent?
21. 24% of what number is 5.76?
22. 36% of 180 is what number?
23. What is 18% of 45?
24. 16 out of 56 is what percent?
25. 150% of what number is 375?
26. What is 4% of 300?
27. What is 8% of 600?
28. $33\frac{1}{3}$% of 282 is what number?

Find the volume.

29.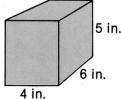

5 in.
6 in.
4 in.

30.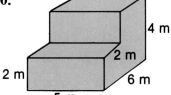

4 m
2 m
2 m
5 m
6 m

31.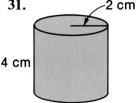

2 cm
4 cm

Write an equation. Solve.

32. The area of a rectangle is 225 m² and one side is 15 m. What are the lengths of the other sides?
33. In a figure skating competition a skater scored twice as many points in the freestyle event as in the school figures. The total of the two events was 14.7 points. How many points did he score in each event?

9

Graphing in the Coordinate Plane

A surveyor looking through a transit, a form of telescope, uses horizontal and vertical lines to map positions. In this chapter we use grids of horizontal and vertical lines to graph points.

9-1 *Collecting Data*

Draw a square and mark a point in it. How many regions are formed if you draw one line through the point? 2 lines? 3 lines? 4 lines?

 1 line
2 regions

 2 lines
4 regions

 3 lines
6 regions

 4 lines
8 regions

The data are shown in the diagrams. We can also show the data in a table.

No. of lines	1	2	3	4
No. of regions	2	4	6	8

+2 +2 +2

Constant change = 2

We see that each time we draw a line, 2 regions are added. We call 2 the *constant change* of the sequence of terms, 2, 4, 6, 8. We also see that the number of regions in each case is twice the number of lines. We can predict the table will continue as follows:

No. of lines	1	2	3	4	5	6	7	8	...	n
No. of regions	2	4	6	8	10	12	14	16	...	$2n$

Example 1   How many regions are formed by 25 lines?

Solution   If $n = 25$, then $2n = 2(25) = 50$. Therefore, 50 regions are formed.

Your Turn   How many regions are formed by the following number of lines?
- 16
- 150
- 1492
- 2001

Here is another experiment. Arrange different numbers of blocks using the following rules:

Rule 1. Blocks must always touch   This , not this
face-to-face.

Rule 2. Blocks must be placed in   This , not this
a straight line.

Our object is to count the number of visible faces in each arrangement and to look for a pattern. Shown below are data collected for the first five arrangements.

No. of blocks	1	2	3	4	5
No. of visible faces	5	8	11	14	17

Each time we add a block, we add 3 visible faces. We have produced the sequence of five terms, 5, 8, 11, 14, 17, with a constant change of 3. We might expect that for n blocks there should be $3n$ visible faces. However, the first block has 5, or 2 *more* than 3, visible faces. So we write the expression $3n + 2$ for the number of visible faces when n blocks are in a row.

No. of blocks	1	2	3	4	5	6	7	. . .	n
No. of visible faces	5	8	11	14	17	20	23	. . .	$3n + 2$

Constant change = 3

+3 +3 +3 +3 +3 +3

Example 2   What is the 18th term in the sequence shown above?

Solution   When $n = 18$, $3n + 2 = 3(18) + 2$
$$= 54 + 2 = 56.$$
The 18th term is 56. So 56 faces are visible when 18 blocks are in a row.

Your Turn   How many faces are visible when the following numbers of blocks are in a row?
- 8   • 10   • 50   • 100

Class Exercises

Find the constant change for each sequence of numbers. Then give the next 5 terms.

	Term							Term				
	1st	2nd	3rd	4th	5th			1st	2nd	3rd	4th	5th
1.	1	3	5	7	9		**2.**	5	6	7	8	9
3.	4	7	10	13	16		**4.**	6	9	12	15	18
5.	8	12	16	20	24		**6.**	6	11	16	21	26

Exercises

Find the constant change for each sequence of numbers. Then find the next 5 terms.

		Term						Term				
		1st	2nd	3rd	4th	5th		1st	2nd	3rd	4th	5th
A	**1.**	12	36	60	84	108	**2.**	23	58	93	128	163
	3.	3	130	257	384	511	**4.**	142	170	198	226	254
	5.	564	540	516	492	468	**6.**	698	646	594	542	490

Find the first five terms of the sequence of numbers produced by the given expression.

Example $6x + 3$

Solution When $x = 1$, $6x + 3 = 6(1) + 3 = 6 + 3 = 9$, the 1st term
When $x = 2$, $6x + 3 = 6(2) + 3 = 12 + 3 = 15$, the 2nd term
When $x = 3$, $6x + 3 = 6(3) + 3 = 18 + 3 = 21$, the 3rd term
When $x = 4$, $6x + 3 = 6(4) + 3 = 24 + 3 = 27$, the 4th term
When $x = 5$, $6x + 3 = 6(5) + 3 = 30 + 3 = 33$, the 5th term

B   **7.** $n + 8$ **8.** $2x + 4$ **9.** $5s - 5$ **10.** $3n + 25$

11. $^-2n + 12$ **12.** $^-3n + 7$ **13.** $^-13z + 6$ **14.** $^-16a +^- 13$

Find the required terms of the given expression.

15. $5n + 8$, 50th and 150th terms   **16.** $12z - 7$, 72nd and 92nd terms

17. $15x - 45$, 84th and 58th terms   **18.** $27a + 80$, 46th and 87th terms

9-2 *Finding Expressions for Data*

In the preceding section we found expressions for any term for two sequences of numbers with constant changes.

$$2, 4, 6, 8, 10, \ldots, 2n \qquad 5, 8, 11, 14, 17, \ldots, 3n + 2$$

We can find an expression for any term of *any* sequence of numbers with a constant change.

Example Find an expression for any term of the sequence 9, 14, 19, 24, ...

Solution 1. Find the constant change by subtracting the first term from the second, the second term from the third, and so on: $14 - 9 = 5$, $19 - 14 = 5$, and so on. The constant change is 5, so $5n$ is one part of the expression.
2. Subtract the constant change from the first term. $9 - 5 = 4$.
3. Write an expression to show that 4 is added to a multiple of 5: $5n + 4$.

Check: Use $5n + 4$ to find the 3rd term.
If $n = 3$, then $5n + 4 = 5(3) + 4 = 19$
The 3rd term is 19, as shown above.

Your Turn • Find an expression for any term of 5, 7, 9, 11, 13, ...

Class Exercises

For the sequence **7, 11, 15, 19, 23, . . . ,** do the following:

1. Find the constant change.

2. Subtract the constant change from the first term.

3. Write an expression for any term: __?__ n + __?__

4. Use the expression to find the 5th term. Does it check with the 5th term given above?

5. Find the 10th and 20th terms.

Exercises

Write an expression for any term of the given sequence. Then find the 10th and 15th terms.

A   **1.** 4, 8, 12, 16, 20, . . .   **2.** 5, 9, 13, 17, 21, . . .

     3. 7, 10, 13, 16, 19, . . .   **4.** 14, 20, 26, 32, 38, . . .

     5. 10, 17, 24, 31, 38, . . .   **6.** 12, 21, 30, 39, 48, . . .

Find the constant change for each sequence of numbers. Then find the next five terms.

B   **7.** 7, 6, 5, 4, 3, . . .   **8.** 4, 2, 0, $^-2$, $^-4$, . . .

     9. $^-25$, $^-20$, $^-15$, $^-10$, $^-5$, . . .   **10.** $^-5$, $^-10$, $^-15$, $^-20$, $^-25$, . . .

     11. $^-9$, $^-6$, $^-3$, 0, 3, . . .   **12.** 9, 6, 3, 0, $^-3$, . . .

Write an expression for any term of the sequence. Then find the 10th and 50th terms.

        12, 10, 8, 6, 4

        Find the constant change: $10 - 12 = {}^-2$
        Subtract constant change from 1st term: $12 - {}^-2 = 14$
        Write an expression for any term: $^-2n + 14$
        10th term: $^-2(10) + 14 = {}^-6$ 50th term: $^-2(50) + 14 = {}^-86$

     13. 10, 9, 8, 7, 6, . . .   **14.** 8, 4, 0, $^-4$, $^-8$, . . .

     15. $^-5$, $^-1$, 3, 7, 11, . . .   **16.** 11, 8, 5, 2, $^-1$, . . .

     17. 3, 8, 13, 18, 23, . . .   **18.** 2, 10, 18, 26, 34, . . .

C   **19.** 8, $10\frac{1}{2}$, 13, $15\frac{1}{2}$, 18, . . .   **20.** 5, $8\frac{3}{4}$, $12\frac{1}{2}$, $16\frac{1}{4}$, 20, . . .

     21. 16, $14\frac{2}{5}$, $12\frac{4}{5}$, $11\frac{1}{5}$, $9\frac{3}{5}$, . . .   **22.** $6\frac{2}{3}$, 4, $1\frac{1}{3}$, $^-1\frac{1}{3}$, $^-4$, . . .

Problems

Solve.

1. On the first week in January, Edna put 75¢ in her piggy bank. On each following week she added 25¢.
   a. How much money was in her piggy bank after 5 weeks? 10 weeks? 50 weeks?
   b. How long did it take Edna to save $20.

The tables in the Seven Seas Restaurant are square-shaped and can seat one person on each side. They can be arranged in rows, as shown at the right, for parties who want to sit together. How many can be seated in a row with the given number of tables?

2. 1   **3.** 2   **4.** 3   **5.** 4   **6.** 5   **7.** 10   **8.** 15

9. A party of 15 persons has made reservations at the Seven Seas for lunch. If the tables are arranged in a row, how many of them are needed?

10. a. Thirty members of the Ecology Club arrived for dinner at the Seven Seas. Only 12 tables were available for them. What is the least number of rows of tables which can seat them exactly if there are the same numbers of tables in each row. How are the tables arranged?
    b. Just as the 30 members of the Ecology Club were about to sit down, 2 more members joined them. How could the tables be rearranged to seat all 32 members exactly? The least number of rows of tables should be used.

9-3 The Coordinate Plane

We can display data like those in the two preceding sections by graphing them on a *coordinate plane*. First we write the data as *ordered pairs* of numbers. The numbers in an ordered pair are called **coordinates.**

The ordered pairs below display the data for the first experiment in Section 9-1.

No. of lines	No. of regions	Ordered pairs
1	2	(1, 2)
2	4	(2, 4)
3	6	(3, 6)
4	8	(4, 8)

In the ordered pair (3, 6), the first coordinate, 3, tells the number of lines drawn through a point in a square. The second coordinate, 6, tells the number of regions formed.

To graph (3, 6) on the coordinate plane, start at 0, move 3 units to the right, then move 6 units up.

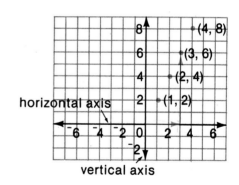

In general, to graph an ordered pair (*a*, *b*):
• Start at 0.
• Move along the horizontal axis the number named by the first coordinate, *a* units.
• Move along the vertical axis the number named by the second coordinate, *b* units.

Example 1   Graph the ordered pairs on a coordinate plane.
a. (6, 3)   Is (6, 3) the same as (3, 6)?
b. (3, −2)   **c.** (⁻4, 1)   **d.** (⁻2, ⁻3)   **e.** (0, 4)   **f.** (⁻3, 0)

Solution   In each case, start at 0.
a. Move 6 right and 3 up.
   No.
b. Move 3 right and 2 down.
c. Move 4 left and 1 up.
d. Move 2 left and 3 down.
e. Do not move on the horizontal axis. Move 4 up.
f. Move 3 left and stop.

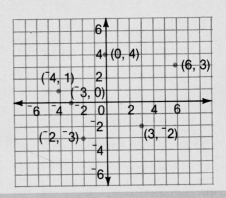

In the example above, we see that the ordered pair (6, 3) is not the same as the ordered pair (3, 6). In general, the ordered pair (a, b) is not the same as (b, a), unless a = b.

Example 2   Write the ordered pair for the point on the coordinate plane.
a. A   **b.** B
c. C   **d.** D

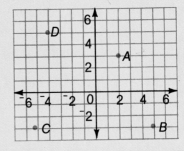

Solution   **a.** (2, 3)   **b.** (5, ⁻3)
c. (⁻5, ⁻3)   **d.** (⁻4, 5)

Your Turn   Graph the ordered pairs on a coordinate plane.
• (3, 5)   • (⁻4, 4)
• (0, 2)   • (1, ⁻4)

Write the ordered pair for the point on the coordinate plane.
• E   • F   • G   • H

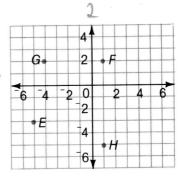

Class Exercises

How far would you move and in what direction for the given coordinate?

1. 2 in (2, 3)   **2.** ⁻3 in (1, ⁻3)   **3.** ⁻5 in (⁻5, 4)   **4.** 0 in (2, 0)

5. ⁻4 in (0, ⁻4)   **6.** 20 in (20, 32)   **7.** 8 in (⁻8, 8)   **8.** 0 in (0, 7)

Exercises

Graph the ordered pairs on a coordinate plane.

A **1.** (7, ⁻2)   **2.** (5, 0)   **3.** (⁻6, 1)   **4.** (⁻3, ⁻3)

5. (0, 8)   **6.** (⁻4, 0)   **7.** (0, 0)   **8.** (0, ⁻10)

9. (⁻4, ⁻1)   **10.** (6, ⁻2)   **11.** (5, 7)   **12.** (⁻2, 4)

Write the ordered pair for the point on the coordinate plane.

13. *M*   **14.** *N*

15. *P*   **16.** *Q*

17. *R*   **18.** *S*

19. *T*   **20.** *U*

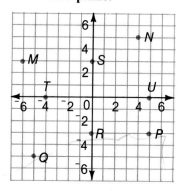

For each exercise, graph the ordered pairs on a coordinate plane, then join the graphs in order, connecting the last graph with the first graph. What geometric figure have you drawn?

B **21.** (3, 3), (⁻3, 3), (⁻3, ⁻3), (3, ⁻3)   **22.** (4, 0), (0, 6), (⁻4, 0)

23. (2, ⁻3), (6, ⁻3), (6, ⁻7)   **24.** (⁻4, 3), (6, 3), (3, ⁻2), (⁻7, ⁻2)

25. (0, 5), (5, 0) (0, ⁻5), (⁻5, 0)   **26.** (2, 6), (5, 3), (⁻3, ⁻5), (⁻6, ⁻2)

27. (⁻1, 3), (5, 3), (3, ⁻1), (⁻3, ⁻1)   **28.** (⁻4, 1), (⁻2, ⁻2), (2, ⁻2), (4, 1)

9-4 *Graphing Data*

In the preceding lesson, we graphed ordered pairs of numbers which showed the number of regions formed by drawing lines through a point. The graph is repeated at the right. Notice that the points graphed lie on a straight line. By extending the line we can predict that 6 lines form 12 regions, 7 lines form 14 regions, and so on.

We can also graph the data we collected in section 9-1 involving the number of visible faces of blocks arranged in a row:

No. of blocks	No. of visible faces	Ordered pairs
1	5	(1, 5)
2	8	(2, 8)
3	11	(3, 11)
4	14	(4, 14)
5	17	(5, 17)

Notice that, like the preceding graph, the plotted points lie on a straight line. By extending the line we can predict that 6 blocks have 20 visible faces, 7 blocks have 23 visible faces, and so on.

In each case above we have graphed a sequence of numbers with a constant change and have found that the graphs lie on a straight line. This is true of any sequence of numbers with a constant change.

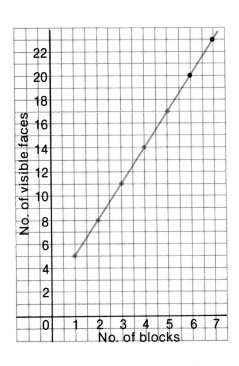

Example What is the constant change for the sequence of numbers shown at the right? Graph the sequence and use the graph to predict the 6th term.

No. of term	Sequence
1	4
2	8
3	12

Solution $4 + 4 = 8$, $8 + 4 = 12$. The constant change is 4. Write the data as ordered pairs and then graph them on a coordinate plane.

Ordered pairs
———————
(1, 4)
(2, 8)
(3, 12)

From the graph, the 6th term is 24.

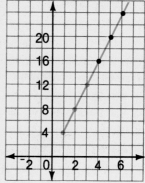

Class Exercises

Use the graph to write the first 5 terms of each sequence.

1.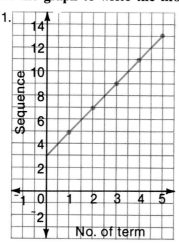

2.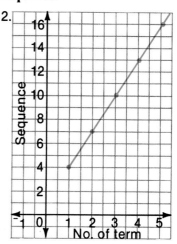

Exercises

Graph the following data. Use your graph to find the 8th term.

A

1. No. of term	Sequence	2. No. of term	Sequence	3. No. of term	Sequence	4. No. of term	Sequence
1	4	1	3	1	1	1	2
2	5	2	6	2	3	2	6
3	6	3	9	3	5	3	10
4	7	4	12	4	7	4	14
5	8	5	15	5	9	5	18

5. No. of term	Sequence	6. No. of term	Sequence	7. No. of term	Sequence	8. No. of term	Sequence
1	0	1	8	1	2	1	15
2	2	2	7	2	5	2	13
3	4	3	6	3	8	3	11
4	6	4	5	4	11	4	9
5	8	5	4	5	14	5	7

B

9. No. of term	Sequence	10. No. of term	Sequence	11. No. of term	Sequence	12. No. of term	Sequence
1	6	1	4	1	$^-5$	1	15
2	5	2	2	2	$^-2$	2	12
3	4	3	0	3	1	3	9
4	3	4	$^-2$	4	4	4	6
5	2	5	$^-4$	5	7	5	3

C

13. No. of term	Sequence	14. No. of term	Sequence	15. No. of term	Sequence	16. No. of term	Sequence
1	$1\frac{1}{2}$	1	$3\frac{1}{2}$	1	$8\frac{1}{4}$	1	$12\frac{1}{4}$
2	$3\frac{1}{2}$	2	5	2	$6\frac{1}{4}$	2	11
3	$5\frac{1}{2}$	3	$6\frac{1}{2}$	3	$4\frac{1}{4}$	3	$9\frac{3}{4}$
4	$7\frac{1}{2}$	4	8	4	$2\frac{1}{4}$	4	$8\frac{1}{2}$
5	$9\frac{1}{2}$	5	$9\frac{1}{2}$	5	$0\frac{1}{4}$	5	$7\frac{1}{4}$

Problems

Solve.

1. A partially full swimming pool contains 6000 gal of water. It is being filled at the rate of 300 gal per h. When full it holds 27,000 gal. Draw a graph to show how much water is in the pool after 10 h, 20 h, and 30 h. Use your graph to predict how long it will take to fill the pool.

2. Comchip Company produced 15,000 computer chips in January. The production manager has set a goal of increasing production by 2000 chips each month for the rest of the year. Draw a graph to show the production figures for January through March. Then use your graph to predict the production figures for each of the remaining months.

Self Test 1

Find the first five terms of the sequence of numbers produced by the given expression.

1. $x - 2$
2. $3y + 5$
3. $^-2n - 4$
4. $^-t + 5$ **9-1**

Write an expression for any term of the sequence. Then find the 10th and 50th terms.

5. 1, 4, 7, 10, 13
6. $^-3$, 0, 3, 6, 9   **9-2**
7. 9, 5, 1, $^-3$, $^-7$
8. 11, 16, 21, 26, 31

Graph the ordered pairs on a coordinate plane.

9. $(3, 5)$
10. $(^-4, 0)$
11. $(6, ^-2)$
12. $(^-5, ^-1)$ **9-3**

Graph the following data. Use your graph to find the 8th term.

13.

No. of term	1	2	3	4	5
Sequence	3	7	11	15	19

9-4

14.

No. of term	1	2	3	4	5
Sequence	8	5	2	$^-1$	$^-4$

Self Test answers and Extra Practice are at the back of the book.

9-5 Equation of a Line

In Section 9-2 you learned that you can write an expression for any sequence of numbers with a constant change. For example, the expression $3x + 2$ produces the sequence of numbers 5, 8, 11, 14, ..., which has a constant change of 3.

In algebra, we often write an expression like $3x + 2$ in an equation with two variables, as $y = 3x + 2$. Then we graph the equation on the coordinate plane. The horizontal axis is called the x-axis and the vertical axis is called the y-axis. We substitute values for x in $y = 3x + 2$ to find values for y. Then we form ordered pairs and graph them.

x	$y = 3x + 2$	(x, y)
$^-2$	$y = 3(^-2) + 2 = {}^-4$	$(^-2, {}^-4)$
0	$y = 3(0) + 2 = 2$	$(0, 2)$
2	$y = 3(2) + 2 = 8$	$(2, 8)$

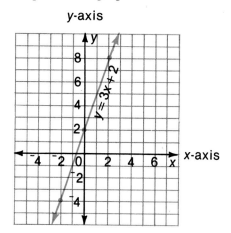

The points all lie on a straight line. The ordered pairs $(-2, -4)$, $(0, 2)$, and $(2, 8)$ are called **solutions** of the equation $y = 3x + 2$. Every point on the line we have graphed has an ordered pair that is a solution of $y = 3x + 2$.

Example Is the ordered pair a solution of the equation $y = 3x + 2$?
  a. $(1, 4)$ **b.** $(^-5, {}^-13)$

Solution **a.** $4 \stackrel{?}{=} 3(1) + 2$

4	$3 + 2$
4	5

$4 \neq 5$, so $(1, 4)$ is not a solution.

b. $^-13 \stackrel{?}{=} 3(^-5) + 2$

$^-13$	$^-15 + 2$
$^-13$	$^-13$

$^-13 = {}^-13$, so $(^-5, {}^-13)$ is a solution.

Your Turn   Is $(10, 32)$ a solution of $y = 3x + 2$?

Class Exercises

Complete the ordered pair so that it is a solution of the given equation. Use the graph.

$$y = 2x - 3$$
$$(x, y)$$

1. $(^-4, ?)$
2. $(^-2, ?)$
3. $(0, ?)$
4. $(?, 1)$
5. $(?, 5)$

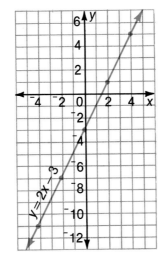

$$y = 3x + 1$$
$$(x, y)$$

6. $(^-3, ?)$
7. $(^-1, ?)$
8. $(0, ?)$
9. $(?, 4)$
10. $(?, 10)$

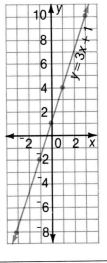

Exercises

Complete the ordered pair so that it is a solution of the given equation.

		x	$y = 2x - 1$	(x, y)
A	**1.**	$^-2$	$y = 2(^-2) - 1 = ^-5$	$(^-2, ?)$
	2.	0	$y = 2(0) - 1 = ^-1$	$(?, ^-1)$
	3.	3	$y = 2(3) - 1 = ?$	$(3, ?)$
	4.	5	$y = 2(5) - 1 = ?$	$(?, ?)$

	x	$y = 4x + 2$	(x, y)
5.	$^-5$	$y = 4(^-5) + 2 = ?$	$(?, ?)$
6.	$^-1$	$y = 4(^-1) + 2 = ?$	$(?, ?)$
7.	0	$y = 4(0) + 2 = ?$	$(?, ?)$
8.	5	$y = 4(5) + 2 = ?$	$(?, ?)$

Find three ordered pairs that are solutions of the given equation, then graph them and draw the line for the equation.

B **9.** $y = x - 6$ **10.** $y = x + 5$ **11.** $y = 2x + 1$

12. $y = 2x - 2$ **13.** $y = ^-x + 3$ **14.** $y = ^-x - 3$

15. $y = 3x - 1$ **16.** $y = ^-2x + 2$ **17.** $y = ^-3x - 1$

18. $y = 4x + 5$ **19.** $y = ^-3x + 1$ **20.** $y = 4x - 8$

C **21.** $y = \dfrac{1}{2}x + 3$ **22.** $y = 2x + \dfrac{1}{3}$ **23.** $y = \dfrac{1}{3}x + \dfrac{2}{3}$

293

9-6 Slope and Intercept

Integers that differ by 1 are called **consecutive integers.** For example, 1, 2, 3, and 4 are consecutive integers, as are ⁻15, ⁻14, and ⁻13.

As shown in the table below, if we substitute the consecutive integers 1, 2, 3, and 4 for x in the equation $y = 2x + 5$, we obtain the values 7, 9, 11, and 13 for y. Thus we produce a sequence of numbers with the constant change of 2, which is the coefficient of x in $y = 2x + 5$.

x	$y = 2x + 5$	(x, y)
1	$y = 2(1) + 5 = 7$	$(1, 7)$
2	$y = 2(2) + 5 = 9$	$(2, 9)$
3	$y = 2(3) + 5 = 11$	$(3, 11)$
4	$y = 2(4) + 5 = 13$	$(4, 13)$

The constant change, or coefficient of x, can also be found by looking at the graph of $y = 2x + 5$. On the graph at the right we can find the constant change by moving from point A to point B horizontally 3 units (the **run**), then vertically 6 units (the **rise**). The ratio of rise to run equals the coefficient of x in the equation $y = 2x + 5$.

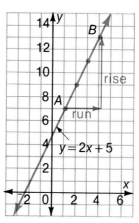

$$\frac{\text{rise}}{\text{run}} = \frac{6}{3} = 2, \text{ the coefficient of } x \qquad \frac{rise}{run} = slope$$

vertical distance

horizontal distance

This ratio of rise to run is the same when moving from any point on the line to any other point on the line and is called the **slope** of the line. The slope is a measure of the steepness of a line.

Notice that the line crosses the y-axis 5 units up. This 5 corresponds to the 5 in the equation $y = 2x + 5$ and is called the **y-intercept.**

An equation of the form $y = mx + b$ is called a **linear** equation. Its graph is a straight line whose slope is m and whose y-intercept is b.

Example 1   Find the slope and y-intercept for the line at the right. Then use them to write an equation for the line.

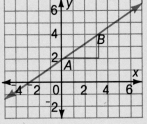

Solution   From point A to point B the run is 3 and the rise is 2.

$$\text{slope} = \frac{\text{rise}}{\text{run}} = \frac{2}{3}$$

The line crosses the y-axis 2 units up, so the y-intercept is 2.

The equation for the line is $y = \frac{2}{3}x + 2$.

Your Turn   • Find the slope and y-intercept for the line at the right. Then use them to write an equation for the line.

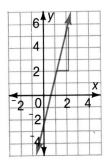

Example 2   What is the slope and y-intercept of the graph of the equation?

a. $y = 3x + 4$ **b.** $y = 5x - 7$ **c.** $y = ^-12x + 8$

Solution   **a.** slope $= 3$ **b.** slope $= 5$ **c.** slope $= ^-12$
    y-intercept $= 4$ y-intercept $= ^-7$ y-intercept $= 8$

Your Turn   What is the slope and y-intercept of the graph of the equation?

• $y = 6x + 9$ • $y = 4x - 3$ • $y = ^-7x - 2$

So far we have shown lines with positive slope. Notice that these lines slant upward from left to right.
Lines that have negative slope slant downward from left
to right. Look at the graph of the equation $y = -3x + 1$.

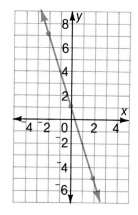

slope $= ^-3$

x	$y = ^-3x + 1$	(x, y)
$^-2$	$y = ^-3(^-2) + 1 = 7$	$(^-2, 7)$
0	$y = ^-3(0) + 1 = 1$	$(0, 1)$
2	$y = ^-3(2) + 1 = ^-5$	$(2, ^-5)$

Class Exercises

Find the rise, the run, and the slope, for each line.

1. 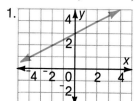 2. 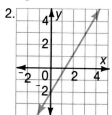 3.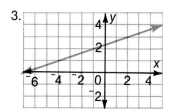

What is the slope and y-intercept of the graph of the equation?

4. $y = 2x + 7$ **5.** $y = ^-3x + 8$ **6.** $y = \frac{3}{5}x - 4$

Exercises

**Find the slope and y-intercept for the line. Then use them to write an
equation for the line.**

A 1. 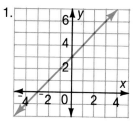 2. 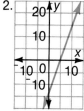 3.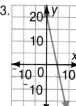

What is the slope and *y*-intercept of the graph of the equation? Does the graph slant upward from left to right, or downward?

4. $y = 6x + 5$

5. $y = \dfrac{3}{5}x - 4$

6. $y = {}^-12x + \dfrac{1}{2}$

7. $y = 1.5x + 7.2$

8. $y = {}^-0.42x - 3.6$

9. $y = 7.3x - 12.6$

For each exercise, what do you notice about the slopes of the graphs of the equations? Graph the equations. What do you notice about the lines?

B **10.** $y = 2x + 5,$ $y = 2x - 3,$ $y = 2x + 7$

11. $y = 5x - 2,$ $y = 5x,$ $y = 5x + 1$

12. $y = {}^-3x + 1,$ $y = {}^-3x - 3,$ $y = {}^-3x + 2$

13. $y = 4x + 1,$ $y = \dfrac{{}^-1}{4}x + 3$

14. $y = \dfrac{2}{3}x + 2,$ $y = \dfrac{{}^-3}{2}x + 4$

Graph each equation. Is the line slanted, vertical, or horizontal?

C **15.** $y = 3$ **16.** $y = {}^-5$ **17.** $x = 2$ **18.** $x = {}^-3$

Problems

Solve.

1. The foot of a ramp is 8 ft from the base of a house. The ramp reaches a doorway that is 18 in. above the ground. What is the slope of the ramp?

2. The foot of a ladder is 3 ft. from the base of a house. Its top rests against a window sill 28 ft from the ground. What is the slope of the ladder?

297

9-7 Linear Inequalities

Inequalities in two variables, such as $y > \frac{1}{2}x + 3$, are called **linear inequalities.** The graph of a linear inequality is not a line but a region.

In the diagram at the right, the following graphs are shown:

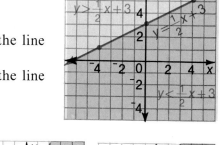

- graph of $y = \frac{1}{2}x + 3$: the line
- graph of $y > \frac{1}{2}x + 3$: the region above the line
- graph of $y < \frac{1}{2}x + 3$: the region below the line

In graphing linear inequalities we use dashed and solid lines in a special way as shown in these two graphs. The graph of $y < 4x - 3$ is shown at the left. In it, the graph of $y = 4x - 3$ is a dashed line, indicating that it is not included in the graph of $y < 4x - 3$. The graph of $y \leq 4x - 3$ is shown at the right. In it, the graph of $y = 4x - 3$ is a solid line, indicating that it is included in the graph of $y \leq 4x - 3$.

We can use a graph to determine whether an ordered pair is a solution of a linear inequality.

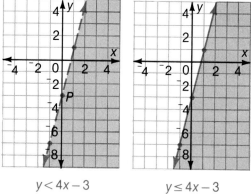

$$y < 4x - 3 \qquad y \leq 4x - 3$$

Example 1 Use the graph of $y < 4x - 3$ above to determine whether or not $(0, {}^-3)$ is a solution of $y < 4x - 3$.

Solution Plotting $(0, {}^-3)$ we find its graph is the point P on the dashed line. Therefore, $(0, {}^-3)$ is not a solution of $y < 4x - 3$.

In Example 2 we determine whether an ordered pair is a solution of a linear inequality by substituting its values in the inequality.

298

Example 2 Use substitution to determine whether $(4, 2)$ is a solution of $y \leq 4x - 3$.

Solution $y \leq 4x - 3$
Is $2 \leq 4(4) - 3$ true?
$2 \leq 13$ is true. Therefore, $(4, 2)$ is a solution.

You can draw the graph of the inequality in Example 2 by first drawing the graph of $y = 4x - 3$; then, of the two possible regions, shade the region containing the known solution $(4, 2)$.

Your Turn
• Use the graph of $y < 4x - 3$ above to determine whether $(^-2, 2)$ is a solution of the inequality.
• Use substitution to determine whether $(3, 4)$ is a solution of $y \geq 5x - 3$.

Class Exercises

Is the point a graph of a solution of $y \geq 2x + 1$?

1. A   **2.** B   **3.** C

4. D   **5.** E   **6.** F

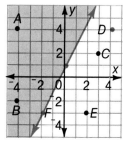

$y \geq 2x + 1$

Exercises

Use the graph at the right to determine whether the ordered pair is a solution of $y < 5x - 4$.

A **1.** $(0, 1)$ **2.** $(1, 0)$ **3.** $(3, 2)$ **4.** $(3, ^-2)$

  5. $(^-2, 3)$ **6.** $(0, ^-4)$ **7.** $(4, ^-6)$ **8.** $(1, 1)$

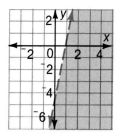

In each exercise, is the ordered pair a solution of the linear inequality?

B   **9.** $(2, 1)$, $y > x - 3$ **10.** $(8, {}^-8)$, $y > 5x - 48$ **11.** $({}^-3, {}^-33)$, $y \leq 12x + 3$

12. $(6, 5)$, $y \leq \dfrac{1}{2}x + \dfrac{1}{4}$ **13.** $(7, 8)$, $y \geq \dfrac{2}{3}x - \dfrac{2}{3}$ **14.** $({}^-16, 0)$, $y > \dfrac{5}{8}x + 4$

Draw the graph of the linear inequality.

15. $y > x + 2$ **16.** $y < 3x - 7$ **17.** $y \geq -2x + 4$ **18.** $y \leq 6x - 9$

Using the Computer

This program will find the slope of the line through two points. RUN
it for these pairs of points:
$(2, 3)$ and $(4, 7)$; $({}^-3, 2)$ and $(1, 8)$; $({}^-3, 1)$ and $(0, {}^-4)$

```
 5  REM: SLOPE OF LINE THROUGH TWO POINTS
10  PRINT "WHAT ARE THE COORDINATES--"
20  PRINT "OF THE FIRST POINT";
30  INPUT X1,Y1
40  PRINT "OF THE SECOND POINT";
50  INPUT X2,Y2
60  PRINT "SLOPE =";Y2-Y1;" /";X2-X1;
70  PRINT " =";(Y2-Y1)/(X2-X1)
80  END
```

Self Test 2

**Find three ordered pairs that are solutions of the given equation. Then
graph them and draw the line for the equation.**

   1. $y = x + 1$ **2.** $y = 3x - 5$ **3.** $y = {}^-2x + 4$ **9-5**

**What is the slope and y-intercept of the graph of the equation? Graph
the equation.**

   4. $y = 4x - 1$ **5.** $y = {}^-x + 2$ **6.** $y = 3x + 4$ **9-6**

Is the ordered pair a solution of the linear inequality?

   7. $(3, 2)$, $y < x + 4$ **8.** $({}^-1, 1)$, $y \geq x - 3$ **9.** $(2, 0)$, $y > 2x + 1$ **9-7**

Algebra Practice

Write an equation. Solve.

1. 49,452 divided by 78 is equal to c.
2. 216 times 980 is equal to f.
3. The sum of 103 and 99.44 is v.
4. The quotient of 0.2313 and 0.09 is m.
5. The sum of $4\frac{2}{3}$ and $4\frac{7}{16}$ is t.
6. $\frac{5}{6}$ of $3\frac{3}{10}$ is equal to b.
7. The perimeter of a square with side $6\frac{7}{8}$ in. is P.
8. The area of a circle with a radius of 2.5 m is A.
9. The volume of a prism with a base area of 20 ft² and a height of 6 ft is V.

Use the distributive property to factor each expression.

10. $2g^2h + g^2h^2 + 5g^2$
11. $3st^2 + s^2t^2 + 2s^2t$
12. $4ab + 8a^2 + 6ab^2$

Solve.

13. $2f + 8 = 4$
14. $5j - 6 = {}^-21$
15. $9s - 4 = {}^-40$
16. $7 > {}^-2w + 3$
17. $6 + {}^-3m < 12$
18. $\frac{a}{2} - 5 > {}^-1$

Find the first five terms of the sequence produced by the given expression.

19. $n + 5$
20. $25a - 12$
21. ${}^-4x + 5$

Write an expression for any term of the sequence. Then find the 12th and 20th terms.

22. ${}^-4, {}^-1, 2, 5, 8, \ldots$
23. $12\frac{5}{7}, 10\frac{3}{7}, 8\frac{1}{7}, 5\frac{6}{7}, 3\frac{4}{7}, \ldots$

Find three ordered pairs that are solutions of the given equation.

24. $y = x + 4$
25. $y = 4x - 3$
26. $y = {}^-2x + 3$

What is the slope and y-intercept of the equation?

27. $y = 5x + 3$
28. $y = 2x - 1$
29. $y = {}^-3x + 6$

Is the ordered pair a solution of the linear inequality?

30. $(3, {}^-5)$, $y > x + 2$
31. $(4, 4)$, $y < 3x - 5$
32. $(2, 0)$, $y \geq {}^-2x + 4$

301

Problem Solving on the Job

Engineering Technician

Are you interested in radio, television, electronics, computers, or scientific research and development? Technicians work in every field of science and engineering. Design technicians must be able to sketch objects and must be creative. Other technicians work closely with production or sales departments.

```
ENGINEERING TECHNICIAN

DESCRIPTION: Analyze and solve en-
gineering and science problems.
Prepare formal reports on experi-
ments, tests, and other projects.
QUALIFICATIONS: 1-4 years training
at a technical school, or on-the-
job training, or a combination.
JOB OUTLOOK: Increase in high tech-
nology jobs and the trend toward
automation in industry will require
more technicians.
```

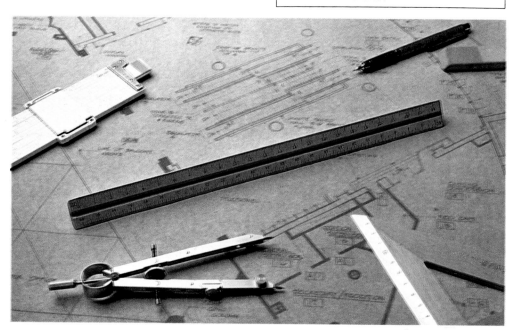

The government employs engineering technicians in many fields. For example, as an aircraft is built a technician checks it for conformance to specifications. If you applied for a job as an engineering technician working for the government, you would have to take a civil service exam. The following problems are similar to those appearing on such an exam. Why don't you see how you would do?

Write a, b, c, or d to show the correct answer.

1. *PQ* is 42 in. long. The ratio of *PR*
   to *RQ* is 5 to 2. How long is *PR*?
   a. 14 in.   **b.** 20 in.
   c. 28 in.   **d.** 30 in.

2. There are *b* boys and *g* girls in a class. What is the ratio of the
   number of boys to the total number of students?
   a. $\dfrac{b}{g}$ **b.** $\dfrac{b}{b+g}$ **c.** $\dfrac{b+g}{b}$ **d.** $\dfrac{g}{b+g}$

3. If 60 ft of wire weighs 80 lb, what is the weight of 141 ft of the wire?
   a. 160 lb   **b.** 162 lb   **c.** 165 lb   **d.** 188 lb

4. If *t* tons of snow falls in *h* hours, how many tons fall at this rate
   in *m* minutes?
   a. $\dfrac{mt}{h}$ **b.** $\dfrac{mt}{60h}$ **c.** $\dfrac{60mt}{h}$ **d.** $\dfrac{60t}{mh}$

5. 0.6927 ft is most nearly how many inches?
   a. $8\dfrac{1}{4}$ in.   **b.** $8\dfrac{9}{32}$ in.   **c.** $8\dfrac{5}{16}$ in.   **d.** $8\dfrac{11}{32}$ in.

6. The distance between two points on the ground is measured with a
   100-ft steel tape and found to be 1012.60 ft. If the tape is 0.04 ft too
   long which measurement is closest to the true distance?
   a. 1012.15 ft   **b.** 1012 ft   **c.** 1012.75 ft   **d.** 1013.00 ft

7. A gallon of paint covers
   350 ft². About how many
   gallons are required to
   paint the inner walls of
   the hall with 1 coat of paint?
   a. 6 gal   **b.** 7 gal
   c. 4 gal   **d.** 5 gal

8. About how many 9″ × 9″
   pieces of tile are needed
   for the hall floor?
   a. 2600   **b.** 2840
   c. 2400   **d.** 2260

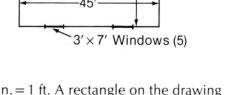

9. The scale of a drawing is ⅛ in. = 1 ft. A rectangle on the drawing
   measures 7 1/8 in. by 6 1/4 in. About how many square feet are there
   in the area of the actual rectangle?
   a. 2725 ft²   **b.** 2580 ft²   **c.** 2675 ft²   **d.** 2850 ft²

Enrichment

Systems of Equations

When we graph two linear equations on the same set of axes, sometimes the graphs intersect. The ordered pair of the point of intersection is a common solution of the two equations. We call this ordered pair a solution of the given **system of equations.**

System of equations
$$x + 2y = 3$$
$$2x - y = {}^-4$$

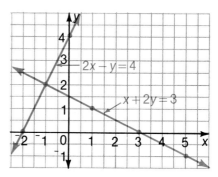

The graphs of the two equations intersect at the point whose coordinates are $({}^-1, 2)$. We check to see that $({}^-1, 2)$ is a solution of the system of equations by substituting in each of the equations.

$x + 2y = 3$	
${}^-1 + 2(2)$	3
${}^-1 + 4$	3
3	3

$2x - y = {}^-4$	
$2({}^-1) - 2$	${}^-4$
${}^-2 - 2$	${}^-4$
${}^-4$	${}^-4$

$({}^-1, 2)$ is a solution of the system of equations.

Sometimes the graphs of two equations in a system are parallel and do not intersect. In this case the system has no solution. Sometimes the graphs coincide, in which case there are an infinite number of solutions.

Is the given ordered pair a solution of the system of equations.

1. $2x - y = 3$; $(3, 3)$
$x + y = 6$

2. $x + 2y = 5$; $(2, 3)$
$x - y = 1$

3. $x + 3y = {}^-4$; $(2, {}^-2)$
$2x - y = 6$

Graph the system of equations. Find and check its solution.

4. $3x - y = 4$
$x + y = 8$

5. $x + 2y = 4$
$x + 3y = 1$

6. $x - 4y = 6$
$2x + y = 3$

7. $2x - y = 1$
$x + 3y = 4$

8. $3x + y = 6$
$x - 2y = 2$

9. $4x - y = 2$
$2x - y = 4$

Chapter Review

Match the expression with the sequence.

1. 3, 7, 11, 15, 19

2. 9, 7, 5, 3, 1

3. 5, 1, $^-$3, $^-$7, $^-$11

4. 1, 6, 11, 16, 21

5. 6, 3, 0, $^-$3, $^-$6

a. $^-3x + 9$

b. $5n - 4$

c. $^-4t + 9$

d. $4d - 1$

e. $^-2y + 11$

Write an expression for any term of the sequence, then find the 15th and 20th terms.

6. 2, 6, 10, 14, 18

7. 1, 6, 11, 16, 21

8. 4, 1, $^-$2, $^-$5, $^-$8

9. $^-$5, $^-$2, 1, 4, 7

Graph the ordered pair on a coordinate plane.

10. $(4, 6)$ **11.** $(4, ^-3)$ **12.** $(2, 0)$ **13.** $(^-7, 1)$ **14.** $(^-2, ^-4)$

Is the ordered pair a solution of the given equation?

15. $y = 2x + 1; (2, 5)$

16. $y = 4x - 1; (0, ^-1)$

17. $y = ^-3x - 2; (1, 5)$

18. $y = 4x + 6; (^-2, ^-4)$

19. $y = ^-x + 7; (3, 10)$

20. $y = ^-4x - 2; (3, ^-14)$

Find three ordered pairs that are solutions of the given equation, then graph them and draw the line for the equation.

21. $y = 3x + 1$ **22.** $y = 2x - 3$ **23.** $y = 4x - 3$

What is the slope and y-intercept of the graph of the equation? Graph the equation.

24. $y = 2x + 6$ **25.** $y = 4x - 5$ **26.** $y = 3x - 3$

Is the ordered pair a solution of the linear inequality?

27. $y < 2x + 3; (1, 6)$ **28.** $y > 3x - 6; (2, 5)$ **29.** $y > x + 8; (1, 3)$

30. $y < 2x - 6; (1, 4)$ **31.** $y < 4x + 2; (2, 10)$ **32.** $y > 5x - 3; (3, 13)$

Chapter Test

Find the first five terms of the sequence of numbers produced by the given expression.

1. $m + 3$ **2.** $3y - 1$ **3.** $^-2x + 5$ **4.** $^-n - 6$ **9-1**

Write an expression for any term of the sequence, then find the 12th and 30th terms.

5. 5, 9, 13, 17, 21   **6.** 12, 9, 6, 3, 0   **9-2**

7. $^-4$, $^-1$, 2, 5, 8   **8.** 6, 1, $^-4$, $^-9$, $^-14$

Graph the ordered pairs on a coordinate plane.

9. $(7, 5)$ **10.** $(4, ^-6)$ **11.** $(^-8, 3)$ **12.** $(2, 0)$ **9-3**

Graph the following data. Use your graph to find the 8th term.

13.

No. of term	1	2	3	4	5
Sequence	7	10	13	16	19

9-4

14.

No. of term	1	2	3	4	5
Sequence	20	16	12	8	4

Find three ordered pairs that are solutions of the given equation, then graph them and draw the line for the equation.

15. $y = 2x + 5$ **16.** $y = 3x - 2$ **17.** $y = 4x + 2$ **9-5**

What is the slope and y-intercept of the graph of the equation. Graph the equation.

18. $y = 3x - 4$ **19.** $y = 2x + 7$ **20.** $y = \frac{1}{2}x + 3$ **9-6**

Is the ordered pair a solution of the linear inequality?

21. $(0, 3)$, $y < 2x - 4$ **22.** $(2, 3)$, $y \geq 4x - 5$ **9-7**

23. $(^-1, 2)$, $y > x - 2$

Cumulative Review (Chapters 1–9)

Add, subtract, multiply, or divide.

1. $3\frac{1}{2} \times 1\frac{1}{8}$

2. $2\frac{1}{3} - 1\frac{5}{8}$

3. $\frac{7}{8} \cdot \frac{1}{3} \cdot \frac{12}{21}$

4. $4.02 - 9.6$

5. $8 - 6.04$

6. $218.7 \div 8.1$

7. $23 - 7 \cdot 2 \div 7$

8. $46 + 28 \div 4$

9. 10×19

10. $(100)(2.6)$

11. $(^-2)(5)$

12. $8 + {}^-5$

13. $^-20 - 7$

14. $14 + {}^-9$

15. $^-8 - {}^-10$

16. $(12)(6)$

17. $(^-8)(^-3)$

18. $^-3 \div {}^-1$

19. $56 \div {}^-7$

20. $(^-9)(^-2) \div (^-3)$

21. $45 \div {}^-9 - 7$

Solve the equation.

22. $a - 6 = 4$

23. $14 + n = 25$

24. $6y = 42$

25. $9f = 81$

26. $\frac{a}{7} = 30$

27. $\frac{1}{3}c = 12$

28. $a + 2a = 60$

29. $3x - 1 = 29$

30. $4n + 17 = 31$

31. $5p - 1 = 3p + 7$

32. $12 + 8s = 20$

33. $6d + 7 = 31 - 2d$

Solve.

34. In a computer game, if a player can score between 450 and 500 points (no more, no less) within 10 minutes, he or she wins. Lois scored the following points in 10 minutes: 80, $^-60$, 90, $^-50$, 110, 70, 40, $^-10$. Did she win?

35. In riding his bike around town Adam went north 2 blocks, then west 5 blocks, then south 10 blocks, then east 6 blocks, then north 8 blocks. How many blocks from his starting point was he then?

36. Eight years ago the monthly rent on the Coopers' apartment was $210. That was $\frac{1}{3}$ of what it is today. What is it today?

37. Here's a formula to relate the surface area (A) of a cube with the length of a side (s).

$$A = 6s^2$$

A cube has sides 12 cm by 12 cm. What's the surface area?

38. On Sunday the temperature at noon in Minneapolis was $^-7°C$. On Monday the noon temperature was 4 degrees warmer. What was the temperature at noon on Monday?

10
Statistics

Statistics helps us know when a new world record for an event is established. In 1980, Juoko Tormanen of Finland achieved the longest ski jump in Olympic history at Lake Placid, NY by jumping 117 m!

10-1 Introduction to Statistics

The traffic safety division of a large city received a request to install a traffic light at the intersection of Thompson and Maynard Streets. The city policy is to install lights at intersections where the average number of vehicles passing through the intersection per day is 1200 or more. Therefore a 7-day electronic vehicle count was conducted at the Thompson and Maynard intersection. The results are shown below.

VEHICLE COUNT—THOMPSON AND MAYNARD INTERSECTION							
Day	1	2	3	4	5	6	7
Number of Vehicles	1254	1189	1261	1304	1295	1083	1285

Numerical facts such as those shown in the chart above are called **data.** The study of the collection, organization, and interpretation of data is called **statistics.**

The traffic safety division used the data to compute the **mean** number of vehicles using the intersection per day. To find the mean of data, divide the sum of the numbers by the total number of addends.

Example 1 Find the mean number of vehicles using the intersection per day.

Solution Let b = the mean.

$$b = \frac{1254 + 1189 + 1261 + 1304 + 1295 + 1083 + 1285}{7}$$

$$b = \frac{8671}{7}, \text{ or about } 1238.7$$

Since the mean number of vehicles using the intersection per day is more than 1200, the traffic light will be installed.

Other terms often used in statistics are **range, mode,** and **median.** The range is the difference between the greatest and least numbers in a set of data. The range of the data above is $1304 - 1083$, or 221.

The mode is the number that occurs most frequently in a set of data. Sometimes a set of data has no mode. The median is the middle number when data are arranged in numerical order. If there is an even number of data, the median is the mean of the two middle numbers.

Example 2   Find the mode of 44 kg, 36 kg, 43 kg, 44 kg, 36 kg, and 44 kg.

Solution   44 kg occurs most frequently so the mode is 44 kg.

Example 3   Find the median of 6.2 m, 0 m, 7 m, 5.7 m, 4.7 m, and 5.9 m.

Solution   Arrange the data in order:
0 m, 4.7 m, 5.7 m, 5.9 m, 6.2 m, 7 m
Find the mean of the two middle numbers:
$$\frac{5.7 + 5.9}{2} = \frac{11.6}{2} = 5.8$$
The median is 5.8 m.

Your Turn   • Find the mean, range, mode and median of 17 L, 12 L, 14 L, 12 L, 17 L, and 12 L.

Class Exercises

Write mean, mode, median, or range to complete the sentence.

1. 18 + 25 + 24 + 25 = 92
92 ÷ 4 = 23
The __?__ is $23.

2. 18, 24, 25, 25
(24 + 25) ÷ 2 = 24.5
The __?__ is $24.50.

SAVINGS				
Month	Jan.	Feb.	Mar.	Apr.
Amount	$18	$25	$24	$25

3. 18, 25, 24, 25
The __?__ is $25.

4. 25 − 18 = 7
The __?__ is $7.

Exercises

Find the mean and mode of the given data.

A **1.** 18, 21, 20, 18   **2.** 0.7, 0.6, 0.6, 0.4   **3.** $\dfrac{5}{6}, \dfrac{1}{8}, \dfrac{1}{8}, \dfrac{7}{12}$

Find the median and range of the given data.

4. 8, 4, 9, 5, 2   **5.** 4.2, 3.6, 3.1, 4.4   **6.** $\dfrac{5}{8}, \dfrac{1}{4}, \dfrac{7}{16}, \dfrac{3}{8}$

Find the mean, mode, median, and range of the given data.

B **7.**

AVERAGE TEMPERATURE							
Day	S	M	T	W	Th	F	S
Degrees Fahrenheit	70°	73°	72°	68°	72°	70°	72°

8.

TEST SCORES							
Week	1	2	3	4	5	6	7
Percent	83%	84%	86%	85%	94%	92%	92%

9.

CALCULATOR PRICES							
Type	A	B	C	D	E	F	G	H
Price	$19.95	$49.95	$69.95	$49.95	$119.95	$49.95	$29.95	$69.95

Problems

Solve.

1. The attendance figures at the first four games of a season were 64,532; 58,711; 62,871; and 64,334. What was the mean attendance?

2. One team's scores for the first five games of a season were 4, 3, 5, 2, and 1. What was the range of the scores?

3. Ticket prices for a home game are as follows: $3.50, $4.50, $5.00, and $3.25. What is the median price of a ticket?

4. The ages of the six youngest players on a team are 24, 26, 25, 24, 23, and 27. What is the mode of the ages of these players?

311

10-2 Circle Graphs

A **circle graph** presents data as a circular region divided into sections. The size of each section corresponds to the part of the data represented.

This circle graph shows the voting status of 72,000 people eligible for voter registration in one city for a recent election.

VOTING STATUS

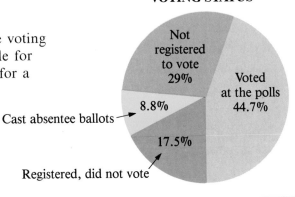

Example About how many people were not registered to vote?

Solution 29% of 72,000 = __?__
0.29 × 72,000 = 20,880

Your Turn • How many people cast absentee ballots?

Class Exercises

The circle graph at the right shows the types of computers sold by a company in one year.

1. What percent of the computer sales were personal computers?
2. What type of computer accounted for the smallest percent of computer sales?

COMPUTER SALES

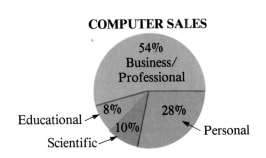

3. Were more scientific or educational computers sold?

4. What type of computer accounted for 54% of the computer sales?

Exercises

The circle graph at the right shows one family's proposed budget. Use the graph to answer the following questions. Write your answers in dollar amounts.

ANNUAL BUDGET ($20,000)

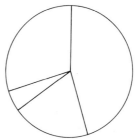

A **1.** About how much do they plan to spend for food?

2. What are their anticipated housing costs for the year?

3. Clothing costs will account for how much of their budget?

B **4.** About how much more do they plan to put in savings than to spend on entertainment?

5. What are they expecting to pay for insurance and taxes altogether?

6. By how much do they expect miscellaneous costs to exceed medical and dental costs?

7. How much money is budgeted monthly for entertainment?

8. How much more is budgeted monthly for clothing than insurance?

A survey of 620 students was conducted to determine the popularity of four radio stations. The chart below shows the results of the survey. Compute the percent of students favoring each station then copy and complete the circle graph, labeling each wedge with the corresponding station and percent.

C **9.**

SURVEY RESULTS	
Station	Student Votes
WROQ	279
WCLS	31
WCAW	124
WANY	186

SURVEY RESULTS

313

10-3 *Pictographs*

A **pictograph** uses a symbol to represent a number. In the graph below, each token represents $40,000. The data in the graph are rounded to the nearest $10,000.

MASON BUS LINES

Quarter	Earnings
1 (Jan – Mar)	$ $
2 (April – June)	$ $ $
3 (July – Sept)	$ $
4 (Oct – Dec)	$ $

$ = $40,000

The first quarter earnings were approximately equal to $40,000 + $40,000, or $80,000.

Amounts less than $40,000 are represented by a fraction of a token.

$40,000 $20,000 $10,000

Example 1   What were the fourth quarter earnings?

Solution $2\frac{1}{4} \times \$40,000 = \frac{9}{4} \times \$40,000 = \$90,000$

The fourth quarter earnings were about $90,000.

Example 2   It is estimated that Mason Bus Lines will earn $112,000 during the first quarter of next year. Draw symbols that represent this amount.

Solution   Round to the nearest $10,000: $112,000 ≈ $110,000

$\$110,000 \div \$40,000 = 2\frac{3}{4}$ ➪ $ $ $

Your Turn Let $\boxed{\$\ \$\ \$\ \$}$ represent $20,000. How much money do these symbols represent?

$\cdot\boxed{\$\ \$\ \$\ \$}\ \boxed{\$\ \$}$ $\cdot\boxed{\$\ \$\ \$}$ $\cdot\boxed{\$\ \$\ \$\ \$}\ \boxed{\$\ \$\ \$\ \$}\ \boxed{\$\ \$\ \$}$

Draw symbols to represent these amounts.

- $50,000
- $75,000
- $102,413.72

Class Exercises

Use the pictograph to tell about how many working women there were in the United States in the given year.

1. 1970   **2.** 1980   **3.** 1960

WORKING WOMEN IN THE U.S.

1960	♀ ♀ ♀
1970	♀ ♀ ♀
1980	♀ ♀ ♀ ♀ ♀

♀ represents 10 million women

Estimates have been made of the number of working women there will be in the United States in the future. How many ♀ symbols are needed to represent the estimates given?

4. 1990: 56 million women   **5.** 2000: 63,900,000 women

Exercises

The pictograph shows the monthly budget for the Pendleton Family. About how much is budgeted for each item? Complete.

A
1. Housing: $4 \times \$100 = \underline{\quad?\quad}$
2. Food: $3 \times \underline{\quad?\quad} = \underline{\quad?\quad}$
3. Insurance: $\frac{1}{2} \times \underline{\quad?\quad} = \underline{\quad?\quad}$
4. Telephone: $\frac{1}{4} \times \underline{\quad?\quad} = \underline{\quad?\quad}$
5. Clothing: $\underline{\quad?\quad} \times \$100 = \underline{\quad?\quad}$
6. Entertainment: $1 \times \underline{\quad?\quad} = \underline{\quad?\quad}$
7. Savings: $\underline{\quad?\quad} \times \$100 = \underline{\quad?\quad}$
8. Miscellaneous: $\underline{\quad?\quad} \times \underline{\quad?\quad} = \underline{\quad?\quad}$

PENDLETON FAMILY MONTHLY BUDGET

Housing	$ $ $ $
Food	$ $ $
Insurance	$
Telephone	▯
Clothing	$ $ $
Entertainment	$
Savings	$ $ ▯
Miscellaneous	$ $

$ represents $100

B
9. About how much more is budgeted for clothing than for entertainment?
10. What is the total amount budgeted for one month?

The data in the table show the number of students enrolled in institutions of higher education in the United States for selected years.

Round the numbers in the table to the nearest 1 million. Let 🙂 represent 2 million students. Draw symbols to represent the data.

Example   1960: $3,789,000 \approx 4,000,000$
$4,000,000 \div 2,000,000 = 2$

大 大

11. 1965: $5,921,000 \approx 6,000,000$
$6,000,000 \div 2,000,000 = \underline{\quad?\quad}$

12. 1970   13. 1975   14. 1980

Enrollment in Institutions of Higher Education (U.S.)	
Year	Enrollment
1960	3,789,000
1965	5,921,000
1970	8,581,000
1975	11,185,000
1980	12,070,000

15. Draw a pictograph to show enrollment in institutions of higher education in the United States for the five years given in the table.

C 16. Make a pictograph from the data in the table. Let ☐ represent 10 million households and ⌐, 5 million households.

U.S. Households with TV Sets (in millions)

1960	Black/White	46.2
	Color	xxxx
1965	Black/White	55.9
	Color	5.5
1970	Black/White	63.2
	Color	27.2
1975	Black/White	72.6
	Color	54.1
1980	Black/White	79.3
	Color	63.4

Self Test 1

Find the mean, mode, median, and range of the given data.

1. 24 mi, 36 mi, 26 mi, 36 mi   **10-1**

2. 9.4 km, 10.1 km, 9.4 km, 9.2 km

Use the circle graph to answer the following questions. Write your answers in dollar amounts.

3. About how much was earned at the car wash?   **10-2**

4. About how much was earned at the craft sale?

CLASS PROJECTS

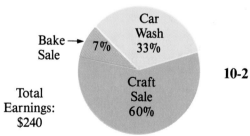

Bake → Sale 7%   Car Wash 33%

Total Earnings: $240   Craft Sale 60%

Use the pictograph to answer the following questions.

5. About how many people attended the 7:00 show on Friday?   **10-3**

6. About how many people attended the 9:20 Saturday show?

THEATER ATTENDANCE

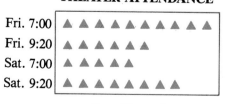

Fri. 7:00 ▲ ▲ ▲ ▲ ▲ ▲ ▲ ▲ ▲ ▲
Fri. 9:20 ▲ ▲ ▲ ▲ ▲
Sat. 7:00 ▲ ▲ ▲ ▲ ▲
Sat. 9:20 ▲ ▲ ▲ ▲ ▲ ▲ ▲

▲ = 20 people

Self Test answers and Extra Practice are at the back of the book.   **317**

10-4 Bar Graphs

A **bar graph** is used to compare data. The bar graph below shows the approximate mean weekly salaries of four different job categories in one city.

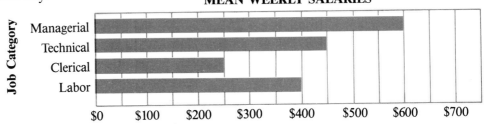

MEAN WEEKLY SALARIES

Example 1	What is the approximate mean weekly salary of the clerical and labor job categories?
Solution	$250 + $400 = $650 $650 ÷ 2 = $325

Your Turn • What is the approximate mean weekly salary of all four job categories?

Multiple bar graphs are used to compare more complex data. The triple bar graph below shows the post-graduation plans of the sophomores, juniors, and seniors at one high school.

POST GRADUATION PLANS

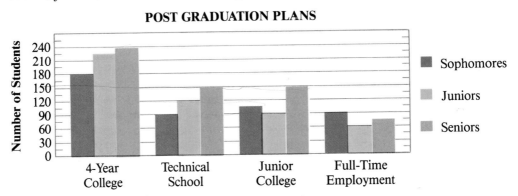

Example 2 How many more seniors than sophomores plan to attend a four-year college?

Solution 60 more seniors

Your Turn • What is the total number of students who plan to attend junior college?

Class Exercises

Use the weekly salaries graph to answer the following questions.

1. What is the approximate mean weekly salary of technical employees?
2. Which job category has the lowest weekly salary?
3. What is the approximate difference between the weekly salary of a clerical worker and that of a laborer?

Exercises

Use the post-graduation plans graph to answer the following questions.

A
1. How many juniors plan to attend junior college?
2. Which class has the greatest number of students planning to attend a four-year college?

B
3. How many seniors participated in the survey?
4. How many students plan to attend a technical school?
5. How many more students plan to attend a four-year college than a junior college?
6. What is the mean number of students who plan to work full time?
7. What is the range in the number of students who plan to attend a four-year college?
8. What is the median number of students who plan to attend a junior college or a four-year college?
9. What percent of the students planning to attend a technical school are sophomores?

A **shaded bar graph** is used to compare individual sets of data as well as combined sets of data. The shaded bar graph below shows the sources of funds for three candidates in a recent election. For example, Thomas received about $10,000 from party funds.

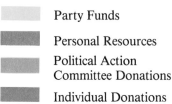

Party Funds

Personal Resources

Political Action Committee Donations

Individual Donations

CAMPAIGN FUND SOURCES

Caldwell   Thomas   Mingle

Use the graph to answer the following questions.

C **10.** Which candidate received the greatest amount of individual donations?

11. Did Mingle receive more campaign contributions from political action committees or from individuals?

12. About how much did Caldwell receive in all?

13. Personal resources accounted for how much of Thomas' campaign funds?

14. What is the range of the amount of funds donated by individuals?

15. What is the approximate median amount of donations from political action committees?

Problems

Construct the graph.

1. Construct a bar graph comparing the swimming speeds given in the table.

SWIMMING SPEEDS

Sailfish	60 mph
Flying fish	40 mph
Dolphin	37 mph
Trout	15 mph
Human	5.19 mph

2. Construct a double bar graph comparing the number of general admission tickets and reserved seat tickets sold for each game.

TICKETS SOLD				
Game	1	2	3	4
General Admission	525	473	681	725
Reserved Seats	54	73	58	104

Challenge

A group of fitness experts rated the benefits of certain activities. Each activity was rated from 0 to 100 relative to physical fitness and general health. Construct a shaded bar graph comparing the benefits of the activities.

|←— Fitness —→|←Health→|
| Total |

	Jogging	Bicycling	Swimming	Basketball	Tennis	Skiing	Walking
Physical Fitness	80	76	78	76	72	70	50
General Health	61	60	55	51	50	50	47
Total	141	136	133	127	122	120	97

10-5 Line Graphs

A **line graph** is used to show changes in data over a period of time. The line graph at the right shows the temperature changes in a city in twelve hours. It shows that the greatest temperature increase occurred between the hours of 10 A.M. and 12 noon.

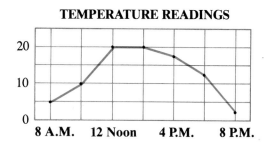

TEMPERATURE READINGS

A **multiple line graph** shows changes in more than one set of data over a period of time. The multiple line graph below shows the changes in the percent of the population that received national political campaign information from various media from 1952 to 1980.

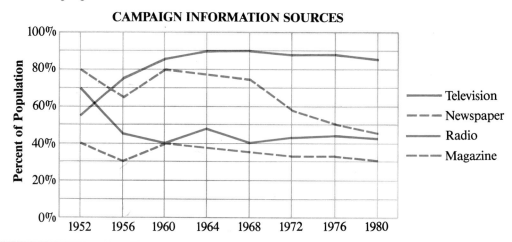

CAMPAIGN INFORMATION SOURCES

Example Which medium declined the most as a source of campaign information from 1968 through 1980?

Solution newspaper

Your Turn • Which medium increased the most as a source of campaign information from 1952 to 1960?

Class Exercises

Use the temperature readings line graph to answer the questions.

1. Did the temperature increase or decrease from 10 A.M. to 12 noon?
2. During what two-hour interval did the temperature decrease 10°?
3. How much did the temperature increase from 8 A.M. to 10 A.M.?
4. How much did the temperature decrease from 2 P.M. to 8 P.M.?
5. During what four-hour interval did the temperature increase 15°?

Exercises

Use the campaign information sources multiple line graph to answer the following questions.

A
1. Did the use of the radio as a source of campaign information increase or decrease from 1952 to 1956?
2. During what four-year interval did the use of magazines as a source of campaign information decrease the most?
3. During what twenty-year interval did the use of television as a source of campaign information stay about the same?
4. Which medium gained the most popularity as a source of campaign information from 1952 to 1968?
5. Which medium experienced the least change as a source of campaign information from 1952 to 1980?

B
6. About how much did the percent of the population using TV as a source of campaign information increase from 1952 to 1960?
7. About how much did the percent of the population using the newspaper as a source of campaign information decrease from 1960 to 1980?
8. What is the approximate difference between the percent of the population using television as a source of campaign information when television was at its highest peak and the percent of the population using radio as a source of campaign information when radio was at its highest peak?
9. What is the approximate difference between the decrease in the percent of the population using the radio as a source of campaign information from 1952 to 1956 and the decrease in the percent of the population using the magazine as a source of campaign information from 1952 to 1956?

323

A **shaded multiple line graph** shows changes in individual sets of data as well as combined sets of data over a period of time. The shaded multiple line graph below shows the population growth of each region in the continental United States as well as the combined regional population growth.

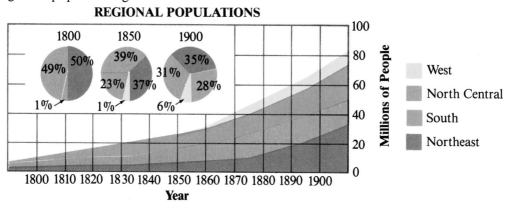

REGIONAL POPULATIONS

Use the regional populations graph to answer the following questions.

Example About 20 million people lived in the north central region in 1895.

C **10.** Which region had the greatest population in 1900?
11. Which region has consistently had the lowest population?
12. When did the population of the west begin to increase?
13. Did the population of the Northeast increase more before or after 1870?
14. About what was the total population in 1900?
15. About what was the total population in 1870?

Problems

Use the data from the chart to construct a double line graph.

1.

CLASS ENROLLMENT								
Year	1956	1960	1964	1968	1972	1976	1980	1984
Science	70%	79%	68%	60%	75%	81%	79%	81%
Math	74%	69%	61%	58%	62%	68%	74%	85%

Using the Computer

The following program will draw a bar graph horizontally. (If you have a microcomputer with graphics capabilities, you will be able to draw vertical lines according to the given instructions.) RUN the program for ten items with values 3, 5, 7, 9, 15, 12, 8, 6, 2, 1.

```
  5  REM: TO DRAW A BAR GRAPH        80  FOR I=1 TO T
 10  PRINT "HOW MANY ITEMS";         90  PRINT "!";
 20  INPUT T                        100  FOR J=1 TO F(I)
 30  PRINT "WHAT IS THE VALUE"      110  PRINT "*";
 40  FOR I=1 TO T                   120  NEXT J
 50  PRINT "FOR ITEM";I;            130  PRINT
 60  INPUT F(I)                     135  PRINT "!"
 70  NEXT I                         140  NEXT I
 75  PRINT                          150  END
```

Self Test 2

Use the multiple bar graph to answer the following questions.

1. About how many more turntable sales than amplifier sales were reported from Store A?

2. About how many recorder sales were reported in all?

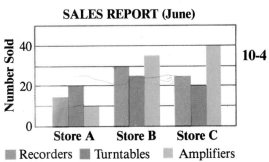

SALES REPORT (June)

10-4

Use the multiple line graph to answer the following questions.

3. Who increased their savings the most from May to June?

4. During which period of time did both Kim and Dave decrease their savings?

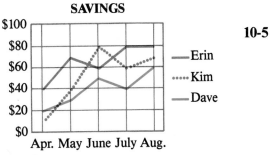

SAVINGS

10-5

Self Test answers and Extra Practice are at the back of the book.

10-6 Problem Solving

A problem usually includes a set of facts and asks a question. Problems in real life are usually decisions you must make or tasks you must do. You usually have to research your own facts and ask your own questions. In this lesson you will write the questions that could help you make a good decision or do the task well.

Example 1   Decision: Whether to rent a telephone or buy a telephone
Facts: You can rent a telephone for $1.50 per month. You can buy a similar telephone for $60.
Write a question that will help you decide whether to buy the telephone or not. Then solve the problem and answer the question.

Solution   "If I rent a telephone at $1.50 per month, how long will it be before rental charges equal the cost of buying the phone?"

$$\$60 \div \$1.50 = 40$$

"At the present rate, I can rent my telephone for 40 months for $60."

Your Turn   Write a question that will help you make the decision and that can be answered with the facts given. Then solve the problem.
- Decision: When to leave home to go to the concert
  Facts: The concert starts at 7:30 P.M. It takes 5 min to walk to the bus stop from your house. Buses leave the corner near your home every 20 min after 6:15 P.M. The bus ride takes 15 min. After you get off the bus, it takes 10 min to walk to the concert hall.

Class Exercises

Ask two questions that can help you make a good decision or do the task.

Example   You want to open a savings account and need to choose a bank.

Solution   • What rate of interest is offered by each bank?
               • How often is interest added to savings accounts?

1. Your class must decide how much to spend on the Spring Dance.
2. You must choose between two job offers for the summer.
3. Choose the most valuable player on the baseball team.
4. You want to change the wallpaper in your bedroom.

Exercises

Suppose you must do the task specified or make the decision. Write a question that can be answered from the facts given. Solve the problem and answer the question.

A   1. Decision: Whether to buy or rent a TV
           Facts: You can rent a TV for $25 a month.
                  You can buy a similar TV for $300.

   2. Task: Tell a friend how your grade average compares with the rest of the class.
      Facts: The mean grade average for the class is 82.6.
             Your grades this term are 78, 85, 86, 80, 77, 85.

B   3. Task: Buy milk and laundry detergent.
           Facts: Milk costs $1.13 for a half gallon or $.59 a quart
                  Detergent costs $1.95 for 32 oz or $2.99 for 64 oz.
                  You have $4.00.

   4. Task: Plan your time after school.

      Facts: School dismissal is at 2:30 P.M. Your job requires $2\frac{1}{2}$ hours. Transportation from school to your job and then home takes about 45 minutes in all. Dinner is served at 6:00 P.M. You must study for 3 hours. Your favorite TV show is from 8:00–9:00 P.M. and bedtime is at 10:30 P.M.

5. Task: Prepare dinner and serve at 6:30 P.M.

Facts: Each dish requires the preparation time and cooking time given.

Dish	Preparation Time	Cooking Time
Meat	10 minutes	2 hours
Potatoes	5 minutes	1 hour
Vegetable	10 minutes	15 minutes
Salad	15 minutes	none
Dessert	30 minutes	45 minutes

6. Task: Eat foods that satisfy the recommended daily allowance (RDA) for iron.

Facts: The RDA for iron is 10 mg. Foods eaten so far today with their iron content are the following:

Orange....0.6 mg   Egg ....0.7 mg
Toast .....0.3 mg   Tuna sandwich ...1.9 mg
Milk.......Trace   Carrot sticks .....0.7 mg

7. Decision: Whether or not to buy a battery charger

Facts: A battery charger costs about $30. You usually spend about $5 a month on batteries for your calculator, flashlights, etc. Rechargeable batteries will cost about $35 and can be recharged many times.

C 8. Decision: Decide which carpet to buy for your bedroom.

Facts: Your bedroom is 10 ft by 12 ft. Royal Plush carpet costs $22.79 per sq yd. Soft Step carpet costs $18.79 per sq yd. Your state has a 5% sales tax. You have $300 to spend.

10-7 *Histograms*

The fuel efficiency rating of an automobile is important to consider when buying a car. The data in the chart below show the fuel efficiency ratings of several different cars.

FUEL EFFICIENCY RATINGS														
Cars	A	B	C	D	E	F	G	H	I	J	K	L	M	N
Rating (mi/gal)	27	33	24	33	30	23	36	26	31	32	28	38	37	36

To analyze the data, you can **tally** the number of times each rating occurs. If the list of data is quite long or if the data covers a wide range of numbers it may be more convenient to group the data into intervals. We can tally the data above using intervals of 5 miles per gallon.

TALLY

Rating (mi/gal)	20–24	25–29	30–34	35–39
Frequency (number of cars)	||	|||	⨯ℋ	||||

A summary of the number of items in each interval is called a **frequency distribution.**

Example 1   Use the tally above to show the frequency distribution of the fuel efficiency ratings.

Solution

FREQUENCY DISTRIBUTION CHART			
Rating (mi/gal)	20–24	25–29	30–34	35–39
Frequency (number of cars)	2	3	5	4

Frequency distribution may be shown on a **histogram.** A histogram is similar to a bar graph; however, there are no spaces between the bars.

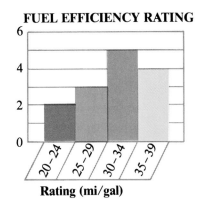

FUEL EFFICIENCY RATING

Rating (mi/gal)

Example 2   Which fuel efficiency rating interval occurs most frequently?

Solution   the 30–34 mi/gal interval

Frequency distribution may also be shown as a **frequency polygon.** A frequency polygon is the broken line connecting the midpoints of the top of each bar of a histogram.

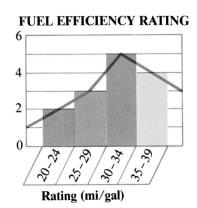

FUEL EFFICIENCY RATING

Rating (mi/gal)

Example 3   Did the frequency increase or decrease from the 20–24 mi/gal interval to the 25–29 mi/gal interval?

Solution   increase

Your Turn   • How much did the frequency decline from the 30–34 mi/gal interval to the 35–39 mi/gal interval?

Class Exercises

The data in the chart below show ten students' absences during the first semester of the school year.

STUDENT ABSENCES—FIRST SEMESTER										
Student	A	B	C	D	E	F	G	H	I	J
Number of absences	2	0	1	2	5	2	4	3	5	3

1. Tally the data from the chart.
2. Make a frequency distribution chart.

Exercises

Tally the data in the chart below in 4-percent intervals. Then use the tally to make a frequency distribution chart.

A **1.**

CLASS TEST SCORES												
Student	A	B	C	D	E	F	G	H	I	J	K	L
Score	77%	86%	78%	93%	76%	92%	96%	88%	81%	98%	87%	99%

Copy and complete the histogram using the data from the frequency distribution chart. Use the completed histogram to answer the following questions.

STUDENT WEEKLY INCOME					
Amount in Dollars	0–4	5–9	10–14 ·	15–19	20–24
Frequency (number of students)	13	15	28	21	16

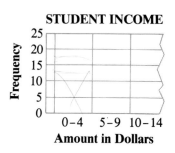

B **2.** Which income interval occurs most frequently?
3. How much greater is the frequency of the $10–$14 income interval than the frequency of the $20–$24 income interval?
4. What is the mean frequency of all of the income intervals?
5. What is the frequency range of all of the income intervals?

The histogram shows the weekly television viewing times of one hundred seventy people. Copy the histogram and make a frequency polygon. Use the frequency polygon to answer the following questions.

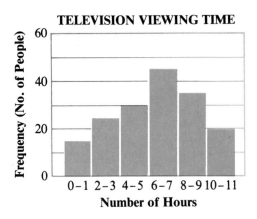

6. Did the frequency increase or decrease from the 0–1 hour interval to the 2–3 hour interval?

7. The greatest decrease in frequency occurred between what two consecutive intervals?

8. How much did the frequency increase from the 0–1 hour interval to the 2–3 hour interval?

A **cumulative histogram** shows the data for each interval added to the data in all preceding intervals. From a cumulative histogram we can more easily answer questions which refer to cumulative data. The cumulative histogram below shows the mean number of miles jogged each week by eighteen joggers.

Use the histogram to answer the following questions.

Example How many people run 7–9 mi each week?

Solution 2 people

C **9.** How many people run 1–3 mi each week?

10. How many people run 6 or fewer miles each week?

11. How many people run fewer than 10 mi each week?

12. How many people run 13–15 mi each week?

13. How many people run more than 15 mi each week?

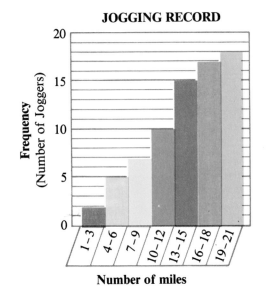

10-8 Scattergrams

A **scattergram** shows the relationship among given data. The scattergram at the right shows the results of a survey about the relationship between the heights of 20 parents and their adult children.

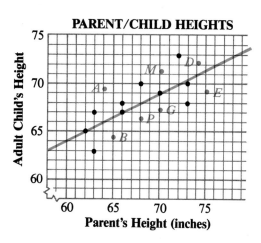

PARENT/CHILD HEIGHTS

Point P represents a parent's height of 68 in. whose child's height is 66 in. Point M represents a parent's height of 70 in. whose child's height is 71 in. Notice that the points slope upward to the right. There is a **positive association** between the heights of the parents and children; the taller parents have taller children. The line to which the scattered points are closest is the **regression** or **trend line.** The trend line best represents the relationship between the parents' and children's heights.

A scattergram may be used to predict future trends. The Parent/Child Height scattergram shows that knowing the height of a parent may help in guessing the height of the child; however, the wide scattering of points shows that there is still room for error.

Example 1 Using the trend line, about what would you expect a child's height to be if the parent's height is 72 in?

Solution about 70 in.

Your Turn • Using the trend line, about what would you expect a child's height to be if the parent's height is 66 in.?

333

The scattergram below shows the relationship between the distance a player was from the target in a baseball toss game and the score.

Point A represents a 6-ft distance from the target and a score of 45. Point B represents a 9-ft distance from the target and a score of 30. Notice that the points slope downward to the right. There is a **negative association** between the distance from the target and the score; the further the distance from the target, the lower the score.

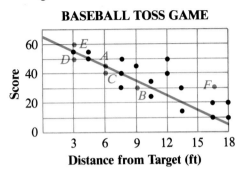

BASEBALL TOSS GAME

Example 2   Using the trend line, what would you expect the score to be if the distance from the target is 15 ft?

Solution   15

Your Turn   • Using the trend line, what would you expect the score to be if the distance from the target is 18 ft?

Class Exercises

Use the Parent/Child scattergram to answer the following questions.

1. Which point represents a parent's height of 74 inches whose child height is 72 inches?
2. Which point represents a parent's height of 65 inches whose child height is 64 inches?

Use the Baseball Toss Game scattergram to answer the following questions.

3. Which point represents a 3-ft distance from the target and a score of 50?
4. Which point represents a 6-ft distance from the target and a score of 40?

Exercises

The scattergram below shows the relationship between the height and weight of several students. Use it to answer the following questions.

A
1. Which points represent students who weigh 140 lb?
2. Which points represent students who are 72 in. tall?
3. Which point represents a student 64 in. tall who weighs 140 lb?
4. Which point represents a student 68 in. tall who weighs 130 lb?

B
5. Does the scattergram show a positive or a negative association?
6. Does the trend line indicate that a tall person will usually weigh more or less than a short person?

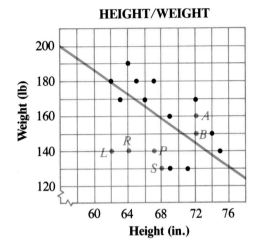

HEIGHT/WEIGHT

7. Using the trend line, how much would you expect a person who is 70 in. tall to weigh?
8. Using the trend line, how much would you expect a person who is 72 in. tall to weigh?

The scattergram below shows the relationship between the world records for the one-mile run and the years elapsed from 1860 to 1980. Use the One-Mile Run scattergram to answer the following questions.

C
9. Does this scattergram show a positive or a negative association between the year and the time?
10. Does the trend line indicate that the time will increase or decrease as the years progress?
11. Using the trend line, what would you expect the time for the one-mile run to be in the year 2000? What does that conclusion indicate to you about predicting future trends from the data on a scattergram?

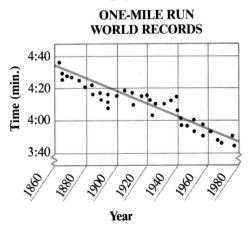

ONE-MILE RUN WORLD RECORDS

335

Problems

The data in the chart below shows the number of students enrolled in industrial arts class at one high school for ten years.

INDUSTRIAL ARTS CLASS ENROLLMENT										
Year	1	2	3	4	5	6	7	8	9	10
Number of students	36	65	51	42	32	68	55	49	64	59

1. Use the data to construct a scattergram.

2. What did you discover about the association between the number of students enrolled in industrial arts class and the progression of years?

Self Test 3

Use the histogram and frequency polygon to answer the following questions.

1. Which age group attended outdoor concerts most frequently?

2. How much did the frequency increase from the 12–15 year old age group to the 15–18 year old age group?

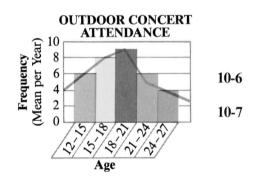

10-6

10-7

Use the scattergram to answer the following questions.

3. Using the trend line, what would you expect the average salary to be for a person with eight years of experience?

4. Using the trend line, about how many years of experience would you expect a person with an average salary of $35,000 to have?

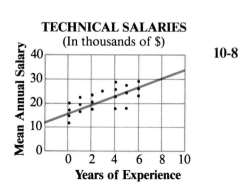

10-8

Algebra Practice

Write an equation. Solve.

1. The product of 385 and 124 is y.

2. 613,718 divided by 826 is a.

3. The sum of 70.54, 8.346, and 21.43 is n.

4. 0.39 times 0.27 is equal to g.

5. The difference between $5\frac{1}{2}$ and $3\frac{4}{5}$ is s.

6. $\frac{4}{9}$ divided by $\frac{5}{18}$ is equal to d.

7. The circumference of a circle with a radius of 6.2 cm is C.

8. The volume of a cone with a radius of 4 in. and a height of 6 in. is V.

Write the LCM of each set of expressions.

9. $2rs$ and r^3

10. c^2, 5, and $2d$

11. $6jk$, $4j^3$, and k^2

Solve.

12. $\frac{j}{4} = {}^-1$

13. $7s - 7 = 7$

14. $5y = 2 + 4y$

15. 80% of 160 is c.

16. $16\frac{2}{3}\%$ of 90 is m.

17. 60 is 80% of r.

18. $6y + 6 = 0$

19. $11n + 7 < {}^-70$

20. $4f - 1 > {}^-21$

Find the first five terms of the sequence produced by the given expression.

21. $c + 5$

22. $5t - 3.8$

23. $6 + 3w$

Write an equation and solve the problem.

24. In Ward Two there are 1614 registered voters.
    There are 44 more women registered than men.
    How many men are registered voters?
    How many women are registered voters?

Problem Solving for the Consumer

Consumer Awareness

Consumer publications provide consumers with information which may be helpful when buying a product.

Suppose you are planning to buy a new 10-speed bicycle. The chart below shows data for six 10-speed bicycles.

BICYCLE						
	A	**B**	**C**	**D**	**E**	**F**
Price ($)	169	198	220	220	190	155
Weight (lb)	$33\frac{1}{2}$	$30\frac{1}{4}$	33	$28\frac{1}{4}$	$34\frac{1}{2}$	$31\frac{3}{4}$

Use the chart to answer the following questions.

1. What is the price range for the six bicycles?
2. If you are interested in purchasing a bicycle that is closest in weight to the average weight of all six bicycles, which bicycle will you choose?
3. What is the average price of bicycles weighing over 31 lb?
4. What is the difference in weight between the two bicycles that are the same price?
5. If you are interested in buying a bicycle that is closest in price to the average price of all six bicycles, which bicycle will you choose?

It is important to compare many different features when purchasing a product. The chart below shows the ratings of four different features for each of the six bicycles.

BICYCLE RATINGS						
	Bicycle					
Features	A	B	C	D	E	F
Ease of Pedaling	3	4	4	4	3	3
Precision of Handling	4	4	4	2	4	5
Ease of Shifting	5	2	5	2	5	4
Braking Control	4	3	4	3	2	4
Overall Rating (Av.)	?	?	?	?	?	?

RATING CODE	
5	Excellent
4	Very Good
3	Good
2	Fair
1	Poor

6. Complete the chart to show the average overall rating for each bicycle.

Use the data from the chart above and the chart on the facing page to answer the following questions.

7. Which bicycles with an overall rating of 3 or more could you purchase for less than $200.

8. Do any of the bicycles weighing less than 31 lb have an average overall rating of 4 or more?

9. What is the average overall rating for all of the bicycles together?

Several students identified the features which were most important to them when purchasing a bicycle. Which bicycle is best suited to each person's needs?

10. Student 1
  - priced less than $200
  - very good or excellent braking control
  - excellent precision of handling

11. Student 2
  - weight 33 lb or less
  - very good or excellent ease of shifting
  - very good ease of pedaling

Enrichment

Percentiles

Standardized test results often indicate scores within a **percentile**. A percentile shows each student's score in comparison to other students' scores in a particular group. The chart below shows forty students' test scores in order from least to greatest. The 90th percentile is the score at which 90% of the scores are at or below and 10% are at or above.

$$90\% \text{ of } 40 = 36 \qquad 10\% \text{ of } 40 = 4$$

36 scores are at or below the score in the 90th percentile and 4 scores are at or above that score. You can now use the chart to locate the score in the 90th percentile. It is the score of Student 36.

Student Number	1	2	3	4	5	6	7	8	9	10	11	12	13	14	15	16
Score	45	46	47	52	53	56	56	58	60	61	65	67	68	70	73	73

Student Number	17	18	19	20	21	22	23	24	25	26	27	28	29	30	31	32
Score	75	76	78	79	80	82	82	84	85	86	87	91	91	91	92	93

Student Number	33	34	35	36	37	38	39	40
Score	93	93	94	95	96	96	97	98

Student 36 had a score of 95. The score in the 90th percentile is 95.

How many scores are at or above the given percentile?

1. 95th **2.** 20th **3.** 65th **4.** 30th **5.** 70th **6.** 55th

Find the score for the given data.

7. 22 scores at or below; 18 at or above **8.** 8 scores at or below; 32 at or above

Find the score for the given percentile.

9. 25th **10.** 50th **11.** 75th **12.** 40th **13.** 60th **14.** 80th

Chapter Review

Write the letter for the graph that presents the data needed to answer the question. Use the data to answer the question.

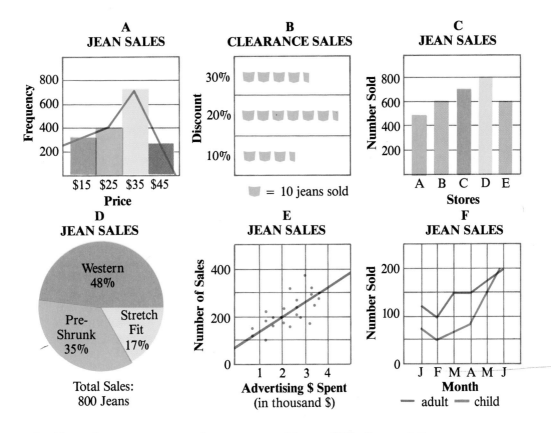

A
JEAN SALES

B
CLEARANCE SALES

= 10 jeans sold

C
JEAN SALES

D
JEAN SALES

Western 48%

Pre-Shrunk 35%

Stretch Fit 17%

Total Sales: 800 Jeans

E
JEAN SALES

F
JEAN SALES

— adult — child

1. About how many more jeans were sold at a 20% discount than a 10% discount?
2. Did the sales of adults' jeans or children's jeans increase more from January to June?
3. What was the price range of jeans?
4. About how many western-look jeans were sold?
5. About how many stretch-to-fit jeans were sold?
6. Does the scattergram show a positive or a negative association?
7. What was the median number of jeans sold at all of the stores?
8. Using the trend line, how many sales would you expect if $4000 was spent on advertising?
9. Use the data in Graphs C and D to write and solve a problem.

Chapter Test

Find the mean of the given data.

1. 87, 92, 88, 96, 87

2. 42.3, 40.6, 42.1, 42.1

10-1

Use the corresponding graph to answer the following questions.

MUSIC FESTIVALS

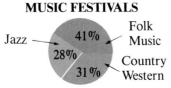

Total Attendance. 4,500

COMMUTING

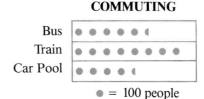

● = 100 people

3. About how many people attended the folk music festival?

4. About how many people commute to work by bus?

10-2, 3

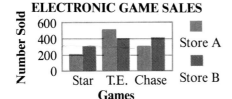

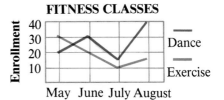

5. About how many more T.E. games than Star games were sold at Store B?

6. How much did enrollment in exercise class decrease from May to July?

10-4, 5

Write a question and solve the problem.

10-6

7. Sandi typed a page 40 lines long in 10 minutes. The mean number of words on each line was 15.

Use the graph to answer the question.

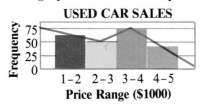

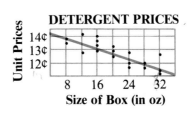

8. Did the frequency increase or decrease from the $2000–$3000 price range to the $4000–$5000 price range?

9. Using the trend line, about how much would you expect the unit price of a 32-oz box of detergent to be?

10-7, 8

Cumulative Review (Chapters 1–10)

Add, subtract, multiply, or divide.

1. $28 - 6 \div 2$

2. $1.183 \div 2.6$

3. 5.6×10

4. 4.3×10^2

5. $27.263 \div 100$

6. $185 + 16 \div 4 \times 0$

7. $\frac{1}{5} \times 6\frac{3}{4}$

8. $2\frac{1}{8} \times 20$

9. $4\frac{1}{3} + \frac{5}{6}$

10. $17 - 3\frac{3}{5}$

11. $24\frac{1}{9} - 5\frac{3}{4}$

12. $21 \div 1\frac{3}{4}$

Solve the equation.

13. $2x + 3 = 9$

14. $25 = 9y - 20$

15. $7(p + 4) = 28$

16. $24 = 6(g - 2)$

17. $56 = 8(2 + w)$

18. $5n = 2n + 9$

19. $4h - 12 = h$

20. $2p - 1 = 4p - 15$

21. $3 + 8r = r + 17$

Solve.

22. What is 20% of 48?

23. 65% of 99 is what?

24. 2% of 74 is what number?

25. What is 5.5% of $840?

26. 150% of 350 is what number?

27. 65 out of 80 is what percent?

Add, subtract, multiply, or divide.

28. $^-2 + 7$

29. $^-7 - 5$

30. $^-21 \div ^-3$

31. $6 \times ^-35$

32. $^-15 - ^-6$

33. $^-26 - ^-4$

34. $^-8 \times ^-12$

35. $45 \div ^-9$

What is the slope and y-intercept of the equation?

36. $y = 2x - 3$

37. $y = ^-5x$

38. $y = ^-x + 9$

39. $y = \frac{1}{2}x + 8$

40. $y = \frac{^-1}{3}x - 1$

41. $y = ^-16x + 4$

Solve.

42. The rent on Jerry's apartment increased 5%. The rent used to be $250 per month. What is the rent now?

43. Jacob Eberhart's father and his father's father both married at 20. Each had their only child when they were 21. If Jacob continues the pattern, how old will Jacob's grandfather be when Jacob's son or daughter is born?

11
Special Triangles

When the sun is at a given angle, the shadows of nearby objects are in proportion to the heights of the objects themselves. You can use this fact to solve a variety of practical problems.

11-1 Similar Triangles

Two triangles are said to be similar if the **corresponding angles** have equal measure and the **corresponding sides** are in proportion. Similar triangles have the same shape but not always the same size.

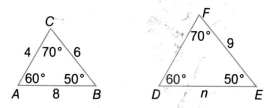

Triangles *ABC* and *DEF* are similar, with vertex *A* corresponding to vertex *D*, vertex *B* corresponding to vertex *E*, and vertex *C* corresponding to vertex *F*. This statement can be written as $\triangle ABC \sim \triangle DEF$. You can identify the corresponding angles by looking for pairs of angles with equal measure. To identify a pair of corresponding sides, notice that both $\overline{AC}$ and $\overline{DF}$ are opposite 50° angles. We say that $\overline{AC}$ corresponds to $\overline{DF}$. The corresponding angles and sides of triangles *ABC* and *DEF* are listed below.

Corresponding Angles

∠ *A* and ∠ *D*
∠ *B* and ∠ *E*
∠ *C* and ∠ *F*

Corresponding Sides

$\overline{AC}$ and $\overline{DF}$ ⟵ *The notation* $\overline{AC}$ *is often*
$\overline{AB}$ and $\overline{DE}$ *used to mean side* AC
$\overline{BC}$ and $\overline{EF}$ *or segment* AC.

You can find an unknown length by writing a proportion using corresponding sides.

Example 1 Find the length of $\overline{DE}$. Use the similar triangles shown above.

Solution $\dfrac{DE}{AB} = \dfrac{EF}{BC}$ ⟵ *The notation EF is used to represent the length of* $\overline{EF}$.

$\dfrac{n}{8} = \dfrac{9}{6}$ ⟹ $6n = 72$ ⟵ *An easy way to solve this equation*
$n = 12$ *is to use cross multiplication.*

The length of $\overline{DE}$ is 12 units.

Sometimes similar triangles are drawn so that one angle is common to both triangles. In the diagram shown at right, notice that $\angle F$ is common to $\triangle FHL$ and $\triangle FGM$. Study the diagram to see that $\overline{HL}$ corresponds to $\overline{GM}$ and $\overline{FL}$ corresponds to $\overline{FM}$.

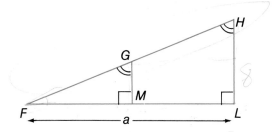

Example 2   Use the similar triangles FHL and FGM above to find a.   Use the following facts:
$$HL = 8, \ GM = 4, \ \text{and} \ FM = 10.$$

Solution $\dfrac{FL}{FM} = \dfrac{HL}{GM} \quad \Rightarrow \quad \dfrac{a}{10} = \dfrac{8}{4}$

$$4a = 80$$
$$a = 20$$

Your Turn   • In the similar triangles FHL and FGM above, $FL = 12$, $HL = 6$ and $GM = 3$. Find the length of $\overline{FM}$.

Class Exercises

In the similar triangles GHK and LMN the corresponding angles are marked alike. Write the corresponding side or angle.

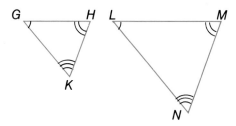

1. $\overline{GH}$ **2.** $\overline{HK}$ **3.** $\overline{LN}$

4. $\angle M$ **5.** $\angle N$ **6.** $\angle G$

Complete the proportion.

7. $\dfrac{GH}{LM} = \dfrac{GK}{?}$

8. $\dfrac{MN}{HK} = \dfrac{LN}{?}$

9. $\dfrac{ML}{HG} = \dfrac{?}{KG}$

Exercises

For each pair of similar triangles, find *n*.

A **1.**

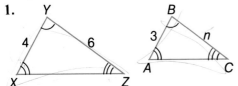

2.

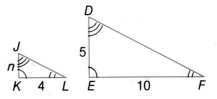

3.

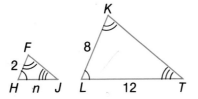

4.

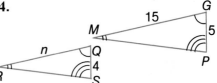

B **5.**

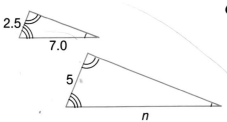

6.

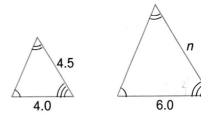

7.

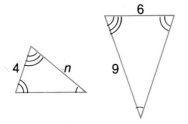

8.

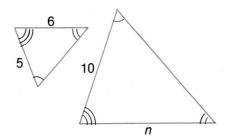

9.

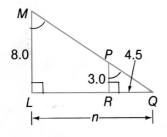

10.

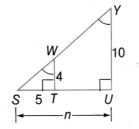

11-2 Similar Right Triangles

You can use similar right triangles to solve a variety of problems. For example, you can find the height of a tree without actually measuring it. The diagram shows a particular tree that casts a 9-ft shadow while a 5-ft pole casts a 3-ft shadow. The triangle formed by the tree and its shadow is similar to the triangle formed by the pole and its shadow.

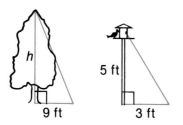

Example   Use the similar triangles in the diagram to find the height of the tree.

Solution $$\frac{\text{height of the tree}}{\text{height of the pole}} = \frac{\text{shadow of the tree}}{\text{shadow of the pole}}$$

$$\frac{h}{5} = \frac{9}{3} \quad \Rightarrow \quad 3h = 45 \quad \Rightarrow \quad h = 15$$

The height of the tree is 15 ft.

Your Turn   • A student 5 ft 4 in. tall casts a shadow that is 5 ft 4 in. long. At the same time, a flag pole casts a shadow that is 45 ft long. What is the height of the flag pole?

Class Exercises

The two triangles are similar. Complete the proportion.

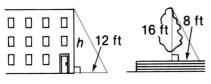

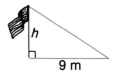

1. $\dfrac{h}{16} = \dfrac{?}{8}$

2. $\dfrac{h}{4} = \dfrac{?}{6}$

Exercises

Use the similar right triangles to find *h* to the nearest tenth.

A 1.
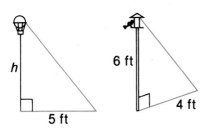

2.
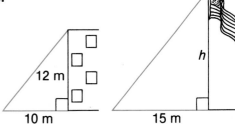

B 3.
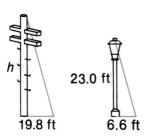

4.

5.
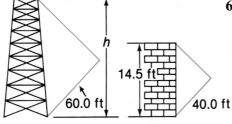

6.
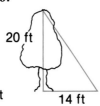

349

Problems

1. To find the length of Sun Lake, Nancy made the measurements shown in the diagram. Triangles *ABC* and *AFG* are similar. How long is Sun Lake?

2. A forest tower is 20 m high and casts a shadow of 4 m. At the same time, a ranger standing nearby casts a shadow of 0.5 m. How tall is the ranger?

3. To find the height of the pole, Alan made use of shadows as suggested by the diagram shown. By measuring, he found that *AB* = 11 ft and *AC* = 20 ft 8 in. If Alan is 5 ft 6 in. tall, how tall is the pole?

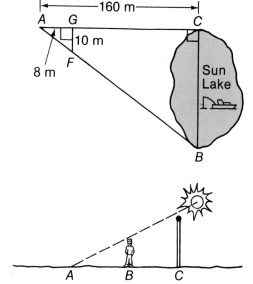

4. A utility pole casts a shadow of 7 m. At the same time, a pedestrian standing next to the pole casts a shadow of 0.6 m. If the pedestrian is 1.8 m tall, how tall is the utility pole?

Using the Computer

This program will calculate the lengths of two sides of a triangle when the length of one side is known, and the lengths of the corresponding sides of a similar triangle are known. RUN it for A = 10, B = 8, C = 12, C1 = 36, then try it for some examples of your own.

```
10  PRINT "TO FIND THE SECOND AND THIRD "
20  PRINT "SIDES OF A SIMILAR TRIANGLE:"
30  PRINT "WHAT ARE A, B, C, C1";
40  INPUT A,B,C,C1
50  PRINT "THE SIDES ARE:"
60  PRINT "A =";A,"A1 =";A*C1/C
70  PRINT "B =";B,"B1 =";B*C1/C
80  PRINT "C =";C,"C1 =";C1
90  END
```

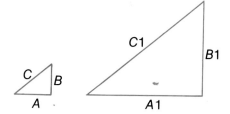

11-3 Pythagorean Theorem

Surely you have seen a ladder set up against a building. The ladder and the building form a right triangle. One side of the triangle is at ground level, from the foot of the ladder to the building. The building forms another side, and the ladder forms the third side. It is interesting to note that the ladder always forms the longest side. In any right triangle, the side opposite the right angle is called the **hypotenuse.** The other two sides are called **legs.**

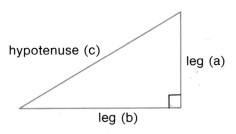

An important property of right triangles is known as the Pythagorean Theorem.

Pythagorean Theorem

For any right triangle, the square of the length of the hypotenuse is equal to the sum of the squares of the lengths of the other two sides.

You can write the Pythagorean Theorem as an equation.

$$c^2 = a^2 + b^2 \qquad \text{or} \qquad a^2 + b^2 = c^2$$

It is also true that if a triangle has the property in which c^2 is equal to $a^2 + b^2$, it is a right triangle.

351

Example 1 Is the triangle shown a right triangle?

Solution

$c^2 = a^2 + b^2$

5^2	$4^2 + 3^2$
25	$16 + 9$
25	25

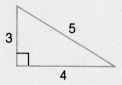

$25 = 25$, so the triangle is a right triangle.

You know that $5^2 = 25$. You also know that $(^-5)^2 = 25$. Since $5^2 = 25$ and $(^-5)^2 = 25$, both 5 and $^-5$ are **square roots** of 25. The positive square root of a number n is shown by the symbol $\sqrt{n}$, and the negative square root is shown by the symbol $^-\sqrt{n}$.

Example 2 Write the positive square root of 18.

Solution 18 is between 16 and 25, so $\sqrt{18}$ is between $\sqrt{16}$ and $\sqrt{25}$. $\sqrt{18}$ is between 4 and 5. From the table of square roots on page 470, $\sqrt{18}$ is approximately 4.243.

If you know the lengths of two sides of a right triangle, you can use the Pythagorean Theorem and the positive square root to calculate the length of the third side.

Example 3 Calculate the length of the hypotenuse.

Solution

$c^2 = a^2 + b^2$
$c^2 = 5^2 + 12^2$
$c^2 = 25 + 144 = 169$
$c = \sqrt{169} = 13$

Your Turn • If $a = 1$ and $b = 2$, what is the length of the hypotenuse?

Class Exercises

Complete.

1. $2^2 + 3^2 = \underline{?}$ **2.** $5^2 + 6^2 = \underline{?}$ **3.** $4^2 + 1^2 = \underline{?}$

4. $10^2 - 6^2 = \underline{?}$ **5.** $12^2 - 8^2 = \underline{?}$ **6.** $13^2 - 12^2 = \underline{?}$

Exercises

Write the positive square root. Use the table of square roots on page 470, if necessary.

A   **1.** $\sqrt{36}$ **2.** $\sqrt{49}$ **3.** $\sqrt{81}$ **4.** $\sqrt{100}$ **5.** $\sqrt{1}$

     6. $\sqrt{12}$ **7.** $\sqrt{6}$ **8.** $\sqrt{10}$ **9.** $\sqrt{13}$ **10.** $\sqrt{14}$

**The lengths of the sides of a triangle are given.
Is the triangle a right triangle?**

B   **11.** $a = 9, b = 12, c = 15$ **12.** $a = 4, b = 5, c = 6$

     13. $a = 10, b = 15, c = 20$ **14.** $a = 10, b = 24, c = 26$

     15. $a = 5, b = 12, c = 13$ **16.** $a = 20, b = 21, c = 29$

The lengths of two sides of a right triangle are given. Find the length of the hypotenuse.

17. 2 km, 3 km   **18.** 3 m, 5 m   **19.** 6 cm, 8 cm

20. 4 mm, 2 mm   **21.** 7 m, 5 m   **22.** 5 mm, 5 mm

A right triangle has sides of lengths a, b, and c, with c the length of the hypotenuse. Find the length of the missing side of the triangle to the nearest thousandth.

Example $a = 7, c = 9$ **Solution** $a^2 + b^2 = c^2$

$$49 + b^2 = 81$$
$$b^2 = 81 - 49 = 32$$

You must find $\sqrt{32}$ to find b. $\longrightarrow$ $b = 5.657$

23. $a = 3, c = 7$ **24.** $a = 4, c = 6$ **25.** $a = 5, c = 9$

26. $b = 1, c = 4$ **27.** $b = 4, c = 9$ **28.** $a = 5, b = 7$

Problems

Solve. Write the answer to the nearest tenth.

1. A rectangle is 5 ft long and 2 ft wide. How long is a diagonal of the rectangle?
2. A square measures 6 yd on one side. What is the length of a diagonal of the square?
3. Marie and Kevin hiked 3 mi east and then 6 mi north. How far were they from their starting point?

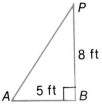

4. A pole is 8 ft high. A wire is attached to the top of the pole and fastened to an anchor in the ground. The anchor is 5 ft from the foot of the pole. What is the length of the wire?
5. A ladder 25 ft long is placed so that its top reaches 24 ft up the side of a house. How far from the house is the foot of the ladder?

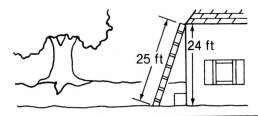

Self Test 1

Triangles *SRT* and *WYZ* are similar. Find the length of $\overline{SR}$.

1.
2.
   11-1

Use the similar triangles to find the height marked *h*.

3.
4.
   11-2

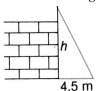 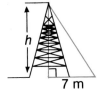

The lengths of the two legs of a right triangle are given. Find the length of the hypotenuse to the nearest tenth.

5. 6 in., 5 in.
6. 4 cm, 6 cm
7. 7 in., 7 in.
   11-3

11-4 Tangent of an Angle

Triangle *ABC* is a right triangle. In relation to ∠ *B*, $\overline{BC}$ is called the adjacent leg and $\overline{AC}$ is called the opposite leg.

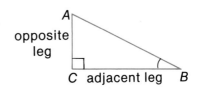

In a right triangle, the ratio of the length of the leg opposite an angle to the length of the leg adjacent to the angle is called the tangent of the angle. We write the tangent of angle *B* as "tan *B*."

$$\tan B = \frac{\text{length of the opposite leg}}{\text{length of the adjacent leg}} = \frac{AC}{BC}$$

Triangles *PQR* and *PTS* shown at the right are similar triangles. Notice that ∠ *P* is common to both triangles. You can find the tangent of ∠ *P* as shown below.

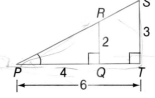

Triangle *PQR*

$$\tan P = \frac{RQ}{PQ} \quad \Rightarrow \quad \frac{2}{4} = \frac{1}{2}$$

Triangle *PTS*

$$\tan P = \frac{ST}{PT} \quad \Rightarrow \quad \frac{3}{6} = \frac{1}{2}$$

For both similar right triangles, tan $P = \frac{1}{2}$. You can see that the tangent of ∠ *P* depends on the measure of the angle itself, not on the size of the triangle.

Example 1   For the given right triangle, find the value of tan *G*.

Solution   Use the Pythagorean Theorem to find *x*.

$$3^2 + x^2 = 5^2$$
$$x^2 = 25 - 9 = 16$$
$$x = \sqrt{16} = 4$$

$$\tan G = \frac{\text{opposite leg}}{\text{adjacent leg}} = \frac{3}{4}$$

Example 2   For the given triangle, find the value of tan C.

Solution   Find the value of y.

$$5^2 + y^2 = 6^2$$
$$y^2 = 36 - 25 = 11$$
$$y = \sqrt{11}$$

$$\tan C = \frac{\text{opposite leg}}{\text{adjacent leg}} = \frac{5}{\sqrt{11}}$$

$$\approx \frac{5}{3.317} \approx 1.507$$

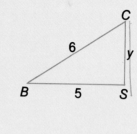

Your Turn   • Find the value of tan B, in the triangle above.

Class Exercises

Use triangle *NBC* to complete.

1. $\overline{NB}$ is opposite $\angle$? .

2. $\overline{BC}$ is opposite $\angle$? .

3. $\overline{BC}$ is ? to $\angle C$.

4. $\overline{BN}$ is ? to $\angle N$.

5. $\tan C = \dfrac{NB}{?}$

6. $\tan N = \dfrac{?}{BN}$

Exercises

Write the value of tan A and tan B in lowest terms.

A   1.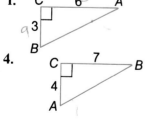

2.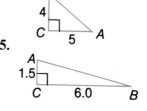

3.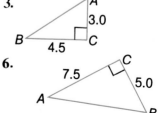

4.

5.

6.

356

For the given triangle, find the value of tan *A* to the nearest tenth.

B **7.**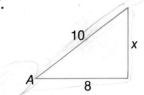

8.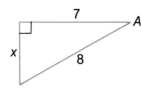

9.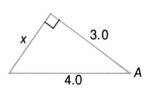

10.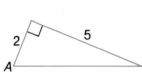

11.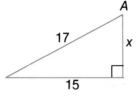

12.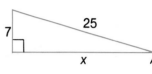

13.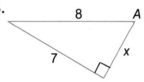

14.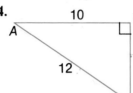

15.

Use the diagram at the right to find the tangent of the angle to the nearest tenth.

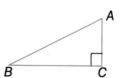

16. ∠A

17. ∠B

18. ∠ACD

19. ∠DCB

20. For any right triangle *ABC* with the right angle at *C*, is it true that tan *A* · tan *B* = 1? Show how you discovered this.

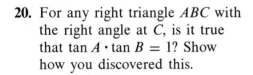

Using the Calculator

Find the value of the symbol in the table on page 470, use a calculator to evaluate the expression, and round to the nearest whole number.

1. $\sqrt{3} \cdot \sqrt{3}$ **2.** $\sqrt{7} \cdot \sqrt{7}$ **3.** $\sqrt{15} \cdot \sqrt{15}$ **4.** $\sqrt{23} \cdot \sqrt{23}$ **5.** $\sqrt{49} \cdot \sqrt{49}$

6. Is the following statement true? $\sqrt{n} \cdot \sqrt{n} = n$

11-5 *Sine and Cosine of an Angle*

In addition to the tangent ratio, you can write two other important ratios, the **sine** ratio and the **cosine** ratio. Notice that $\triangle MLS$ is a right triangle. We write the sine of $\angle M$ as sin M and the cosine of $\angle M$ as cos M.

$$\sin M = \frac{\text{length of opposite leg}}{\text{length of hypotenuse}} = \frac{SL}{MS}$$

$$\cos M = \frac{\text{length of adjacent leg}}{\text{length of hypotenuse}} = \frac{ML}{MS}$$

In the diagram at the right, triangles *ADE* and *AFG* are similar right triangles. You can compute sin *A* and cos *A* for both triangles.

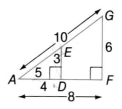

Triangle *ADE*

$$\sin A = \frac{ED}{AE} = \frac{3}{5}$$

$$\cos A = \frac{AD}{AE} = \frac{4}{5}$$

Triangle *AFG*

$$\sin A = \frac{GF}{AG} = \frac{6}{10} = \frac{3}{5}$$

$$\cos A = \frac{AF}{AG} = \frac{8}{10} = \frac{4}{5}$$

You can see that sin *A* and cos *A* depend on the size of $\angle A$, not on the size of the triangle.

Example   Find sin *T* and cos *T* in the given triangle.

Solution   Find the value of *x*.

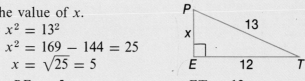

$$12^2 + x^2 = 13^2$$
$$x^2 = 169 - 144 = 25$$
$$x = \sqrt{25} = 5$$

$$\sin T = \frac{PE}{PT} = \frac{5}{13} \qquad \cos T = \frac{ET}{PT} = \frac{12}{13}$$

Your Turn • Find sin *P* and cos *P*. • Find sin *M* and cos *M*.

Class Exercises

Use the triangles shown to complete.

1. $\sin A = \dfrac{BC}{?}$

2. $\sin C = \dfrac{?}{AC}$

3. $\cos A = \dfrac{AB}{?}$

4. $\cos C = \dfrac{?}{AC}$

5. $\sin F = \dfrac{?}{6}$

6. $\cos D = \dfrac{\sqrt{11}}{?}$

7. $\sin \underline{\ ?\ } = \dfrac{5}{6}$

8. $\underline{\ ?\ } F = \dfrac{5}{6}$

Exercises

For each triangle, find the value of sin *A*, cos *A*, sin *B*, and cos *B*. Write the ratio in lowest terms.

A　**1.**

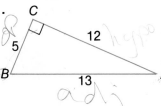

2.

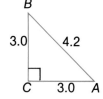

3.

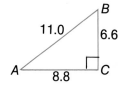

4.

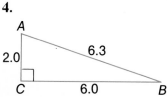

5.

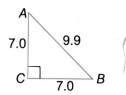

6.

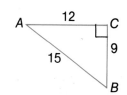

Find the value of x, then find $\sin A$.

B   **7.**
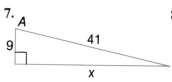

8.
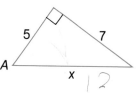

9.
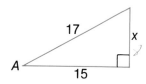

Find the value of x, then find $\cos A$.

10.
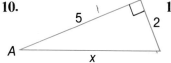

11.
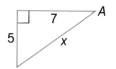

12.
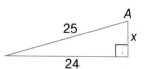

For the triangles shown at the right, find the value of the ratio.

13. $\sin \angle BCD$ **14.** $\cos \angle BCD$

15. $\sin \angle ACD$ **16.** $\cos \angle ACD$

17. $\sin A$ **18.** $\cos A$

19. $\cos B$ **20.** $\sin B$

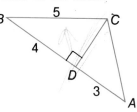

Refer to rectangle $ABCD$ to find the value of the ratio.

C **21.** $\tan \angle 2$ **22.** $\sin \angle 1$

23. $\cos \angle 1$ **24.** $\tan \angle 3$

25. $\sin \angle 4$ **26.** $\cos \angle 4$

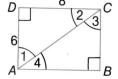

27. Find the values of $\sin M$ and $\cos N$ in $\triangle MNP$. What do you notice? Is this result always true for every right triangle?

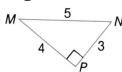

Challenge

To express $\dfrac{5}{\sqrt{2}}$ without the symbol in the denominator, you can multiply the numerator and the denominator by $\sqrt{2}$. Write the following expressions without the symbol in the denominator.

1. $\dfrac{1}{\sqrt{6}}$ **2.** $\dfrac{1}{\sqrt{3}}$ **3.** $\dfrac{2}{\sqrt{5}}$ **4.** $\dfrac{3}{\sqrt{8}}$ **5.** $\dfrac{7}{\sqrt{3}}$

11-6 Using Trigonometry Tables

In order to solve some practical problems involving right triangles, you must have access to the values of certain sine, cosine, and tangent ratios. A few of these can be easily computed using the properties of special triangles and the Pythagorean Theorem. Study triangle ABC shown below to see how the statement sin 30° = 0.5000 is derived.

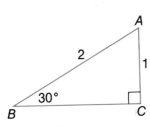

Angle	Sine	Cosine	Tangent
26°	.4384	.8988	.4877
27°	.4540	.8910	.5095
28°	.4695	.8829	.5317
29°	.4848	.8746	.5543
30°	.5000	.8660	.5774

$\sin 30° = \dfrac{1}{2} = 0.5$

$\sin 30° = 0.5000$ ⟵ —— *Remember that 0.5000 has the same value as 0.5.*

Most values of trigonometric ratios have to be computed by advanced methods. The table on page 471 gives approximate values of the sine, cosine, and tangent of angles with measures 1° to 90°. A portion of the table is shown above. You can use the table to find the lengths of sides of right triangles.

Example Use the tangent ratio to find the length of $\overline{XE}$ in the given triangle.

Solution $\tan 30° = \dfrac{XE}{AE}$

Substitute the value of tan 30° from the table.

$0.5774 = \dfrac{XE}{5}$ ⟵ —— *Use multiplication to solve the equation.*

$5 \times 0.5775 = XE$

$2.8875 = XE$

Your Turn
- If $AE = 4$, use the tangent ratio to find the length of $\overline{XE}$ in the triangle above.
- If $XE = 3$, use the sin ratio to find the length of $\overline{AX}$.

Class Exercises

Use the table on page 471 to write the measure of the angle.

1. $\sin \underline{} = .1736$ **2.** $\sin \underline{} = .2588$ **3.** $\cos \underline{} = .9703$

4. $\cos \underline{} = .9962$ **5.** $\tan \underline{} = 11.4301$ **6.** $\tan \underline{} = 6.3138$

Exercises

Use the tangent values from the table on page 471 to find the length of a to the nearest tenth.

A   **1.**   **2.** 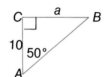   **3.**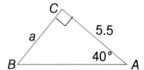

Use the table on page 471 to find the length of x to the nearest tenth.

B   **4.** 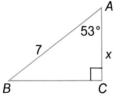   **5.** 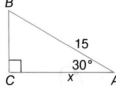   **6.**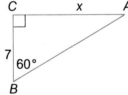

7. 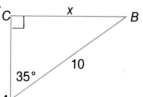   **8.** 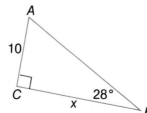   **9.**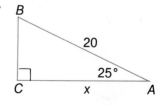

Problems

Use the table on page 471 to solve.

1. What is the distance across the lake to the nearest meter?

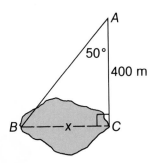

2. What is the height of the tree to the nearest foot?

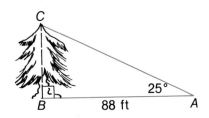

3. A trail bike is driven along a straight path from point *D* to point *C*. At point *C* the driver makes a 90° turn and travels to point *A*. The bike is then driven along a straight path from point *A* back to point *D*. Find the total distance the bike is driven to the nearest mile.

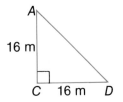

4. Lisa and her brother Eric are flying their kite. Eric is standing at point *E*. The kite is directly over his head at point *K*. The kite string makes an angle of 54° with the ground. To the nearest meter, how high is the kite above Eric's head?

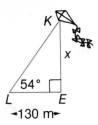

11-7 Finding Angle Measures

You can use trigonometry tables to find the approximate measure of an angle in a right triangle if you know the lengths of two sides.

Example   Find the measure of $\angle A$ to the nearest degree, then find the measure of $\angle B$.

Solution $\sin A = \dfrac{BC}{BA} = \dfrac{5}{8}$

$\dfrac{5}{8} = 0.625$ ⟵ *Write $\frac{5}{8}$ as a decimal since the entries in the table are decimals.*

Use the table on page 471 to estimate the measure of the angle with a sine of 0.6250.

Sin 38° = 0.6157 and sin 39° = 0.6293.

Since 0.6250 is closer to 0.6293 than it is to 0.6157, $\angle A = 39°$ to the nearest degree.

$\angle B = (180° - 90°) - 39° = 51°$ ⟵ *The sum of the measures of the angles of a triangle is 180°.*

Your Turn   • Find the measure of $\angle A$ to the nearest degree, then find the measure of $\angle B$.

Class Exercises

In triangle ABC, $\angle C = 90°$. The measure of one other angle is given below. Find the measure of the angle that is not given.

1. $\angle A = 53°$ 2. $\angle B = 88°$ 3. $\angle A = 71°$ 4. $\angle B = 63°$

5. $\angle B = 47°$ 6. $\angle A = 39°$ 7. $\angle B = 29°$ 8. $\angle A = 55°$

Exercises

Use the table on page 471 to find the measure of the angle to the nearest degree.

A 1. $\sin A = 0.5625$ 2. $\cos B = 0.7049$ 3. $\cos A = 0.44$

   4. $\sin B = 0.3846$ 5. $\sin A = 0.875$ 6. $\tan B = 1.0758$

Find the measures of $\angle A$ and $\angle B$ to the nearest degree.

7. 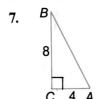   8.   9.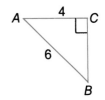

B 10.   11. 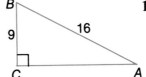   12.

 In triangle ABC, $\angle C = 90°$. Find the measures of $\angle A$ and $\angle B$ to the nearest degree.

13. $AB = 5$ and $AC = 2$ 14. $AB = 6$ and $AC = 4$

15. $BC = 3$ and $AB = 6$ 16. $AC = 7$ and $AB = 9$

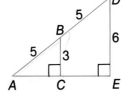

Triangles ABC and ADE are similar. Find the measure of the angle to the nearest degree.

17. $\angle A$ 18. $\angle ABC$ 19. $\angle D$

Find the measures of the angles and sides that are not given. Round each measure to the nearest unit.

C 20. 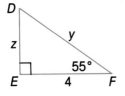   21.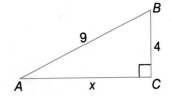

Problems

1. The picture at the right shows a section of the roof of a house. What is the approximate measure of the angle that the rafters form with the attic floor? (Hint: The rafters and the attic floor form the angle with vertex at *R*.)

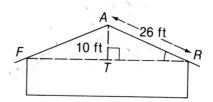

2. Engineers at point *P* would like to get to point *T*, 15 m below ground level. They can get to point *T* by tunnelling at an angle of 40°. How long will the tunnel be to the nearest meter?

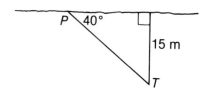

Self Test 2

For the given triangle, find the value of tan *A*.

1.
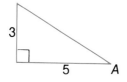

2.
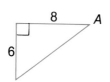

3.
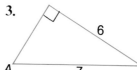

11-4

Find the value of the ratio for the triangle shown.

4. sin *B*

5. cos *B*

6. cos *A*

7. sin *A*

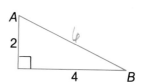

11-5

Use the table on page 471 to find *x*. Round to the nearest tenth.

8.

9.

10.
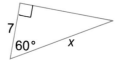

11-6

In △*ABC*, ∠*C* = 90°. Find the measures of ∠*A* and ∠*B* to the nearest degree.

11. *AB* = 7.5 and *AC* = 4.5

12. *BC* = 2 and *AC* = 1.5

11-7

Algebra Practice

Write an equation. Solve.

1. The quotient of 293,675 and 425 is m.

2. 768 times 392 is equal to k.

3. 700 minus 469.89 is equal to c.

4. 510.2 divided by 46.8 is equal to y.

5. The sum of $12\frac{3}{5}$ and $5\frac{9}{10}$ is g.

6. The product of $11\frac{1}{5}$ and $1\frac{3}{7}$ is n.

7. The area of a circle with a diameter of 8.4 in. is A.

8. The surface area of a box 8 cm wide, 9 cm high, and 14 cm long is S.

Use the distributive property to factor each expression.

9. $2g^2 - 2g + 6g^2$ 10. $6ab - 2a^2b + ab^2$ 11. $12c^2d + 3c^2d^2 - 3cd^3$

Solve.

12. $\frac{1}{2}d = 5\frac{1}{2}$ 13. $3 - 2c = 15$ 14. $8r = 7r - 3$

15. 15% of 60 is m.   16. $\frac{3}{4}$% of 800 is y.   17. 39 is 12.5% of h.

18. $10 = 20 + {}^-5x$ 19. $10v - 12 > {}^-42$ 20. ${}^-14 - 3q > 7$

Write an expression for any term of the sequence. Then find the 15th and 30th terms.

21. ${}^-7, {}^-2, 3, 8, 13$ 22. ${}^-10.7, {}^-9.4, {}^-8.1, {}^-6.8, {}^-5.5$

Write an equation and solve the problem.

23. The base of a triangle is 2.8 cm more than its height.
The height of the triangle is 5.3 cm.
What is the base of the triangle?

24. In triangle ABC, the measure of $\angle A$ is 92°.
The measure of $\angle A$ is twice the measure of $\angle B$.
What is the measure of $\angle B$?

Problem Solving on the Job

Medical Records Administrator

Many hospitals, health insurance companies and health equipment manufacturers require the service of medical record administrators.

> MEDICAL RECORD ADMINISTRATOR
>
> Description: Develop systems for recording, storing and retrieving patients' medical information—compile medical statistics—supervise the medical record staff
>
> Qualifications: Bachelor's degree in medical record administration
>
> Job Outlook: Openings are expected to grow faster than average for all occupations

The medical record department at a large hospital compiled the following statistics on 470 male and 530 female surgery patients over a three-year period. Statistics such as these are useful when plans are being made for patient turn-over and room assignments.

Recovery Time			
Age Group	**Male**	**Female**	**Days**
0–15	35	45	12
16–30	90	85	11
31–45	105	110	13
46–60	135	155	15
61+	105	135	18

1. What percent of the patients in each age group were male?
2. What percent in each age group were female?
3. What percent of the patients were over 60 years old?
4. What percent were under 16 years old?
5. Of the patients under 46 years old, what percent was female?
6. Of the patients under 31 years old, what percent was male?

7. How much greater was the average recovery time for the 61+ age group than for the 31–45 age group?

8. Copy and complete the chart to find the average recovery time for each male patient.

Age Group	Number of Male Patients	Number of Days	Total Recovery Time
0–15	35	12	420
16–30	90	11	?
31–45	105	13	?
46–60	135	15	?
61+	105	18	?
Total	?		?

$$\text{Average Recovery Time} = \frac{\text{total recovery time}}{\text{total number of male patients}}$$

9. What was the average recovery time for each female patient?

10. A full-time medical record administrator usually works 40 hours per week. One part-time medical record administrator works 21 hours per week. What percent of the week does she work?

11. Salaries are often based on training and experience. Six medical records administrators received these salaries in a recent year: $14,500, $15,200, $19,500, $18,000, $23,600, $30,000. What was the range in salary? What was the mean salary?

12. In a recent year 15,000 medical record administrators were employed in hospitals. Of these about 6500 were RRA's (Registered Record Administrators). What fraction was RRA's?

13. During the summer Bill McLean worked 12 hours per week as a medical record administrator intern at City Hospital. The table shows the number of hours he worked at each task in one week.

How many hours credit will Bill have in each section after 15 weeks?

Task	Hours
Verify information	2
Code forms	$4\frac{1}{2}$
Transcribe information to patients charts	4
Maintain files	$1\frac{1}{2}$

Enrichment

Special Right Triangles

Right triangles are so important in construction and in many other areas of work that it is worthwhile to study them in detail. One special right triangle has two 45° angles. Also, the two legs are equal in length. In the right triangle shown below, each leg is 1 unit long. Through the Pythagorean Theorem, we know that the hypotenuse is $\sqrt{2}$.

$$c^2 = 1^2 + 1^2$$
$$c^2 = 2$$
$$c = \sqrt{2}$$

45°-45° Right Triangle Property:
If each leg of a 45°-45° right triangle has length x, then the hypotenuse has length $x\sqrt{2}$.

Triangle SPQ is drawn so that each side is 2 units long. The 30°-60° right triangle, $\triangle PRQ$, can be thought of as half of $\triangle SPQ$. Since $\overline{RQ}$ is half of $\overline{SQ}$, then $\overline{RQ}$ is 1 unit long.

$$(RQ)^2 + (RP)^2 = (PQ)^2 \longleftarrow \text{Pythagorean Theorem}$$
$$1^2 + (RP)^2 = 2^2 \longleftarrow \text{Subtract } 1^2 \text{ to solve for } (RP)^2.$$
$$(RP)^2 = 3$$
$$RP = \sqrt{3}$$

30°-60° Right Triangle Property:
If the shorter leg of a 30°-60° right triangle has length x, then the other leg has length $x\sqrt{3}$, and the hypotenuse has length $2x$.

Find the length of each side labeled with a variable.

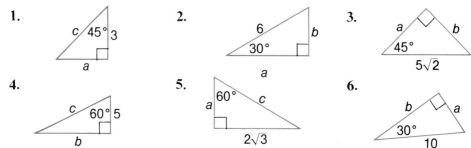

1.

2.

3.

4.

5.

6.

Chapter Review

How high is the tree? Write a, b, or c.

1. An 8-ft pole casts a shadow 5.5 ft long. At the same time a tree casts a shadow 16.5 ft long. **a.** 16 ft **b.** 24 ft **c.** 10.5 ft

The lengths of the two legs of a triangle are given. Find the length of the hypotenuse to the nearest tenth.

2. 6 ft, 8 ft **3.** 4 m, 6 m **4.** 7 in., 5 in.

Find the tangent of the angle. Write a, b, or c.

5. ∠A
   a. 1.6 **b.** 16 **c.** 0.6

6. ∠B
   a. 10 **b.** 1.6 **c.** 0.625

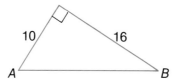

7. Triangles *ABC* and *DEF* are similar. Choose the correct measure for *n*.
   a. 28 **b.** 50 **c.** 7

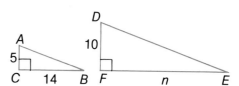

Use the table on page 471 to find the length of *n* to the nearest tenth.

8.
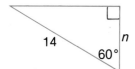

9.
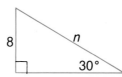

10.

In triangle *RNP*, ∠N = 90°. Find the measures of ∠R and ∠P to the nearest degree.

11. $RP = 15$, $RN = 12$ **12.** $NP = 10$, $RP = 12.5$ **13.** $RN = 12$, $RP = 16$

Find the value of *x*, then find the value of the ratio.

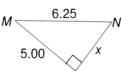

14. sin *M* **15.** cos *M*

16. sin *N* **17.** cos *N*

Chapter Test

For each pair of similar triangles, find *n*.

1.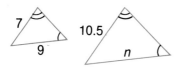

2.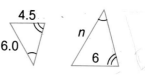

11-1

Use the similar right triangles to find the missing height.

3.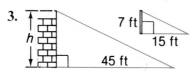

4.

11-2

A right triangle has sides of lengths *a*, *b*, and *c*, with *c* the length of the hypotenuse. Find the length of the missing side to the nearest tenth.

5. $a = 6$, $c = 10$ **6.** $c = 12$, $b = 10$ **7.** $b = 11$, $c = 13$ **11-3**

Find the value of tan *B*.

8.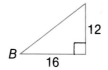

9.

10.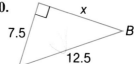

11-4

In $\triangle MRP$ $\angle R = 90°$, $MR = 6$, $RP = 8$, and $MP = 10$. Find the value of the ratio.

11. $\sin M$ **12.** $\sin P$ **11-5**

13. $\cos P$ **14.** $\cos M$

Use the table on page 471 to find the length of *x* to the nearest tenth.

15.

16.

17.

11-6

Find the measure of $\angle A$ and $\angle B$ to the nearest degree.

18.

19.

20.

11-7

Cumulative Review (Chapters 1–11)

Write the fraction in lowest terms.

1. $\dfrac{4}{18}$
2. $\dfrac{7}{49}$
3. $\dfrac{12}{48}$
4. $\dfrac{35}{70}$

5. $\dfrac{18}{32}$
6. $\dfrac{9}{27}$
7. $\dfrac{14}{27}$
8. $\dfrac{16}{40}$

9. $\dfrac{18}{45}$
10. $\dfrac{15}{35}$
11. $\dfrac{56}{72}$
12. $\dfrac{24}{56}$

13. $\dfrac{42}{48}$
14. $\dfrac{20}{32}$
15. $\dfrac{27}{81}$
16. $\dfrac{32}{72}$

Solve.

17. What percent is 9 out of 15?
18. 12 out of 50 is what percent?
19. 64 percent of 86 is what number?
20. What is 93% of 82?
21. 24% of what number is 8.4?
22. $\dfrac{1}{2}\%$ of 44 is what number?

23. How much simple interest will $1000 make at $7\dfrac{1}{2}\%$ for 2 years?

24. If you invest $500 and the money increases by $4\dfrac{1}{2}\%$, how much money do you gain?

Add, subtract, multiply, or divide.

25. $^-2 + 0$
26. $^-5 + {}^-7$
27. $6 + {}^-13$
28. $^-4 + 25$
29. $^-9 - 2$
30. $6 - {}^-4$
31. $0 - {}^-8$
32. $^-2 \cdot {}^-7$
33. $4 \cdot {}^-12$
34. $^-5 \div {}^-1$
35. $^-35 \div 7$
36. $^-80 \div {}^-10$

Draw the graph of the equation.

37. $y = x + 7$
38. $y = x + 3$
39. $y = 4x - 6$
40. $y = 2x + 1$
41. $y = 3x - 2$
42. $y = \dfrac{1}{2}x - 3$

43. Record albums range from $4.95 to $10.95 at Dave's Record City. What is the minimum and maximum you could spend for 5 albums?

12
Probability

A package of certified wheat seed must be 96% "pure live seed." That is, the probability that a seed in the package will be a wheat seed, and that it will sprout under proper conditions must be 96%.

12-1 Probability

One student from Granville High School will be interviewed on a TV program called "Students Speak Out." To select the student, names are written on cards and placed in a box. One card will be drawn at **random.** That is, the outcome of the drawing is strictly a matter of chance.

There are 520 students at Granville High School. We say that there are 520 **outcomes.** A set of one or more outcomes is called an **event.** The outcomes are **equally likely** if each card is as likely to be chosen as any other card. Stephen hopes that the card with his name will be drawn. As far as he is concerned, there is only 1 **favorable outcome.** The **probability** that Stephen's name will be drawn is 1 out of 520, or $\frac{1}{520}$. We write $P(Stephen) = \frac{1}{520}$.

If all outcomes are equally likely, the probability of an event is the ratio of the number of favorable outcomes to the total number of outcomes.

$$P(\text{event}) = \frac{\text{number of favorable outcomes}}{\text{total number of outcomes}}$$

Since a ratio may be written as an equivalent decimal or as a percent, the probability of an event may also be written as a decimal or a percent.

Example 1   What is *P*(even number) on the spinner? Write the probability as a ratio, as a decimal, and as a percent.

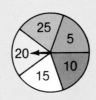

Solution   There are 2 favorable outcomes: 10 and 20.
There are 5 outcomes in all: 5, 10, 15, 20, and 25.

$$P(\text{even number}) = \frac{2}{5} = 0.4 = 40\% \longleftarrow \quad \begin{array}{l} 2 \div 5 = 0.4 \\ 0.4 = 0.40 = 40\% \end{array}$$

Example 2   What is *P*(6)? What is *P*(whole number)?

Solution   There is no 6 on the spinner. We say that 6 is an **impossible event.** Every number on the spinner is a whole number. Getting a whole number is a **certain event.**

$$P(6) = \frac{0}{5} = 0 \qquad P(\text{whole number}) = \frac{5}{5} = 1, \text{ or } 100\%$$

Your Turn   There are 100 beads in a bag. Of these, 50 are red, 30 are blue, and 20 are yellow. You draw one bead without looking. State the probability as a ratio, as a decimal, and as a percent.
  • *P*(red)   • *P*(blue or yellow)   • *P*(colored bead)   • *P*(green)

Examples 1 and 2 show some useful properties of probabilities.

- The probability of an impossible event is 0.
- The probability of a certain event is 1, or 100%.
- The probability of an event that is neither certain nor impossible is a number between 0 and 1.

Class Exercises

A game cube has six faces numbered 0, 3, 6, 9, 12, and 15. The cube is tossed once. State the probability as a ratio.

1. $P(9)$

2. $P(0 \text{ or } 15)$

3. $P(\text{odd number})$

4. $P(\text{not } 6)$

5. $P(1)$

6. $P(\text{number} < 16)$

State the probability as a percent.

7. $P(A) = \dfrac{1}{2} = \underline{\ ?\ } \%$

8. $P(B) = \dfrac{3}{5} = \underline{\ ?\ } \%$

9. $P(C) = \dfrac{7}{10} = \underline{\ ?\ } \%$

Exercises

A record store is giving away a randomly selected album free with each purchase of a new album. The free albums are selected from a group of 500 albums. They include 250 rock albums, 150 country albums, and 100 classical albums. Complete to determine the probability of the album selected for the first customer.

A

1. $P(\text{rock album}) = \dfrac{250}{500} = \dfrac{?}{2}$

2. $P(\text{not a rock album}) = \dfrac{250}{?} = \dfrac{?}{?}$

3. $P(\text{classical album}) = \dfrac{?}{500} = \dfrac{?}{?}$

4. $P(\text{country album}) = \dfrac{?}{?} = \dfrac{?}{?}$

5–8. State the probabilities in Exercises 1–4 as percents.

At a school carnival students can buy chances to spin the wheel for the prizes shown. State the probability as a ratio.

Outcome	Prize
Red	Free lunches for 1 week
White	Free spin

B

9. $P(\text{yellow})$

10. $P(\text{blue})$

11. $P(\text{yellow or blue})$

12. $P(\text{free lunches})$

13. $P(\text{free spin})$

14. $P(\text{prize})$

15. $P(\text{no prize})$

16. $P(\text{white})$

17. $P(\text{not white})$

18. $P(\text{red})$

19. $P(\text{not red})$

20. $P(\text{purple})$

377

The table shows the percent of students at Rockbury School whose ages are 15, 16, 17, or 18. No student is older than 18, but some are younger than 15. If a student is selected at random, write the probability as a percent.

Age at Last Birthday	Percent
15	20%
16	30%
17	25%
18	18%

21. P(age 16) **22.** P(age at least 15)

23. P(age 19) **24.** P(age less than 15)

25. P(not 15) **26.** P(age greater than 15)

On the spinner, $P(A)$ equals $\frac{1}{3}$, and $P(\text{not } A)$ equals $1 - \frac{1}{3}$, or $\frac{2}{3}$. Event A and Event (not A) are called **complementary events.**

$$P(\text{not } A) = 1 - P(A)$$

Let $P(R) = \frac{1}{3}$, $P(S) = \frac{3}{4}$, and $P(T) = 0.4$. Determine the probability.

C **27.** P(not S) **28.** P(not T) **29.** P(not R)

Problems

Solve. To the nearest whole percent, what is the probability?

1. Nathan randomly answered a true-false question on a quiz. What is the probability that he answered correctly?
2. Lisa randomly selected 1 of 5 choices on a multiple choice question. What is the probability that she selected the one correct answer?
3. One student from Valleyview Central School will be chosen at random to be "Principal For a Day." If 52% of the students are girls, what is the probability that a boy will be chosen?
4. At Dee's Bike Shop 50% of the customers pay for their purchases with cash and 30% use a credit card. The rest pay with a check. What is the probability that a customer will pay with a check?

12-2 Sample Spaces

Ace Tours offers passengers a choice of three routes from Los Angeles to St. Louis and two routes from St. Louis to New York. The sketch shows the possible routes from Los Angeles to New York through St. Louis.

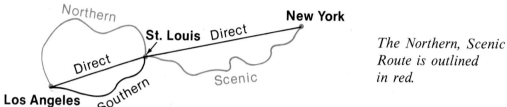

The Northern, Scenic Route is outlined in red.

The first prize in a contest is a trip from Los Angeles to New York. We write *P(Northern, Scenic)* to represent the probability that the Northern Route *and* the Scenic Route will be chosen.

To determine the probability of an event, you first need to know the number of possible outcomes. The set of all outcomes is called the **sample space.** One way to find the number of outcomes in a sample space is to list them. A **tree diagram** can help you do this.

Los Angeles-St. Louis	St. Louis-New York	*Outcomes*
Northern	Direct	*Northern, Direct*
	Scenic	*Northern, Scenic*
START ──── Direct	Direct	*Direct, Direct*
	Scenic	*Direct, Scenic*
Southern	Direct	*Southern, Direct*
	Scenic	*Southern, Scenic*

Example 1   What is *P(Northern, Scenic)*?

Solution   There are 6 outcomes in all; only 1 is favorable.

$$P(\text{Northern, Scenic}) = \frac{1}{6}$$

Your Turn In addition to the cities on the preceding page, Ace Tours offers a choice of two routes from New York to Boston, a direct route and an historical route.
- Draw a tree diagram to list all the routes from Los Angeles to Boston through St. Louis and New York.
- If you select one route from Los Angeles to Boston at random, what is the probability that part of the trip will be via a direct route?

Class Exercises

Suppose you spin the spinner and then toss a coin. The outcomes of the two-stage activity are shown in the tree diagram.

1. Complete the list of outcomes.

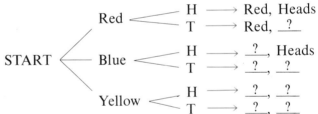

Outcomes

Red ⟨ H ⟶ Red, Heads
     T ⟶ Red, ?

START ⟨ Blue ⟨ H ⟶ ?, Heads
          T ⟶ ?, ?

Yellow ⟨ H ⟶ ?, ?
       T ⟶ ?, ?

What is the probability?

2. P(Red, Heads) **3.** P(Red or Heads) **4.** P(Green, Tails)

Exercises

For his birthday, Jason's parents will buy tickets to a movie or a game for him and a friend. Jason will choose one of four friends: Jack, Bill, Tim, or Chris. Copy and complete the tree diagram. Use the tree diagram to determine the probability.

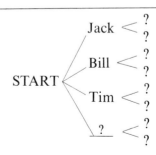

A **1.** P(Jack, movie) **2.** P(Bill, game) **3.** P(Tim or Chris, game)

    4. P(not Harry, movie) **5.** P(Bill, not movie) **6.** P(not Bill or Chris, game)

Jason decides to go with Jack, Tim, or Chris to either a football game or a baseball game. Use a tree diagram to determine the probability.

B 7. *P*(Jack, football)

8. *P*(Tim, game)

9. *P*(Chris, soccer)

10. *P*(Jack or Chris, baseball)

11. *P*(not Tim, football)

12. *P*(not Jack, not football)

A coin is tossed. Then a game cube is rolled. The game cube has faces numbered 1 to 6. Use a tree diagram to determine the probability.

13. *P*(H, 4)

14. *P*(*T*, 6)

15. *P*(*T*, 8)

16. *P*(H, even number)

17. *P*(H or T, 6)

18. *P*(not H, not 5)

19. *P*(T, prime number)

20. *P*(H or T, odd number)

21. *P*(H, $n > 4$)

A chart is sometimes used to give additional information about a two-stage activity. For example, the Tourist-Day flight costs $110. Use the chart to determine the probability. Let *r* represent the rate.

Jetaway Airline
Carlton to Oxbow Rate Chart

	Day	Night
1st Class	$125	$100
Tourist	$110	$90
Standby	$80	$70

C 22. *P*(least expensive)

23. *P*($r < 100)

24. *P*($r > 100)

25. *P*(Tourist or Night)

Problems

Draw a tree diagram and solve.

1. Susette plans a trip from Chicago to an island off the coast of Maine. She can go to the coast by plane, by car, by train, or by bus. She then has a choice of going by plane or by boat from the coast to the island. What is the probability that she will fly on both parts of the trip if she chooses at random?

2. Daily specials at a restaurant offer customers these choices:

      Meat: roast beef, chicken, turkey, ham
      Potato: baked, mashed
      Vegetable: salad, squash, peas

If a customer selects a meal at random, what is the probability that the customer will not select either ham or peas?

381

12-3 *Counting Principle*

It is not always necessary to list all the outcomes in a sample space to determine the number of possible outcomes. You can use the **counting principle** to find the number of outcomes for an activity with two or more stages.

The number of outcomes for an activity with two or more stages equals the product of the number of outcomes for each stage.

Suso is buying a telephone. He has a choice of these colors, styles, and dials.

Colors	Styles	Dials
black	antique	dial
white	modern	push-button
yellow	futura	
red	standard	
tan		
blue		
green		

Example 1 From how many different telephones must Suso choose?

Solution This is a three-stage activity.

Stage 1	Stage 2	Stage 3
Color	Style	Dial
7 choices	4 choices	2 choices

There are $7 \times 4 \times 2$, or 56 different outcomes possible.

Example 2   If Suso chooses a telephone at random, what is the probability that he will select a yellow or red, futura, push-button phone?

Solution   There are 2 favorable outcomes:

yellow, futura, push-button
red, futura, push-button

There are 56 possible outcomes.

P(yellow or red, futura, push-button) $= \dfrac{2}{56}$, or $\dfrac{1}{28}$.

Your Turn   • Jan has won a vacation trip for two to Bermuda, Hawaii, Jamaica, or Puerto Rico. She will go with either Sue, Pamela, or Fran. She can go by plane all the way or she can take a boat cruise. If she chooses at random, what is P(Hawaii, Fran, boat cruise)?

Class Exercises

How many different paths are there?

1. from A to B 2. from B to C
3. from A to C 4. from C to D
5. from B to D 6. from A to D

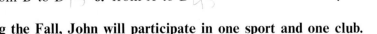

During the Fall, John will participate in one sport and one club.

Sports: soccer, football, swimming
Clubs: drama, photography, French, computer

If John selects at random from the lists above, what is the probability?

7. P(football, drama)   8. P(baseball, drama)   9. P(swimming, computer)

10. How many 2-digit numerals are there if there are 9 digits to choose from for the tens' place, and 10 digits for the ones' place? How many 3-digit numerals are there?

Exercises

Four activities are used for Exercises 1–20. They are tossing a coin, rolling a game cube, spinning the spinner shown, and drawing one of three cards marked *A, B, C*. The faces of the game cube are numbered 1 through 6.

  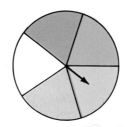

Complete to determine the number of outcomes for each set of activities.

A
1. Toss the coin; then spin the spinner. $2 \times 5 = \underline{?}$

2. Roll the game cube; then draw a card. $6 \times \underline{?} = \underline{?}$

3. Toss the coin twice; then spin the spinner. $2 \times \underline{?} \times \underline{?} = \underline{?}$

4. Roll the game cube twice; then toss the coin. $6 \times \underline{?} \times \underline{?} = \underline{?}$

Determine the number of outcomes for each set of activities.

B
5. Spin the spinner 3 times.

6. Spin the spinner twice; then draw a card.

7. Draw card; spin the spinner; then roll the game cube.

8. Toss the coin 3 times; roll the game cube; then draw a card.

Suppose you spin the spinner and draw a card. What is the probability?

9. *P*(red, *B*)   10. *P*(yellow, *D*)   11. *P*(white, *A* or *C*)

Suppose you toss the coin, roll the game cube, and draw a card. What is the probability?

12. *P*(H, 5, *B*)   13. *P*(T, 4, *A* or *C*)   14. *P*(H, even number, *A*)

Suppose you roll the game cube, spin the spinner, and draw a card. What is the probability?

15. *P*(1, blue, *B*)   16. *P*(prime number, red, *C*)   17. *P*(even number, red, *B*)

Suppose you toss the coin, roll the game cube, spin the spinner, and draw a card. What is the probability?

18. $P(H, 1, red, A)$ **19.** $P(H$ or $T, 3, red, A)$ **20.** $P(H, 1$ or $2, red, B)$

Two bowling teams, the Strikers and the All-Stars, are tied for the league championship. The four bowlers on each team are ranked 1, 2, 3, and 4. For the playoff, one person, selected at random, from each team will bowl. Answer the question.

C **21.** How many possible combinations of players are there for the playoff?

22. What is the probability that Player 1 of the Strikers will bowl against Player 3 of the All-Stars?

23. What is the probability that both players will have even numbers?

Problems

Solve.

1. A club has 5 seniors, 3 juniors, and 4 sophomores. How many senior-junior-sophomore combinations are possible for a three-person committee?

2. A baseball team has 5 pitchers and 2 catchers. How many pitcher-catcher combinations are possible?

3. A car dealer offers 15 body colors and 10 interior colors. How many color combinations are possible?

4. Barbara has 8 mystery books. 6 science fiction books, and 5 animal books. She wants to take one of each on a camping trip. From how many combinations can Barbara choose?

5. A clothing manufacturer makes dresses in 10 styles. Each style is made in 4 different colors and 6 sizes. A different code number is used for each style-color-size combination. How many code numbers are there?

6. A department store sells TVs from 5 different manufacturers. Each manufacturer offers a choice of screen in 8 different sizes. All TVs are available in either color or black and white. If one TV is chosen at random as a prize in a contest, what is the probability that it will be a Brand X color TV in the largest size available?

12-4 Independent Events

Suppose you spin the spinner then toss the coin. The outcome on the spinner in no way affects the outcome on the coin. A *4* on the spinner and *heads* on the coin are **independent events.** If two events are independent, you can use the following rule to determine the probability of both events occurring.

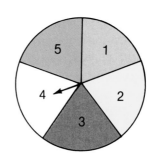

If Event *A* and Event *B* are independent, then *P(A and B)* equals *P(A)* times *P(B)*.

$$P(A, B) = P(A) \cdot P(B)$$

In general, the probability of two or more independent events occurring is the product of the probabilities of the individual events.

Example 1   If you spin the spinner then toss a coin, what is *P*(odd number, H)?

Solution   First, determine the probability of each event separately.

$$P(\text{odd number}) = \frac{3}{5} \qquad P(\text{H}) = \frac{1}{2}$$

Then multiply: $P(\text{odd number, H}) = \frac{3}{5} \cdot \frac{1}{2}$

$$= \frac{3}{10}$$

Your Turn   If you spin the spinner above two times, what is the probability?
- *P*(even number, odd number)
- *P*(not 4, 4)

Example 2   If a coin is tossed three times, what is $P(H, H, T)$?

Solution $P(H, H, T) = P(H) \cdot P(H) \cdot P(T)$

$$= \frac{1}{2} \cdot \frac{1}{2} \cdot \frac{1}{2}$$

$$= \frac{1}{8}$$

Your Turn   A game cube has faces numbered 1 through 6. If the cube is rolled three times, what is the probability?
- $P(1, 1, 6)$ • P(even number, odd number, 6)

Class Exercises

Suppose you spin the spinner shown twice. Complete to determine the probability.

1. $P(7, 7) = \frac{1}{5} \cdot \underline{\ ?\ } = \underline{\ ?\ }$

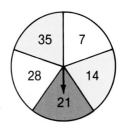

2. P(even number, 7) $= \frac{2}{?} \cdot \frac{1}{5} = \underline{\ ?\ }$

3. P(prime number, composite number) $= \frac{?}{5} \cdot \frac{?}{5} = \underline{\ ?\ }$

4. P(multiple of 3, odd number) $= \frac{?}{5} \cdot \frac{?}{5} = \underline{\ ?\ }$

Exercises

Two game cubes each have faces numbered 1 through 6. Determine the probability.

A **1.** $P(2, 5)$ **2.** $P(6, 6)$ **3.** P(even number, 3)

   4. P(odd number, 1)   **5.** $P(n < 6, 6)$ **6.** $P(4, n > 4)$

One boy and one girl from each class at Taylor High School will be chosen at random for an exchange program. The table shows the number of boys and the number of girls in each class as well as the number of girls named Mary and the number of boys name Bill in each class. Use the table for Exercises 7-21.

What is *P*(Bill, Mary) in each class?

Taylor High School Enrollment

B 7. Freshman   **8.** Sophomore
   9. Junior   **10.** Senior

Class	Boys	Girls	Bill	Mary
Freshman	28	32	7	4
Sophomore	30	35	6	7
Junior	25	27	5	9
Senior	24	30	6	5

What is *P*(neither Bill nor Mary) in each class?

11. Freshman   **12.** Sophomore
13. Junior   **14.** Senior

Determine the probability.

15. *P*(Mary in all 4 classes)   **16.** *P*(Bill in all 4 classes)

17. *P*(no girl named Mary is chosen)   **18.** *P*(no boy named Bill is chosen)

19. *P*(Mary in all 4 classes, Bill in all 4 classes)

Solve.

C **20.** In which of the classes is it most likely that a boy named Bill and a girl named Mary will be chosen? (Hint: Before comparing the probabilities, write them as decimals.)

21. In which class is it least likely that a boy named Bill and a girl named Mary will be chosen?

Problems

Solve.

1. The probability that a healthy child will contract a particular disease after exposure is 0.8. What is the probability of two children who have been exposed to it getting the disease?

2. The probability that a person is allergic to a new drug is 0.004. What is the probability that the first three persons for whom a doctor prescribes this drug will all be allergic?

Using the Calculator

Statistics about past events can sometimes be used to estimate the probability of future events if conditions remain unchanged. Recent statistics about child births in the United States show that for every 100 girls that are born 105 boys are born. Let $P(G, B)$ represent the probability that the first child is a girl and the second child is a boy.

What is the probability, rounded to the nearest thousandth?

1. $P(B)$ **2.** $P(G)$ **3.** $P(B, G)$ **4.** $P(B, B, B)$ **5.** $P(G, G, B, B)$

Self Test 1

The letters of the word "CALCULATOR" are each printed on a separate card. A game cube has faces numbered 1 through 6. Use these cards and this game cube for Exercises 1–11.

One card is selected at random. Write the probability as a ratio.

1. $P(C)$ **2.** $P(\text{vowel})$ **3.** $P(\text{not A})$ **4.** $P(\text{L or A})$ **12-1**

Draw a tree diagram to list the outcomes.

5. Toss a coin; then roll the game cube. **12-2**
6. Choose a card; then toss a coin.

How many outcomes are there for each set of activities?

7. Choose a card; then roll the game cube. **12-3**

8. Toss a coin twice; then roll the game cube.

Suppose you choose a card, and then toss a coin. What is the probability?

9. $P(\text{A, Heads})$ **10.** $P(\text{vowel, Tails})$ **11.** $P(\text{L or A, Tails})$ **12-4**

Self Test answers and Extra Practice are at the back of the book. **389**

12-5 Dependent Events

Mary and Marty are twins. The 25 members of their class have decided to draw names for a gift swap at a party. If Mary draws first and Marty second, what is the probability that each will draw the other's name?

Mary draws 1 name out of 25 and does not replace the name she has drawn. Then Marty draws 1 name out of 24. This is called *drawing without replacement.* The occurrence of the first event affects the occurrence of the second. Such events are called **dependent events.**

To calculate the probability of two dependent events occurring you first determine the probability of each event separately in the order in which it occurs. Then multiply the probabilities.

Example 1   What is P(Marty, then Mary)? ← *This means Marty on the first draw, and Mary on the second.*

Solution   On the first draw $P(\text{Marty}) = \dfrac{1}{25}$.

On the second draw $P(\text{Mary}) = \dfrac{1}{24}$.

$P(\text{Marty, then Mary}) = \dfrac{1}{25} \cdot \dfrac{1}{24} = \dfrac{1}{600}$

Your Turn   Two of the seven cubes shown are drawn at random, one after the other, without replacement. What is the probability?
  • P(R, then R)   • P(B, then R)

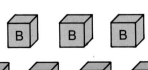

Class Exercises

A box contains 4 green, 5 red and 6 yellow marbles. If a green marble is drawn first and not replaced, what is the probability for the second draw?

1. P(green)   **2.** P(red)   **3.** P(yellow)

If a red marble is drawn first and not replaced, what is the probability for the second draw?

4. P(green)   **5.** P(red)   **6.** P(yellow)

If a yellow marble is drawn first and a green marble is drawn second without replacement, what is the probability for the third draw?

7. P(green)   **8.** P(red)   **9.** P(yellow)

Exercises

Events A and B are dependent events. The probability of A is given, as well as the probability of B after the occurrence of A. Complete the steps to determine $P(A,\ \text{then}\ B)$.

A   **1.** $P(A)$: $\frac{1}{4}$ **2.** $P(A)$: $\frac{1}{3}$ **3.** $P(A)$: 0.2

$P(B)$ after A: $\frac{2}{3}$ $P(B)$ after A: $\frac{6}{7}$ $P(B)$ after A: 0.1

$P(A,\ \text{then}\ B) = \frac{1}{4} \cdot \underline{\ ?\ }$ $P(A,\ \text{then}\ B) = \frac{1}{3} \cdot \underline{\ ?\ }$ $P(A,\ \text{then}\ B) = \underline{\ ?\ }$

A drawer contains 6 black socks, 4 blue socks, and 2 white socks. Two socks are drawn at random without replacement. If a black sock is drawn first, determine the probability for the second draw.

4. P(black)   **5.** P(blue)   **6.** P(white)

What is the probability of the given sequence of events?

B   **7.** P(black, then black)   **8.** P(black, then white)

9. P(white, then white)   **10.** P(blue, then blue)

11. P(blue, then black)   **12.** P(white, then blue)

Two students from the class executive committee are to be president and vice president. The committee includes 5 girls (Kim, Jennifer, Lisa, Maria, and Alice) and 3 boys (John, Charles, and Nathaniel). The names are placed in a container and two are drawn at random without replacement. The first person whose name is drawn is president and the second is vice president. Find the probability.

13. P(boy, then boy)

14. P(girl, then girl)

15. P(boy, then girl)

16. P(girl, then boy)

17. P(Kim, then Charles)

18. P(Lisa, then Nathaniel)

19. P(John, then Charles)

20. P(Maria, then Alice)

Twelve cards are numbered 1 to 12 and placed in a container. Two cards are drawn at random, one after the other, without replacement. What is the probability of each sequence of draws? Let m and n represent whole numbers from 1 to 12.

21. P(10, then 3)

22. $P(n$, then $2n)$

23. $P(n$, then $n + 1)$

C **24.** $P(2n$, then $2m)$

25. $P(n < 6$, then $m > 6)$

26. $P(2n$, then $2n + 1)$

Problems

Six-digit numbers are used for license plate numbers in some states. Suppose zero is never the first digit, and that license numbers are chosen at random. Solve the problem.

1. What is the probability that no two digits will be the same?
2. What is the probability that each digit is 1 more than the preceding digit?
3. What is the probability that all digits will be the same?
4. What is the probability that all digits will be even?
5. What is the probability that no two digits will be the same and that all digits will be even?
6. Some states use a combination of letters and numbers for license plates. Suppose license plates are to have a three-digit number followed by three letters and that plates are assigned at random. What is the probability that an individual will be assigned *100-WOW*?

12-6 Odds

A civic association is sponsoring a raffle to raise money for a children's summer sports camp. They plan to sell 2000 tickets to a local concert and will award door prizes by drawing names at random.

Summer Camp Raffle

1 First Prize: Bicycle
5 Second Prizes: Radio
10 Third Prizes: Calculator

If you buy 1 ticket, there is only 1 chance that you will win first prize and 1999 chances that you will not win first prize. We say that the **odds in favor** of your winning first prize are 1 to 1999, or $\frac{1}{1999}$. The **odds against** your winning first prize are 1999 to 1, or $\frac{1999}{1}$.

If for an event there are a favorable outcomes and b unfavorable outcomes, then the following is true.

Odds in favor: $\frac{a}{b}$ Odds against: $\frac{b}{a}$

Example 1   A total of 16 prizes will be awarded in all. If you buy only 1 ticket, what are the odds of winning a prize?

Solution   Number of favorable outcomes: 16
Number of unfavorable outcomes: $2000 - 16 = 1984$

Odds in favor of winning: $\frac{16}{1984}$, or $\frac{1}{124}$

Example 2 If you do not win the first prize, what are the odds against your winning another prize?

Solution Number of favorable outcomes: 15 ◄——— *5 second prizes + 10 third prizes*

Number of unfavorable outcomes: 1984

Odds against winning second or third prize: $\dfrac{1984}{15}$

Your Turn • After the first and second prizes have been awarded, what are the odds in favor of winning a third prize? What are the odds against winning a third prize? State the answers in lowest terms.

Class Exercises

There are 10 beads in a bag. Of these, 5 are blue, 3 are red, and 2 are green. You draw one bead at random. What are the odds in favor of each draw? What are the odds against each draw?

1. red bead **2.** green bead **3.** blue bead **4.** red or green bead

The odds in favor of an event are given. What are the odds against this event occurring?

5. 4 to 1 **6.** 3 to 10 **7.** 2 to 7 **8.** 4 to 3

Exercises

If you spin the spinner shown once, what are the odds in favor of each event? State the answers in lowest terms.

A **1.** blue **2.** red **3.** yellow

4. 3 **5.** 1 or 4 **6.** even number

7. blue or even number

8. blue and even number

9. yellow or odd number

10. yellow and odd number

What are the odds against the event?

11. blue   12. odd number   13. 1 or red   14. 1 and red

The odds in favor of an event are given. Write the odds against the event.

B 15. $\dfrac{2}{3}$ 16. $\dfrac{2}{1}$ 17. $\dfrac{4}{5}$ 18. $\dfrac{7}{10}$ 19. $\dfrac{1}{n}$ 20. $\dfrac{p}{q}$

The number given is the number of favorable outcomes. If there are 100 possible outcomes in all, what are the odds in favor of each event?

21. 30   22. 5   23. 80   24. 50   25. n 26. $2a$

The odds against four events are given in the table.

27. Which event is least likely to occur (has the worst odds)?

28. Which event is most likely to occur (has the best odds)?

Event	Odds Against
A	10 to 1
B	7 to 3
C	5 to 4
D	9 to 2

The number given is the probability that an event will occur. If the total number of possible outcomes is 120, what are the odds in favor of each event?

C 29. $\dfrac{4}{5}$ 30. $\dfrac{2}{3}$ 31. $\dfrac{1}{3}$ 32. $\dfrac{1}{5}$ 33. $\dfrac{1}{n}$ 34. $\dfrac{a}{b}$

Solve.

35. The odds in favor of an event are 3 to 5. The total number of outcomes is 10 more than the number of favorable outcomes. How many outcomes are there? What is the probability that the event will occur?

36. The odds against an event are 2 to 3. The total number of outcomes is 1 less than 3 times the number of unfavorable outcomes. How many outcomes are there? What is the probability that the event will occur?

12-7 Experimental Probability

The results of an experiment repeated many times can be used to get a good idea of the probability of an event. We call the result the **experimental probability** of the event.

Experimental Probability $= \dfrac{\text{Number of favorable outcomes}}{\text{Number of experiments}}$

Experimental Probability $\approx P(\text{event})$

You can approximate the probability of an event by finding the experimental probability. A greater number of experiments usually gives a closer approximation than a lesser number of experiments.

Example 1 Pedro and his 24 classmates each tossed a coin 20 times. Pedro got 8 Heads out of 20 tosses. Altogether, the entire class counted 251 Heads out of 500 tosses. Determine $P(\text{H})$ experimentally for Pedro and for the entire class.

Solution $P(\text{event}) \approx \dfrac{\text{Number of favorable outcomes}}{\text{Number of experiments}}$

Pedro: $P(\text{H}) \approx \dfrac{8}{20} = \dfrac{2}{5}$, or 0.4

Class: $P(\text{H}) \approx \dfrac{251}{500}$, or 0.502

Your Turn Suppose you spin a spinner 300 times. The spinner stops on red 200 times, on blue 40 times, and on yellow 60 times. What is the approximate probability?
 • $P(\text{red})$ • $P(\text{blue})$ • $P(\text{yellow})$

Example 1 illustrates an important property of experimental probability. You know that if a coin is "fair", then $P(H) = \dfrac{1}{2}$, or 0.5. In Example 1 the greater number of experiments (tosses) gave the closer approximation to 0.5. Try tossing a coin yourself 10 times, then 100 times to approximate $P(H)$. Which set of experiments gives you the better approximation?

Statistics about past events are often considered as experiments to approximate the probability of a future event. A baseball player's batting average is based on statistics of past performances. It is often used to approximate the probability of a hit the next time at bat.

Example 2 A baseball player hit the ball 48 times out of the last 150 times at bat. What is the probability that that player will hit the ball next time at bat?

Solution $P(\text{hit}) \approx \dfrac{\text{Number of hits}}{\text{Number of times at bat}}$ ⟵ *This ratio is called the "batting average."*

$P(\text{hit}) \approx \dfrac{48}{150}$, or 0.320

Your Turn • A baseball player hit the ball 72 times out of 320 times at bat. What is the approximate probability that that player will hit the ball next time at bat?

Class Exercises

If a bottle cap is tossed in the air, it will land either "up" or "down." Use the results of these experiments to determine $P(\text{up})$ and $P(\text{down})$.

1. 50 tosses
   Up: 33
   Down: 17

2. 100 tosses
   Up: 67
   Down: __?__

3. 200 tosses
   Up: 130
   Down: __?__

Exercises

A game cube is weighted so that not all outcomes are equally likely. The faces of the cube are each painted a different color. The cube is rolled 200 times and the results recorded in the table. About what is the probability of each event?

Red	Blue	White	Green	Yellow	Orange
21	30	43	76	15	15

A **1.** $P(\text{red})$ **2.** $P(\text{blue})$ **3.** $P(\text{white})$ **4.** $P(\text{green})$ **5.** $P(\text{orange})$

When tossed, a paper cup can land up, down, or on its side. Four friends experimented to determine the probability of each event. Use their results for Exercises 6–20.

	Up	Down	Side
Pam	3	5	8
Max	4	6	14
Sue	2	3	7
Lee	6	9	15

What is the approximate probability based on each person's experiments?

B **6.** Pam, $P(\text{up})$ **7.** Max, $P(\text{up})$ **8.** Sue, $P(\text{up})$ **9.** Lee, $P(\text{up})$

10. Pam, $P(\text{down})$ **11.** Max, $P(\text{down})$ **12.** Sue, $P(\text{down})$ **13.** Lee, $P(\text{down})$

14. Pam, $P(\text{side})$ **15.** Max, $P(\text{side})$ **16.** Sue, $P(\text{side})$ **17.** Lee, $P(\text{side})$

What is the approximate probability of each event based on the total of all the experiments?

18. $P(\text{up})$ **19.** $P(\text{down})$ **20.** $P(\text{side})$

Problems

During the first two baseball games of the season, a player hit the ball 4 times out of 10 times at bat. During the entire season, that same player hit the ball 157 times out of 600 times at bat. Solve.

1. What is the batting average for that player for the first two games?
2. What is the batting average for that player for the season?
3. Which average is a better approximation of $P(\text{hit})$ for that player? Why?

12-8 *Using Samples*

You can approximate the probability of an event by examining a **random sample.** A sample is a relatively small set of outcomes which we analyze and from which we draw conclusions.

Characteristics of a random sample:
- Each element of the sample space is as likely to be examined as any other element of the sample space.
- The sample is large enough to reflect the variety of the sample space.

Example 1 Heidi closed her eyes and opened her Math book at random 10 times. She found photos on 5 of the 20 pages. About what is the probability of finding a photo on a given page?

Solution $P(\text{photo}) \approx \dfrac{5}{20}$, or 0.25

Your Turn • Suppose that out of 20 pages of this book, chosen at random, you find photos on 7 pages. Based on your sample, about what is the probability of finding a photo on any given page?
• John used his computer to obtain a random set of 50 page numbers of his textbook. He found photos on 10 of the 50 pages. Based on John's sample, about what is the probability of finding a photo on a particular page?

The way you choose a sample is very important. Suppose you looked only at the first page of each chapter of this book. You would find a photo on each page. You might then falsely conclude that $P(\text{photo})$ is $\frac{14}{14}$, or 100%. The first page of each chapter is not a random sample.

We often use a random sample to draw conclusions about the entire sample space. Surveys and opinion polls are frequently conducted for this purpose.

Example 2　A new school district conducted a survey to determine the needs of its students for bike racks, bus service, and parking. The question "How do you plan to get to and from school?" was asked of 200 students. Here are the results.

Walk	55	Ride a bus	85
Ride in a car	40	Ride a bike	15
Drive a car	5		

The school district has 10,000 students. Approximately how many students plan to drive a car?

Solution　Based on the survey, the probability that a student will drive is about $\dfrac{5}{200}$, or 0.025.

$$0.025 \times 10,000 = 250$$

About 250 students will probably drive.

Your Turn　Approximately how many students plan to use each method of transportation to and from school?
- Walk
- Bus
- Bike

Class Exercises

A box contains 1000 marbles. A random sample of 20 marbles contains 10 red, 2 blue, 5 yellow, and 3 green marbles. Complete to approximate the number of each color marble.

1. Number of red marbles
$P(\text{red}) \cdot 1000 = \underline{\ ?\ } \cdot 1000$
$= \underline{\ ?\ }$

2. Number of blue marbles
$P(\text{blue}) \cdot 1000 = \underline{\ ?\ } \cdot 1000$
$= \underline{\ ?\ }$

Exercises

Opinions on 4 brands of toothpaste were obtained from a random sample of toothpaste users. The results are shown in the table. Based on the poll, what is the probability that an individual will prefer the brand specified?

Brand	Number Favoring
A	240
B	320
Y	160
Z	80
Total	800

A   **1.** Brand A   **2.** Brand Z   **3.** Brand B or Y

The Midvale High School Student Council asked 50 randomly chosen students to vote on the name of the school mascot. The results are shown in the table. Based on the poll, what is the probability that a student will favor the name specified?

Tigers	15
Cougars	21
Bears	6
Muskrats	8

B   **4.** Tigers   **5.** Cougars   **6.** Bears or Muskrats

Use the results of the Midvale Student Council survey to estimate how many of the 1200 Midvale students favor the given mascot name.

     7. Tigers   **8.** Cougars   **9.** Bears or Muskrats

The Committee to Re-elect the Mayor randomly surveyed 60 of the 4000 Hardyville eligible voters. Of the 60, 28 favored re-electing the mayor, 20 favored Opponent A, and 12 favored Opponent B. About how many eligible voters favor each candidate?

     10. Mayor   **11.** Opponent A   **12.** Opponent B

Solve. Assume that each sample is random.

13. When 50 ninth graders were questioned, 18 said they participate in a sport. About how many of the 450 ninth graders participate in a sport?

14. A survey indicated that 75% of Midville High School students plan to attend college. How many of the 120 students interviewed said they planned to attend college?

15. A grower chose 30 trees in an orchard of 1450 trees and found that fruit on 24 trees was ripe enough to pick. About how many trees in the orchard are ripe enough to pick?

401

Using the Computer

You can use something called the random number function to simulate a coin tossing experiment on a computer. The following program may need to be adjusted for use on your computer to provide different results on every RUN. RUN the program several times for N = 20.

```
  5  REM: SIMULATING TOSSING A COIN
 10  PRINT "HOW MANY TOSSES";
 20  INPUT N
 30  FOR I=1 TO N
 40  LET A=RND(1)  ←——— To get different tosses on every RUN,
 50  IF A<.5 THEN 90     you may need to alter line 40. Check the
 60  LET T=T+1           manual for your computer.
 70  PRINT "  T"
 80  GOTO 110
 90  LET H=H+1
100  PRINT "H"
110  NEXT I
120  PRINT "H =";H,"T =";T,"H/N =";H/N
130  END
```

Self Test 2

A seed package contains seeds for 6 red flowers, 6 blue flowers, 8 white flowers, and 4 yellow flowers. What is the probability for the first two seeds planted?

1. P(red, then red)   **2.** P(red, then blue)   **3.** P(red, then white)   **12-5**

What are the odds in favor of the event for the first flower planted?

4. red   **5.** red or white   **6.** red, white, or blue   **12-6**

A survey of 200 people showed that 60 read the paper only on Sunday and 40 read the paper every day. Approximate the probability.

7. P(read paper only on Sunday)   **8.** P(read paper every day)   **12-7**

The newspaper survey above was conducted in an area with about 12,000 people. About how many people in the area read the paper?

9. only on Sunday   **10.** every day   **12-8**

Algebra Practice

Write an equation. Solve.

1. The product of 304 and 465 is b.

2. 327,012 divided by 916 is v.

3. The sum of 5.84, 9.473, and 6.958 is j.

4. 0.208 times 0.41 is equal to q.

5. The difference between $20\frac{3}{8}$ and $12\frac{5}{6}$ is h.

6. $8\frac{1}{3}$ divided by $\frac{5}{9}$ is equal to c.

7. The perimeter of a rectangle 6.2 m long and 5.9 m wide is P.

8. The volume of a cylinder with a radius of 6 ft and a height of 20 ft is V.

Use the distributive property to factor each expression.

9. $48x^3 - 42x^2$ 10. $5fg + f^2g - f^2g^2$ 11. $2r^3 - 6r^2s + 4rs^3$

Solve.

12. $0 = \dfrac{m}{3}$ 13. $4 - 8r = 20$ 14. $b = 12 - 2b$

15. 4% of 96 is j 16. 300% of 21 is w 17. 13 is $16\frac{2}{3}$% of t

18. $^-3h + 5 = ^-4$ 19. $4 - \dfrac{3}{2}k > ^-5$ 20. $7d + 3 < ^-6$

Find three ordered pairs that are solutions of the given equation.

21. $y = x - 6$ 22. $y = 3x + 2$ 23. $y = 8 - 5x$

Write an equation and solve the problem.

24. The probability of snow today is 30%.

   The probability of snow is $\frac{2}{3}$ the probability of rain.

   What is the probability of rain today?

25. Last month the Lees spent $58.42 on clothing.
   They spent $18.79 less on clothing than on gasoline.
   How much did they spend on gasoline?

Problem Solving for the Consumer

Polls and Surveys

Many ads claim that surveys have been made and that people prefer a particular product. All surveys are not necessarily good ones. Consumers should be aware of what makes a survey good and what makes a survey poor or even worthless. Here's a typical ad.

"Four out of five doctors recommend Clean Gel for deep cleaning oily skin."

There are two ways the survey could have been conducted.

<table>
<tr><td align="center">One Survey</td><td align="center">Another Survey</td></tr>
<tr><td>Doctors surveyed: 10</td><td>Doctors surveyed: 200</td></tr>
<tr><td>Doctors recommending Clean Gel: 8</td><td>Doctors recommending Clean Gel: 160</td></tr>
</table>

Study the two possible surveys.

1. In each survey, what is the ratio of doctors recommending Clean Gel to those not recommending it?
2. Which survey stands the greater chance of representing the opinion of doctors in general?
3. If 1000 doctors had been surveyed and the results of the survey had been the same, how many doctors would have recommended Clean Gel?

Let's consider an opinion poll.

> Issue: Whether or not members of the senior class at Stamford High School should be given any special privileges in buying school play or school concert tickets.

Here are results of five different polls.

	Description of Group Polled	No. of Persons Polled	"Yes" Votes	"No" Votes	"Maybe" Votes
Poll 1	members, senior class	90	82	3	5
Poll 2	members, freshman class	85	0	80	5
Poll 3	members, sophomore class	82	0	78	4
Poll 4	members, junior class	86	6	78	2
Poll 5	randomly selected students	95	24	66	5

4. If you want a poll to represent the general opinion of the school body, which poll does that best?

5. Suppose all five polls are conducted. Which will be more reliable to use, the result of Poll 5, or the sum of the results from Polls 1–4?

6. If 50 more seniors are asked their opinion, about how many of them will most likely vote "yes"?

Barnes, a candidate for mayor, comes from the south part of town. His campaign workers conducted two independent surveys. In the first survey, 120 people from the south part of town were polled. A second survey selected 150 people at random from the telephone book.

7. If you were Barnes, which survey would you trust more?

The results of the poll of 150 people selected at random from the telephone book are given in the table. Twelve people are undecided.

8. Set up another table to show the results of this political poll. In the last column of your table show the percent of people polled who said they were going to vote for the candidate.

9. Make a circle graph to represent the results of the poll.

Candidate	Number Favoring
Barnes	45
Lipinsky	48
Nathan	33
Stowe	12

Enrichment

Logical Thinking

Let's look at three key statements from a sports broadcast.

> 1: "That's right, folks, if Jones steals only one more base, then he will hold the new world record for stolen bases."
> 2: "He's done it! Al Jones has just stolen another base."
> 3: "Al Jones holds the new world record for stolen bases."

A statement in the form "If . . ., then . . ." is called a **conditional statement.** Sometimes the word "then" is understood. A conditional statement has two parts.

"If Jones steals one more base, (then) he will break the record."

         Condition   Conclusion

The three statements above are in a special form called a **syllogism.** Syllogisms are often used in logical thinking.

    Syllogism
1: If A, then B.   ⟵   *A conditional statement*
2: A ⟵   *States that the condition is true*
3: Therefore, B.   ⟵   *Draws the conclusion*

Complete the syllogism.

1. If it is snowing, then it is cold. It is snowing. Therefore, __?__ .

2. If $a + 6 < b + 6$, then $a < b$. $a + 6 < b + 6$ Therefore, __?__ .

3. If a number has only two factors, itself and 1, then it is a prime number. The number 13 has only two factors, 13 and 1. Therefore, __?__ .

4. If riddles are fun, then we'll talk in riddles. Riddles are fun. Therefore, __?__ .

5. If a number is a multiple of 12, it is also a multiple of 4. 48 is a multiple of 12. Therefore, __?__ .

6. If two triangles are similar, corresponding sides are proportional. $\triangle ABC \sim \triangle XYZ$. Therefore, __?__ .

7. If you win the race, you will receive $1000. If you win the race, you will be on TV. If you have money and you are on TV, you will be a celebrity. You win the race. Therefore, __?__ .

Chapter Review

A bus has 30 orange seats and 20 green seats. If passengers choose seats at random, what is the probability for the first passenger?

1. P(orange)
2. P(green)
3. P(red)

4. What are the odds that the first passenger will choose a green seat?

Suppose the first passenger gets off before the second passenger gets on. What is the probability that the first two passengers will select the seat colors in the order specified?

5. P(orange, orange)
6. P(green, orange)
7. P(green, green)
8. P(orange, green)

If the second passenger gets on before the first passenger gets off, what is the probability?

9. P(orange, then green)
10. P(green, then green)
11. P(orange, then orange)
12. P(green, then orange)

Jim, Joan and Jane are the first three passengers on the bus.

13. Draw a tree diagram to show all the possible combinations of passenger and seat color.
14. How many outcomes are there in the sample space?

A random sample of 500 passengers shows that 300 use tokens and 200 pay with exact change. For any given passenger, what is the probability?

15. P(token)
16. P(exact change)

17. If a bus driver counts 1200 passengers a day, about how many pay with a token?
18. What are the odds that a passenger will use exact change?

One-half of the green seats and one-half of the orange seats are next to a window. The other seats are on the aisle.

19. What is the probability that the first passenger will choose an orange seat next to a window?
20. What is the probability that the first two passengers will each choose a green seat on the aisle?

Chapter Test

A game cube has faces numbered 10, 20, 30, 40, 50, 60. Each letter of MATH is printed on one of four cards. Use this game cube, these cards, and the spinner shown to do the exercises on this page.

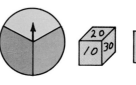

The game cube is rolled once. What is the probability, written as a ratio?

1. $P(10)$ **2.** P(even number)   **3.** P(odd number)   **12-1**

Draw a tree diagram to list the outcomes.

4. Choose a card; then spin the spinner.
5. Spin the spinner; then toss a coin.   **12-2**

How many outcomes are there for each set of activities?

6. Roll the game cube; then spin the spinner.
7. Spin the spinner twice; then choose a card.   **12-3**

If you roll the game cube; then choose a card, what is the probability?

8. $P(10, M)$ **9.** $P(10$ or $20, A)$ **10.** $P(60,$ consonant)   **12-4**

Suppose you choose two cards, one after the other, without replacement. What is the probability?

11. $P(M,$ then $A)$ **12.** $P(A,$ then T or H)   **12-5**

What are the odds for each event?

13. Odds in favor of choosing the card marked A   **12-6**

14. Odds against a multiple of 3 on the game cube

You spin the spinner 12 times and it stops on red 6 times, on blue 4 times, and on yellow 2 times. What is the experimental probability?

15. P(red)   **16.** P(blue)   **17.** P(yellow)   **12-7**

A survey of 120 voters shows that 60 afe in favor of a new mall, 40 oppose it, and 20 have no opinion. What is the probability?

18. P(favor)   **19.** P(oppose)   **20.** P(no opinion)   **12-8**

21. If there are 10,000 voters, about how many can be expected to vote in favor of the new mall?   **12-9**

Cumulative Review (Chapters 1–12)

Add, subtract, multiply, or divide.

1. $4\dfrac{3}{4} - \dfrac{1}{3}$

2. $6\dfrac{7}{8} - 1\dfrac{3}{5}$

3. $2\dfrac{1}{9} - \dfrac{3}{5}$

4. $14.06 - 0.32$

5. $24.9 \div 0.3$

6. 6.7×10^3

7. $63.92 \div 0.02$

8. $^-6 \cdot {}^-27$

9. $^-1 - {}^-18$

10. $^-12 \div 4$

11. $^-35 \div {}^-5$

12. $28 - {}^-9$

Find the measure of the missing angle in triangle ABC.

13. $\angle A = 27°$, $\angle B = 41°$

14. $\angle C = 18°$, $\angle B = 45°$

15. $\angle C = 90°$, $\angle A = 40°$

16. $\angle A = 104°$, $\angle C = 12°$

17. $\angle C = 25°$, $\angle A = 56°$

18. $\angle B = 80°$, $\angle C = 38°$

Find the area.

19. rectangle
    $b = 12.0$ cm
    $h = 2.7$ cm

20. triangle
    $b = 14$ in.
    $h = 3$ in.

21. square
    $s = 4.2$ m

Solve.

22. An \$80 pair of ice skates is on sale at 30% off. How much will the skates cost on sale?

23. A child's plastic swimming pool is a cylinder in shape. It is 6 ft in diameter and 1 ft deep. What is its volume?

24. A certain subway car used to be able to carry 80 passengers. When the seats were rearranged, there was a 15% increase in the number of passengers the car could carry. How many passengers could the subway car then carry?

Write an equation. Solve the problem.

25. Emily plans to drive 725 mi to visit a relative. If she drives 420 mi the first day, how far will she have to drive the second day?

26. The sum of two numbers is 693. One number is twice as great as the other. What are the two numbers?

27. The Keep Our City Green Club raised money to plant trees and shrubs. Team A raised \$160 more than Team B. Team C raised \$70 less than Team B. Together the teams raised \$1860. How much money did each team raise?

13
Geometry

The art of quiltmaking often relies on the use of geometric shapes. Look carefully and you'll see squares, triangles, and windmills of various sizes hiding in the myriad shapes.

13-1 Points, Lines, and Planes

Did you know that the photos you see reproduced in books, magazines, and newspapers are made up of thousands of closely packed dots? The dots are so small and so close together that they can only be seen when the photo is enlarged.

In geometry, dots and capital letters are used to represent **points.** **Lines** are made up of an infinite number of **collinear points** (points in a line) and have no starting or ending point. We often work with parts of lines that do have endpoints; these parts are called line **segments.**

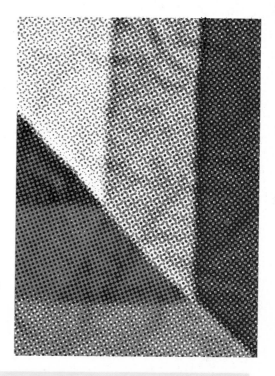

Point *A*	Line *AB*	Segment *AB*
•A	A B	A B
A	$\overleftrightarrow{AB}$	$\overline{AB}$

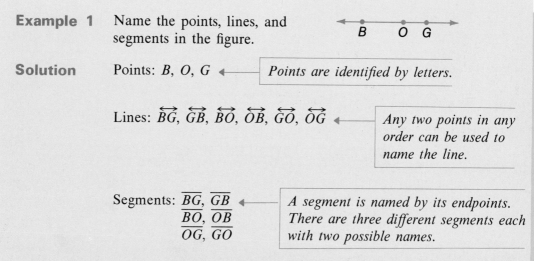

Example 1   Name the points, lines, and segments in the figure.

B   O   G

Solution   Points: *B, O, G* ← *Points are identified by letters.*

Lines: $\overleftrightarrow{BG}, \overleftrightarrow{GB}, \overleftrightarrow{BO}, \overleftrightarrow{OB}, \overleftrightarrow{GO}, \overleftrightarrow{OG}$ ← *Any two points in any order can be used to name the line.*

Segments: $\overline{BG}, \overline{GB}$ ← *A segment is named by its endpoints. There are three different segments each with two possible names.*
$\overline{BO}, \overline{OB}$
$\overline{OG}, \overline{GO}$

411

A **plane** is a set of points that extends forever in all directions. We usually deal with parts of a plane as suggested by a sheet of paper or pane of glass. An actual plane does not have sides, but is often represented by a four-sided figure.

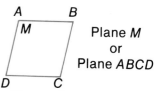

Plane *M*
or
Plane *ABCD*

When two planes intersect, their intersection is a line. If two planes do not intersect they are parallel.

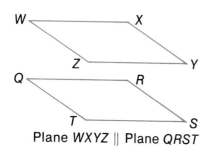

Plane *WXYZ* ∥ Plane *QRST*

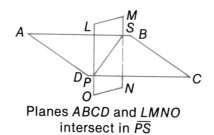

Planes *ABCD* and *LMNO*
intersect in $\overleftrightarrow{PS}$

Example 2   Name the parallel planes in the figure.

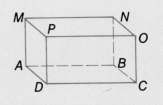

Solution   Plane *MNOP* ∥ Plane *ABCD*
Plane *MPDA* ∥ Plane *NOCB*
Plane *MNBA* ∥ Plane *POCD*

Your Turn   Use the figure in Example 2 to complete.
• Plane *MNBA* intersects plane *NOCB* in line ___?___.

For lines to be parallel, they must be in the same plane. Look at lines $\overleftrightarrow{AB}$ and $\overleftrightarrow{CD}$. They will never intersect but they are not parallel because they are not in the same plane. $\overleftrightarrow{AB}$ and $\overleftrightarrow{CD}$ are called **skew** lines.

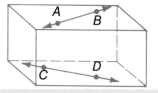

Example 3   Write skew or parallel to describe the lines.
$\overleftrightarrow{TY}$ and $\overleftrightarrow{QN}$ $\overleftrightarrow{LT}$ and $\overleftrightarrow{QN}$

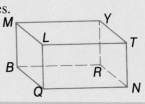

Solution $\overleftrightarrow{TY}$ and $\overleftrightarrow{QN}$ are skew.
$\overleftrightarrow{LT}$ and $\overleftrightarrow{QN}$ are parallel.

Your Turn   Use the figure in Example 3. Write skew or parallel to describe the lines.

• $\overleftrightarrow{BR}$ and $\overleftrightarrow{QN}$ • $\overleftrightarrow{MB}$ and $\overleftrightarrow{LT}$

A **transversal** is a line that intersects two lines in two different points. If the two lines are parallel, then the transversal forms several pairs of equal angles. Angles 3 and 5, called **alternate interior** angles, are equal as are angles 4 and 6. Angles 2 and 6, called **corresponding angles,** are equal. Other pairs of equal corresponding angles are angles 1 and 5, 4 and 8, and 3 and 7.

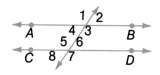

Example 4   In the figure, $\overleftrightarrow{LM} \parallel \overleftrightarrow{NO}$, and $\angle 4 = 140°$. What are the measures of angles 8, 2, and 6?

Solution

$\angle 8 = 140° \longleftarrow$ *Angles 4 and 8 are equal since they are corresponding angles.*

$\angle 2 = 140° \longleftarrow$ *Angles 8 and 2 are equal since they are alternate interior angles.*

$\angle 6 = 140° \longleftarrow$ *Angles 6 and 2 are equal since they are corresponding angles.*

Your Turn   • Use the figure in Example 4. Suppose $\angle 1 = 50°$. What are the measures of angles 3, 7, and 5?

Class Exercises

1. Give three examples of models that could represent each of the following: point, line, segment, plane.

Use the figure on the right.

2. Name five different segments in the figure.
3. Name three different planes.
4. Name a pair of parallel lines.
5. Name a pair of skew lines.

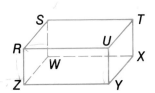

Exercises

True or false? Write *T* or *F*.
For Exercises 1–5 use the figure on the right. Assume that $\overleftrightarrow{AB} \parallel \overleftrightarrow{CD}$.

A 1. Angles 1 and 5 are alternate interior angles.
  2. Angles 6 and 13 are corresponding angles.
  3. $\overleftrightarrow{AB}$ is a transversal.
  4. Angles 10 and 14 are equal.
  5. $\angle 5$ measures 110° so $\angle 1$ measures 110°.

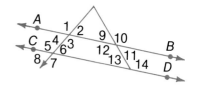

For Exercises 6–10 use the figure on the right.

  6. Plane *ABCD*‖Plane *EFGH*.
  7. Planes *ABCD* and *ADHE* intersect in $\overleftrightarrow{EH}$.
  8. Lines $\overleftrightarrow{AD}$ and $\overleftrightarrow{BC}$ are skew.
  9. Planes *BCGF* and *ABFE* intersect in $\overleftrightarrow{BF}$.
  10. $\angle ADC = \angle EHG$.

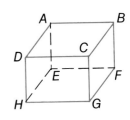

For Exercises 11–18 use the figure on the right. $\overleftrightarrow{WX} \parallel \overleftrightarrow{YZ}$.
Give the special names for the following pairs of angles.

B 11. $\angle 3$ and $\angle 9$ **12.** $\angle 6$ and $\angle 16$
  13. $\angle 5$ and $\angle 13$ **14.** $\angle 7$ and $\angle 15$
  15. Name four pairs of alternate interior angles.
  16. Name eight pairs of corresponding angles.
  17. $\angle 8 = 125°$ so $\angle 16 = \underline{\ ?\ }$.
  18. $\angle 2 = 70°$ so $\angle 12 = \underline{\ ?\ }$.

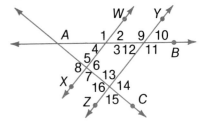

Use the figure at the right for Exercises 19–24. If two lines are cut by a transversal and the resulting corresponding angles are equal, then the lines are parallel. If two lines are cut by a transversal and the resulting alternate interior angles are equal, then the lines are parallel.

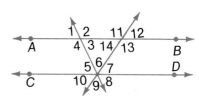

Write *Yes* or *No* to tell if the information given is enough to prove that $\overleftrightarrow{AB} \parallel \overleftrightarrow{CD}$.

C 19. $\angle 14 = \angle 7$ **20.** $\angle 1 = \angle 5$ **21.** $\angle 5 = \angle 8$
  22. $\angle 12 = \angle 14$ **23.** $\angle 10 = \angle 7$ **24.** $\angle 11 = \angle 5 + \angle 6$

13-2 Polygons

A **polygon** is a closed, plane figure formed when three or more line segments are connected at their endpoints. Polygons can be concave or convex. To decide if a polygon is convex you can think of stretching an elastic band around its perimeter. If the band fits snugly, the polygon is convex. Our discussion will focus on convex polygons.

Convex Polygon   Concave Polygon

A polygon is classified by the number of its sides. The table shows some of the polygons with which you are familiar.

Name	Number of sides
Triangle	3
Quadrilateral	4
Pentagon	5
Hexagon	6
Octagon	8

A polygon in which all the sides are the same length and all the angles are equal is called a **regular polygon.**

Regular Polygons

A **vertex** is a point where two adjacent sides of a polygon intersect. When referring to a polygon, we name the vertices (plural of vertex) in order. The polygon shown could be called *LMNO* but not *OMLN*. A **diagonal** is a segment that joins two non-consecutive vertices.

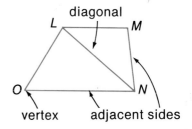

diagonal

vertex   adjacent sides

Example 1 Write *Yes* or *No* to tell whether the segment is a diagonal of polygon *ABCD*.

1. $\overline{AC}$
2. $\overline{AB}$
3. $\overline{CD}$
4. $\overline{DB}$

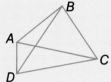

Solution   1. Yes   2. No   3. No   4. Yes

If you could arrange the angles of a triangle so they were side-by-side, you would see they form a straight angle which measures 180°. This is true of any triangle, no matter what its size. You can draw a diagonal in a quadrilateral to form two triangles. Since the sum of the angles in a triangle is 180°, you can see that the sum of the angles in a quadrilateral is twice as large or 360°. A pentagon can be divided into three triangles. So the sum of the angles of a pentagon is 3 × 180° = 540°.

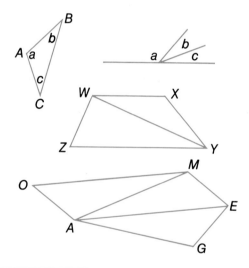

In general, the sum of the angles of a polygon with *n* sides is 180°(*n* − 2).

Example 2 What is the sum of the angles of the polygon?

Solution   Let *s* = the sum of the angles
          Then *s* = 180°(*n* − 2) ⟵—— *Since the polygon has*
              = 180°(6 − 2)   *6 sides,* n *= 6.*
              = 180°(4)
              = 720°
          The sum of the angles is 720°.

Class Exercises

Is the figure a polygon? Write *Yes* or *No*. If *Yes* is it a regular polygon?

1.   **2.**   **3.**

Exercises

True or false? Write *T* or *F*.

A **1.** A polygon has six sides.
    2. A diagonal connects consecutive vertices in a convex polygon.
    3. A triangle has two possible diagonals.
    4. In a regular polygon all the sides are of equal length.
    5. The sum of the angles of a triangle is 180°.

Use the figure on the right for Exercises 6–9.

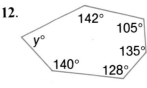

B **6.** Is figure *LAUNCH* a hexagon?
    7. Do the letters *CHLNUA* describe the figure?
    8. How many diagonals can be drawn from vertex *C*? from vertex *U*?
    9. Are *C* and *N* consecutive vertices? *A* and *N*?

For Exercises 10–12 find the sum of the angles in the polygon and the measure of $\angle y$.

10.
$y°$ 104°
102°   97°
124°

11.
112° $y°$
96°   82°

12.
142°   105°
$y°$ 135°
140°   128°

For Exercises 13–15 find the measure of each angle of the polygon.

C **13.**
B
$2x$ $3x+20$
A
100° x
D C

14.
A B
$21x$ $30x$ $29x$ C
F $31x$ $13x$
$20x$ D
E

15.
108° B
A $7x+4$
E
$18x+6$ 90°
D C
$17x-4$

13-3 *Similar, Congruent Polygons*

You know from Chapter 11 that if two triangles are similar, their corresponding angles are equal and their corresponding sides are in proportion. They are the same shape but not necessarily the same size. This relationship is true for all similar polygons. We use the symbol ∼ to mean "is similar to."

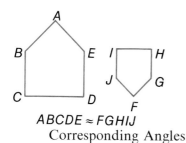

Just think of FGHIJ *as being upside down. Imagine flipping it over. Then it's easier to find the corresponding sides and angles.*

ABCDE ≈ FGHIJ

Corresponding Angles		Corresponding Sides	
∠A and ∠F	∠D and ∠I	$\overline{AB}$ and $\overline{FG}$	$\overline{DE}$ and $\overline{IJ}$
∠B and ∠G	∠E and ∠J	$\overline{BC}$ and $\overline{GH}$	$\overline{EA}$ and $\overline{JF}$
∠C and ∠H		$\overline{CD}$ and $\overline{HI}$	

Since the sides of similar polygons are in proportion, you can solve a proportion to find the length of a missing side.

Example 1   QRSTU ∼ ABCDE
Find the length of $\overline{AE}$.

Solution

$$\frac{QR}{AB} = \frac{QU}{AE}$$ ← *When you want to refer to the length of a segment, just write the letters without a bar above them.*

$$\frac{16}{8} = \frac{12}{n}$$

$$(12)(8) = 16n$$
$$96 = 16n$$
$$6 = n$$ ← $\overline{AE}$ *is 6 units long.*

Your Turn • $\triangle ABC \sim \triangle XYZ$. Find the length of side AC.

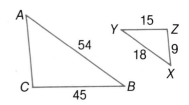

Two polygons are **congruent** if they have the same size and the same shape. We use the symbol $\cong$ to mean "is congruent to." In congruent polygons, corresponding sides are equal and corresponding angles are equal. To show congruence, corresponding parts are marked in the same way. In referring to congruent polygons it is important to pair corresponding sides in order and to list corresponding vertices in order. For the polygons on the right, the corresponding angles and sides are listed below.

Corresponding Angles   Corresponding Sides

$$\angle A = \angle W \qquad \overline{AB} \cong \overline{WX}$$
$$\angle B = \angle X \qquad \overline{BC} \cong \overline{XY}$$
$$\angle C = \angle Y \qquad \overline{CD} \cong \overline{YZ}$$
$$\angle D = \angle Z \qquad \overline{DA} \cong \overline{ZW}$$

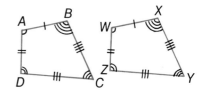

To show that two polygons are congruent, you must show that corresponding parts are equal. When the polygons in question are triangles, it is only necessary to show that certain parts are equal to prove that the triangles are congruent.

Side-Side-Side (SSS)

If the three sides of one triangle are equal to the corresponding sides in another triangle, the triangles are congruent.

Side-Angle-Side (SAS)

If two sides and the angle between them in one triangle are equal to the corresponding parts in another triangle, the triangles are congruent.

Angle-Side-Angle (ASA)

If two angles and the side included between them in one triangle are equal to the corresponding parts in another triangle, the triangles are congruent.

If two triangles share a common segment, the segment is congruent to itself.

Example 2 Are triangles *MAD* and *HDA* congruent? Give a reason for your answer.

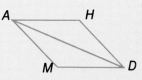

Solution
$$\angle 1 = \angle 4$$
$$\angle 2 = \angle 3$$
$$AD = AD \longleftarrow$$ | *Notice that* $\overline{AD}$ *is included between angles 1 and 3, and 2 and 4.*

The triangles are congruent by ASA $\longleftarrow$ | *Angle-side-angle*

Class Exercises

Complete. Assume *MNOP* ~ *EFGH*.

1. $\dfrac{MN}{EF} = \dfrac{MP}{\underline{\ ?\ }}$ 2. $\dfrac{PM}{HE} = \dfrac{\underline{\ ?\ }}{GF}$

3. $\dfrac{\underline{\ ?\ }}{PM} = \dfrac{GH}{HE}$ 4. $\dfrac{MP}{\underline{\ ?\ }} = \dfrac{EH}{HG}$

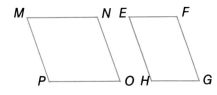

Complete. Assume $\triangle ABC \cong \triangle XYZ$.

5. $AC = \underline{\ ?\ }$ 6. $\angle B = \underline{\ ?\ }$ 7. $XY = \underline{\ ?\ }$

8. $BC = 15$, so $YZ = \underline{\ ?\ }$

9. $\angle C = 85°$ so $\angle \underline{\ ?\ } = 85°$

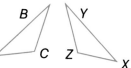

Exercises

Write *Yes* or *No* to tell if the two polygons are similar, and give a reason for your answer.

A 1.

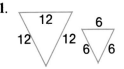

2.

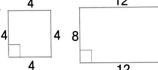

3.

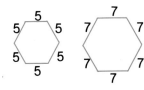

The figure shown is made up of congruent polygons. Write *Yes* or *No* to tell if the statements correctly name congruent polygons.

4. $ABKL \cong EFGH$ 5. $BMK \cong EHM$

6. $ABMKL \cong FEMHG$ 7. $EFGH \cong KHIJ$

8. $BEM \cong KHM$ 9. $KEDC \cong KEIJ$

10. $AFGL \cong CDIJ$ 11. $CDEB \cong EFHG$

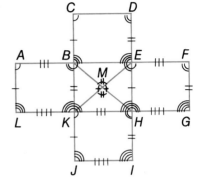

In Exercises 12-14 the pairs of polygons are similar. Find the lengths of the missing sides.

12.

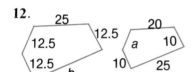

13.

14.

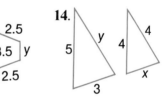

In Exercises 15-17 the pairs of polygons are congruent. List the corresponding sides and angles.

B 15.

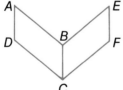

16.

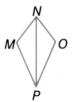

17.

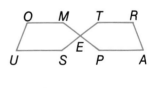

Are the triangles congruent? Give a reason for your answer.

18.

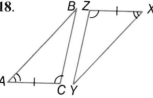

19.

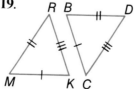

20.

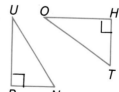

13-4 Problem Solving

Many problems are difficult to solve because the words are unfamiliar, the numbers are intimidating, or because they are long and therefore just look hard.

It sometimes helps to rewrite a problem in your own words.

Example 1   On the average, the earth's temperature increases 25°C for every kilometer of depth below the earth's surface. If an oil well were drilled to a depth of 9160 m, about how much hotter would you expect the temperature at the bottom to be than at the surface?

Solution   Rewrite the problem in your own words. Then change the measures of depth to the same units.
Simpler Problem—An oil well is 9.16 km deep. The temperature in it rises 25°C per 1 km of depth.
    How much hotter is it at the bottom of the well?
Computation—25° × 9.16 = 229°

You might also choose to reorganize the information to make it easier to use.

Example 2   Four of the elements of which the human body is made up are present in the following amounts: oxygen—65%, carbon—18%, hydrogen—10%, nitrogen—3%. What would the weight be of each in a 178 lb person?

Solution   Reorganize the information to make it easier to use.

Element	Oxygen	Carbon	Hydrogen	Nitrogen
Percentage	65	18	10	3
Decimal	0.65	0.18	0.10	0.03
178 lb × %	115.7 lb	32.0 lb	17.8 lb	5.3 lb

Your Turn • Some elements are present in the body in very small amounts. They include potassium—0.35%, sulfur—0.25%, sodium—0.15%, chlorine—0.15%, magnesium—0.05% and iron—0.004%. What would be the weight of each element in a 122 lb person?

Class Exercises

Rewrite the problem in your own words. Write *add, subtract, multiply,* or *divide* to tell how you would solve the problem.

1. The lowest land spot on earth lies 8100 ft below sea level in Marie Byrd, Antarctica. The lowest known spot in the ocean is 36,198 ft below sea level in the Pacific Ocean's Marianas Trench. About how many times deeper is the Marianas Trench?

2. The Helena's hummingbird, at about $2\frac{1}{4}$ in. from bill tip to tail, is the smallest bird on earth. A North African ostrich, the largest bird in the world, can reach a height of 9 ft. How many hummingbirds would it take lined up from bill tip to tail to reach the height of a 9 ft ostrich?

Exercises

Write a simpler problem in your own words; then solve the problem.

A 1. A chameleon can extend its tongue to a distance equal to its body length. If Cammie the chameleon is $7\frac{5}{8}$ in. long, how long would he be from tongue tip to tail, if his tongue were fully extended?

2. Though the name centipede means "100 feet", these creatures may actually have from 28 to 354 legs. Each body segment has one pair of legs and the number depends on the species. Suppose a centipede had 37 body segments. How many legs would it have?

B 3. Automobiles contribute heavily to pollution of the atmosphere. For every liter of gas burned, a car traveling at 40 km/h might give off many substances. They include 284.9 g of carbon monoxide, 22.2 g of nitrogen oxides, 14.8 g of hydrocarbons, 1.3 g of solid particles, and 1.0 g of sulfur oxides. If a car uses 76 L of gas in a week, how much of each type of pollutant is released into the atmosphere?

4. In a sample of 1,000,000 parts of seawater the following dissolved minerals were found: 25,416 parts of sodium chloride, 3986 parts of magnesium chloride, 1475 parts of magnesium sulfate, and 1282 parts of calcium sulfate. What percent of the seawater does each substance account for?

Self Test 1

For Exercises 1–3 refer to the figure on the right. $\overleftrightarrow{AB} \parallel \overleftrightarrow{CD}$.

1. Name two pairs of alternate interior angles.
2. $\angle 4 = 146°$ so $\angle 6 =$? .
3. $\angle 3 = 34°$ so $\angle 7 =$? .

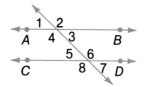

13-1

For Exercises 4 and 5 find the sum of the angles in the polygon and the measure of $\angle x$.

4.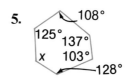

5.

13-2

In Exercises 6 and 7 the pairs of polygons are similar. Find the length of the missing side.

6.

7.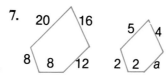

13-3

Solve.

8. A water truck holds 650 gal of water. Due to a leak, it loses about 18 gal of water each time it makes a trip from the hydrant to a construction site. About how much water is delivered in 8 trips?

13-4

13-5 Triangles

Triangles can be classified by their angles or by their sides.

Triangles Classified by Angles

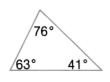

Acute Triangle
All angles acute

Right Triangle
One right angle

Equiangular Triangle
All angles equal

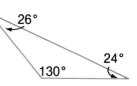

Obtuse Triangle
One angle obtuse

Triangles Classified by Sides.

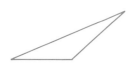

Scalene Triangle
No two sides equal

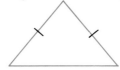

Isosceles Triangle
At least two sides equal

Equilateral Triangle
All three sides equal

If a triangle is equiangular it is also equilateral.

Example 1 Name an acute, an isosceles, and a scalene triangle in the figure.

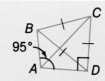

Solution

Acute triangle—*BCD* ⟵ *All angles are less than 90°.*
Isosceles triangle—*BCD* ⟵ *Sides BC and BD are equal.*
Scalene triangle—*ABD* ⟵ *No two sides are equal.*

Your Turn Use the figure in Example 1.
- Name a right triangle.
- Name an obtuse triangle.
- Name an equilateral triangle.

You can use the properties of special triangles to solve problems.

Example 2   The perimeter of the triangle is 30.
Find the length of each side.

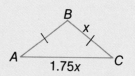

Solution   Write an equation to show that the perimeter is the
sum of the lengths of the sides.
$$x + x + 1.75x = 30 \longleftarrow \boxed{\textit{The two sides marked are equal.}}$$
$$3.75x = 30$$
$$x = 8$$
The lengths of the sides are:
$AB = 8$, $BC = 8$, $AC = (1.75)(8) = 14$

Your Turn   • The perimeter of $\triangle XYZ$ is 23.
Find the length of each side.

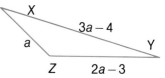

• $\triangle MTA \cong \triangle BUS$. Is $\triangle BUS$ an
acute or obtuse triangle?

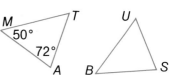

Class Exercises

True or false? Write *T* or *F*.

1. A right triangle can also be an obtuse triangle.
2. An isosceles triangle can also be an equilateral triangle.
3. An isosceles triangle can also be a right triangle.
4. An obtuse triangle can have more than one obtuse angle.
5. A scalene triangle can also be an equilateral triangle.
6. If $\triangle RUN \cong \triangle HOP$, both triangles must be right triangles.
7. If isosceles $\triangle SAM \cong \triangle TED$ and side $SA = 17$, then side
   $ED = 17$.
8. In right $\triangle SAP$, $SP = AP$. $\triangle SAP$ is an isosceles triangle.

Exercises

Classify the triangle as acute, right, obtuse, or equiangular.

A 1.

2.
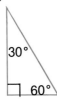

3.

4.
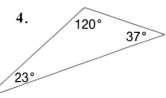

Classify the triangle as scalene, isosceles, or equilateral.

5.
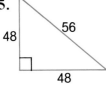

6.
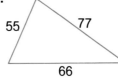

7.

8.
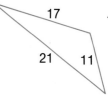

Find the lengths of the sides of the triangle.

9.

The perimeter is 44.

10.
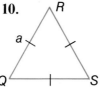
The perimeter is 108.

11.

The perimeter is 66.

12.

The perimeter is 18.

13. $\triangle QED \cong \triangle MLS$. What is the measure of $\angle SML$?
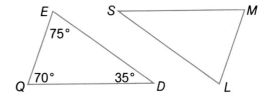

14. $\triangle MOP \cong \triangle BRD$. Is $\angle DBR$ acute, obtuse, or right?
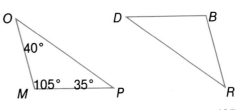

13-6 Quadrilaterals

Many convex quadrilaterals have special properties. Some of the most common are described below.

Parallelogram—
Quadrilateral with
opposite sides
parallel.

Rectangle—
Parallelogram with
four right angles.

Square—
Parallelogram with
four right angles
and all sides equal.

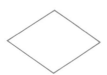

Rhombus—
Parallelogram with
all sides equal in
length

Trapezoid—
Quadrilateral with
exactly one pair of
opposite sides parallel.

You can use what you know about congruent triangles and alternate interior angles to show that opposite sides and opposite angles of a parallelogram must be equal.

Quadrilateral *ABCD* is a parallelogram. To show that $AB = CD$ and $AD = CB$ you can go through a simple proof. Though many of the steps seem obvious, they are all necessary to form a logical conclusion.

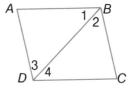

Proof: You can prove that $AB = CD$ and $AD = CB$ by showing that
        $\triangle ABD \cong \triangle CDB$.
        Step 1. $\overline{AB} \parallel \overline{CD}$ and $\overline{BD}$ is a transversal. Therefore, $\angle 1 = \angle 4$.
        Step 2. $\overline{BC} \parallel \overline{AD}$ and $\overline{BD}$ is a transversal. Therefore, $\angle 2 = \angle 3$.
        Step 3. $BD = BD$
        Step 4. Therefore, $\triangle ABD \cong \triangle CDB$ by ASA.

Quadrilateral *PORK* is a parallelogram.
Show that ∠*KPO* ≅ ∠*ORK* and ∠*POR* ≅ ∠*RKP*.

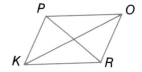

Proof: Step 1. $\overline{KO}$ is a diagonal that forms congruent triangles *KPO* and *ORK*.

Step 2. ∠*KPO* ≅ ∠*ORK* since corresponding parts of congruent triangles are congruent.

Step 3. $\overline{PR}$ is a diagonal that forms congruent triangles *POR* and *RKP*.

Step 4. ∠*POR* ≅ ∠*RKP* since corresponding parts of congruent triangles are congruent.

   The preceding proofs allow us to make some statements about parallelograms.

A diagonal of a parallelogram divides the parallelogram into congruent triangles.

Opposite sides of a parallelogram are congruent.

If both pairs of opposite sides of a quadrilateral are congruent, the quadrilateral is a parallelogram.

Opposite angles of a parallelogram are congruent.

If two sides of a quadrilateral are both parallel and congruent, the quadrilateral is a parallelogram.

Class Exercises

True or false? Write *T* or *F*.

1. A rhombus can never be a square.
2. All squares are rectangles.
3. A trapezoid always has one pair of equal opposite sides.
4. A parallelogram always has at least one right angle.
5. Opposite sides of a parallelogram are equal.
6. A trapezoid can be a parallelogram.
7. A rhombus is a regular polygon.

Exercises

For Exercises 1–9 refer to parallelogram *TUNA*.

A   **1.** $\triangle TUA \cong \triangle$ __?__   **2.** $\triangle ANT \cong \triangle$ __?__

   3. $TA = 50$ so $NU =$ __?__   **4.** $AN = 78$ so $UT =$ __?__

   5. $\angle 3 = 40°$ so $\angle 2 =$ __?__

   6. $\angle TUN = 85°$ so $\angle NAT =$ __?__

   7. $\angle 1 = 45°$, $\angle 3 = 40°$, so $\angle TAN =$ __?__

   8. $\angle 2 = 50°$, $\angle 4 = 55°$, so $\angle ANU =$ __?__

   9. $AT = 16$, $TU = 33$, so the perimeter of *TUNA* is __?__

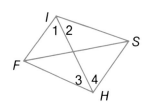

For Exercises 10–12 refer to parallelogram *FISH*. Find the value of *x*.

B   **10.** $FI = 3x + 7$, $HS = 16$

   11. $FH = \dfrac{26}{x}$, $IS = 13$

   12. $\angle 3 = (x - 4)°$, $\angle 2 = 48°$

Self Test 2

Find the lengths of the sides of the triangle.

1.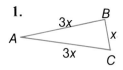

The perimeter is 35.

2.

The perimeter is 51.

3. 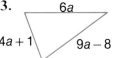   13-5

The perimeter is 88.

Exercises 4–6 refer to parallelogram *LAMP*.

   4. $\triangle PAM \cong \triangle$ __?__   13-6

   5. $\angle 2 = 60°$ so $\angle 3 =$ __?__

   6. $LP = 15$ so $AM =$ __?__

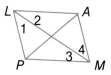

Algebra Practice

Write an equation. Solve.

1. 1,336,064 divided by 136 is p.

2. 3462 multiplied by 863 is f.

3. The difference between 152 and 107.98 is z.

4. The quotient of 0.1716 and 0.33 is b.

5. The sum of $6\frac{1}{4}$ and $1\frac{2}{3}$ is m.

6. $\frac{1}{6}$ of $7\frac{1}{8}$ is equal to g.

7. The circumference of a circle with a radius of 5.2 m is c.

8. The surface area of a cylinder with a radius of 3 in. and a height of 10 in. is s.

Write the LCM of each set of expressions.

9. $4x$ and y^2

10. 7, gh, and h^2

11. a^2, 10, and $4b^2$

Solve.

12. $\frac{1}{4}n = {}^-8$

13. $\frac{2}{5}x - 1 = 3$

14. $1.4m = 0.4m - 3$

15. 30% of 72 is f.

16. 0.75% of 1000 is t.

17. 2.9 is 116% of c.

18. $\frac{{}^-3}{8}a - 2 = \frac{1}{4}$

19. $6 - 3p > \frac{{}^-3}{4}$

20. $7 < {}^-5w + 3$

What is the slope and y-intercept of the equation?

21. $y = 2x + 7$

22. $y = 4x - 3$

23. $y = 12 - 5x$

Find the lengths of the sides of the triangle.

24. Perimeter 70, sides $3m - 4$, $2m + 2$, $4m$

25. Perimeter 82, sides $6d - 14$, $4d$, $3d + 5$

26. Perimeter 59, sides $j + 12$, $3j - 7$, $2j$

Write an equation and solve the problem.

27. Last year Riverside Motors sold 815 cars and trucks.
    They sold 129 more cars than trucks.
    How many trucks did they sell last year?
    How many cars did they sell last year?

Problem Solving on the Job

Insurance Broker

An insurance broker may help you choose your car insurance. There are four basic types of car insurance that most people buy.

1. *Bodily injury*—insures against accidental injuries to people.
2. *Property Damage*—insures against damage to the property of others.
3. *Collision*—insures against damage to your car.
4. *Comprehensive*—insures against loss due to fire, vandalism, or theft of property.

```
INSURANCE BROKER

DESCRIPTION: Helps individuals and
businesses choose best insurance to
suit their needs.
QUALIFICATIONS: College work is
very desirable but many agencies
will train qualified high school
graduates. Must pass a written test
to obtain a license.
JOB OUTLOOK: Demand expected to
grow as fast as the average for all
jobs.
```

The rating group depends on the make and model of the car.

If you choose a $50 deductible, you must pay the first $50 of each loss.

The age group depends on the age of the car.

		Insurance Rating Group					
		1–5	**6**	**7**	**8**	**9**	**10**
AGE		Full comprehensive					
	A	$19.60	$22.00	$26.00	$32.40	$45.60	$58.40
	B	16.40	18.40	22.00	27.60	38.80	49.60
	C	14.80	16.40	19.60	24.40	34.40	44.00
	D	12.80	14.40	16.80	21.20	29.60	38.00
		Comprehensive $50 deductible					
	A	$12.80	$15.60	$19.60	$26.40	$39.20	$52.00
	B	10.80	13.20	16.80	22.40	33.20	44.40
	C	9.60	11.60	14.80	20.00	29.60	39.20
	D	8.40	10.00	12.80	17.20	25.60	34.00
		Collision $50 deductible					
	A	$92.00	$97.20	$108.00	$124.40	$140.40	$156.80
	B	78.40	82.80	92.00	105.60	129.20	133.20
	C	69.20	72.80	81.20	93.20	10.20	117.60
	D	55.20	58.40	64.80	74.80	84.40	94.00
		Collision $100 deductible					
	A	$74.00	$83.20	$92.40	$106.30	$124.80	$143.20
	B	62.80	70.80	78.40	90.40	106.00	121.60
	C	55.60	62.40	69.20	80.00	93.60	109.60
	D	44.40	50.00	55.60	64.00	76.80	86.00

PROPERTY DAMAGE LIMITS	BODILY INJURY LIMITS			
	15/30	25/100	50/100	100/300
$10,000	$ 93.20	$104.00	$109.20	$120.00
$25,000	$ 97.20	$108.00	$113.20	$124.00
$50,000	$100.00	$111.20	$116.00	$127.20

◄— *A choice of 15/30 limits means the insurance company would pay a maximum of $15,000 for injuries to one person in an accident or $30,000 maximum for injuries to several people.*

Complete the table.

	Customer Name | Car	Age Group	Rating Group	Bodily Injury (Limits)	Property Damage (Limits)	Comprehensive	Collision	Total
1. Plan 1 — Smith Pronto		B	7	25/100	25,000	full	$ 50 ded.	
Premium				$108		$ 22	$ 92	?
2. Plan 2 — Smith Pronto		B	7	100/300	50,000	$ 50 ded.	$ 50 ded.	
Premium					?	$16.80	$92	?
3. Plan 1 — Barkley Excella		A	10	25/100	10,000	full	$ 100 ded.	
Premium					?	?	?	?
4. Plan 2 — Barkley Excella		A	10	50/100	50,000	$ 50 ded.	$ 100 ded.	
Premium					?	?	?	?
5. Plan 1 — Chou Mesa		D	3	15/30	10,000	$ 50 ded.	$ 50 ded.	
Premium					?	?	?	?
6. Plan 2 — Chou Mesa		D	3	100/300	50,000	$ 50 ded.	$ 100 ded.	
Premium					?	?	?	?

Use the completed table to solve.

7. The owner of the Pronto decided to purchase Plan 1. The broker asked for a 25% deposit. How much was the deposit?

8. The owner of the Excella chose Plan 2. The broker informed her that she was entitled to a good driver discount of 15% on the BI/PD portion of her insurance. What was her total premium?

9. The owner of the Mesa chose Plan 2. He also chose to pay in 12 monthly installments. The broker told him that with interest, his monthly payments would be $17.35. How much interest is he paying?

Enrichment

Arrangements

Peaches, Muffy, and Buster are finalists in the competition for best behaved dog at the Tip Top Dog Training Center. In how many different ways can the three places be chosen?

You could list all the possibilities, but that would be pretty tedious. Instead, let's make a tree diagram to show the possible choices.

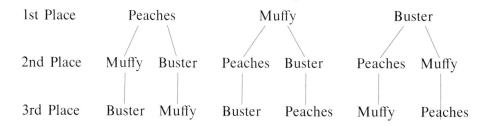

There are three possible choices for first place. Once that has been decided there are two possible choices left for second place and then one choice for third place. So there are $3 \times 2 \times 1 = 6$ possible ways to choose the three places.

1. Five contestants are entered in the Broadway Bowl-a-thon. In how many orders can the bowling lineup be chosen?
2. The Quincy High School Essay Contest attracted 22 entrants. In how many ways can the three prizes be awarded?
3. Suppose you are buying a system of stereo components. You can choose from three different turntables, four sets of speakers, four cassette players, and five tuners. How many different four-piece systems are possible?
4. You just installed an electronic burglar alarm system that can only be disarmed by punching in a 4-digit code. You can choose any 4-digit code using the numerals from 0 through 9 with no repetitions. How many choices do you have?
5. If 8 runners compete in a race, in how many ways can prizes be awarded for first, second, and third place?
6. In some states automobile license plates consist of 3 letters followed by 2 or 3 digits. How many possibilities are there for those plates with 2 digits? for those with 3 digits?

Chapter Review

True or false? Write *T* or *F*.

1. The sum of the angles of a polygon is always 180°.
2. A transversal intersects two lines in two different points.
3. A diagonal of a parallelogram cuts the parallelogram into two congruent triangles.
4. Parallel lines can also be skew lines.
5. When two planes do not intersect they are parallel.
6. An equilateral triangle is a regular, three-sided polygon.
7. If you know that the three sides of △*ARK* are equal to the corresponding sides of △*RED*, you know the two triangles are congruent.
8. You can use a compass to measure an angle.

The triangles are congruent. Write a, b, or c to indicate the correct reason.

9.
   a. ASS
   b. SSS
   c. SAS

10.
    a. ASA
    b. SSS
    c. SAS

The polygons are congruent. Complete to show the corresponding parts.

11. $AB = $ ___?___

12. $\angle BCD = \angle$ ___?___

13. $BC = $ ___?___

Complete.
Polygon *MIST* is a rhombus.

14. $\overline{MI} \cong$ ___?___

15. $\overline{MT} \cong$ ___?___

16. $\angle T \cong \angle$ ___?___

17. $\angle I = $ ___?___

Polygon *MINK* is a parallelogram.

18. $\angle IMN \cong \angle$ ___?___

19. $\angle IMN = $ ___?___

20. $\angle MKN = $ ___?___

21. $\angle MIN = $ ___?___

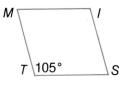

435

Chapter Test

For Exercises 1–3 refer to the figure on the right. $\overleftrightarrow{WX} \parallel \overleftrightarrow{YZ}$.

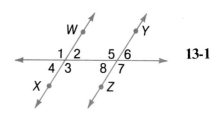

1. Name four pairs of corresponding angles.   **13-1**

2. $\angle 2 = 50°$ so $\angle 8 = \underline{\ ?\ }$ and $\angle 4 = \underline{\ ?\ }$.

3. $\angle 1 = 130°$ so $\angle 5 = \underline{\ ?\ }$ and $\angle 3 = \underline{\ ?\ }$.

For Exercises 4 and 5 find the sum of the angles in the polygon and the measure of $\angle t$.

4.   **13-2**

5.

In Exercises 6 and 7 the polygons are similar. Find the length of the missing side.

6. 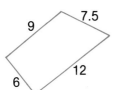   **13-3**

7.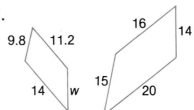

Solve.

8. In a recent year it cost 39.59¢ per mile to run a compact car. The next year the cost rose 1.09¢. What percent increase was that?   **13-4**

Find the lengths of the sides of the triangle.

9.   **13-5**

Perimeter: 108

10.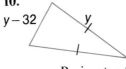

Perimeter 268

11.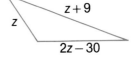

Perimeter 79

Exercises 12–13 refer to parallelogram MATH.

12. $\overline{MA} \parallel \underline{\ ?\ }$ 13. $MH = 25$, so $AT = \underline{\ ?\ }$ **13-6**

Cumulative Review (Chapters 1–13)

Choose the best estimate.

1. The difference between 7648 and 895   **a.** 5000   **b.** 7500   **c.** 6700

2. The product of 38 and 195   **a.** 8000   **b.** 6000   **c.** 6500

3. The sum of $4.85 and $9.25   **a.** $13.50   **b.** $14.00   **c.** $15.00

4. $2250 minus $1105   **a.** 1300   **b.** 1000   **c.** 1100

5. The difference between $16\frac{3}{5}$ and $3\frac{1}{2}$ **a.** 13   **b.** 10   **c.** 14

6. The sum of 8427, 3592, and 1600   **a.** 13,000   **b.** 12,000   **c.** 14,000

For each pair of similar triangles, find n.

7.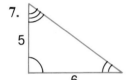

8.

The lengths of the sides of a triangle are a, b, and c. Use the Pythagorean Theorem to check to see if the triangle is a right triangle or not. Write *Yes* or *No*.

9. $a = 2$
$b = 4$
$c = 6$

10. $a = 3$
$b = 4$
$c = 5$

11. $a = 5$
$b = 7$
$c = 8$

State the probability as a fraction in lowest terms and as a percent.

12. $P(A) = \dfrac{6}{24}$ **13.** $P(B) = \dfrac{18}{36}$ **14.** $P(X) = \dfrac{8}{40}$

Write an equation for the problem. Solve it.

15. A coach ordered jerseys for the field hockey team. She ordered twice as many size M as size S, and four more size L than size S. Altogether she ordered 16 jerseys. Let n = the number of size S jerseys. How many of each size did she order?

16. When you double a number and add 35, the result is 23. What is the original number?

437

14
Programming in BASIC

Fasten your seatbelts! Though this picture looks real, it was actually generated by a computer without a single pen stroke by an artist.

14-1 Symbols and Line Numbers

Sets of instructions called programs are used to communicate with computers. You will be learning a language called BASIC. There are many similar though distinctly different versions of BASIC. The language demonstrated in this chapter covers many common statements and uses common punctuation.

The BASIC symbols for arithmetic operations are given below.

Operation	BASIC Symbol	Expression	BASIC Expression
Addition	$+$	$4 + 3$	$4+3$
Subtraction	$-$	$10 - 6$	$10-6$
Multiplication	$*$	$8 \cdot 8$	$8*8$
Division	$/$	$12 \div 4$ or $\frac{12}{4}$	$12/4$
Exponentiation	$\uparrow$, or $\wedge$, or $**$	5^2	$5\uparrow2$, or $5\wedge2$, or $5**2$

Programs are written one line at a time. Each line is numbered and usually contains only one instruction. Line numbers can be any number from 1 to 99999. PRINT commands are used to tell the computer to perform calculations and print the results.

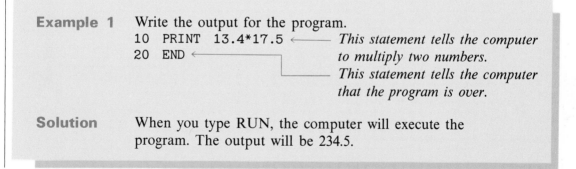

Example 1   Write the output for the program.
   10   PRINT 13.4*17.5 ⟵———— *This statement tells the computer to multiply two numbers.*
   20   END ⟵
   This statement tells the computer that the program is over.

Solution   When you type RUN, the computer will execute the program. The output will be 234.5.

Your Turn   Write a program to compute the following:
   • $26.7 \div 15.25$ •4^6

Programs usually have many lines of instructions. They are usually numbered in multiples of 10 or 100 to allow the programmer to insert new lines in the correct sequence without renumbering the lines.

Example 2   Write the output for the program.
```
10   PRINT   1+2+3
20   PRINT   2*57
30   END
15   PRINT   1.75/25
```

Solution   6
0.07 ◄——————   *The computer calculates line 15 in*
114   *order between lines 10 and 20.*

Your Turn   Write the output for the program.
```
● 10   PRINT   15-9
  20   END
  15   PRINT   18/10
   5   PRINT   16*3
```

When a statement contains more than one operation a computer follows these rules:
1. Perform all operations within parentheses.
2. Perform all exponentiations.
3. Do all multiplications and divisions in order from left to right.
4. Do all additions and subtractions in order from left to right.

Example 3   Write the output for the program.   **Solution**   3
```
10   PRINT   5-6/3                                            9
20   PRINT   3*(8-5)                                          0.25
30   PRINT   (2+3)/(4*5)
40   END
```

Class Exercises

Write the expression using BASIC symbols.

1. 3×4 **2.** 16^3 **3.** $a \div 6$ **4.** $(17 - 2) \times 3$

Write the output for the program.

5.
```
10  PRINT  15+32+61
20  PRINT  18/6
30  END
15  PRINT  11*3*10
```

6.
```
10  PRINT  48-29
20  END
 5  PRINT  47*3
15  PRINT  19+51
16  PRINT  8↑3
```

Exercises

Write the expression using BASIC symbols.

A

1. $2x - 7$

2. $\dfrac{2}{a^3}$

3. $\dfrac{1}{2}bh$

4. $\dfrac{a}{2} + b^5$

5. $\dfrac{a}{2} + \dfrac{b}{3}$

6. $3.14r^2$

7. $\dfrac{xy}{z}$

8. $\dfrac{1}{2} + 4\dfrac{1}{5}y$

Write the output for the program.

9.
```
10  PRINT  3+6/2
20  PRINT  3+5*6.1
30  PRINT  1-7/28+4*2
40  PRINT  -5+9*4
50  END
```

10.
```
10  PRINT  (3+6)/2
20  PRINT  (3+5)*6.1
30  PRINT  (9-1)/(28+4)*2
40  PRINT  -(5+9)*4
50  END
```

Write a BASIC program to compute the value of the expression.

B **11.** $1 \times 2 \times 3$

12. $\dfrac{18}{36}$

13. $512 - 367$

14. $396 + 405 + 271$

15. Write a program that computes the circumference of a circle with a radius of 7.2. Use 3.14 to approximate π.

Challenge

Write the output for the following program.

```
10  PRINT 8+4↑2/2*3-5
20  END
```

Show the different ways in which one set of parenthesis could be added to line 10 to produce the following outputs: 211, -8, 24

14-2 REM, LET, and PRINT

You can use a REMARK (REM) statement to state what a program or part of a program is supposed to do. These statements are numbered within the program and are printed only when the program is listed. They do not interfere with or have any effect on the execution of the program. LET statements are used to assign a value to a variable.

Example 1   Write a program to calculate the circumference of a circle with a radius of 2.75.

Solution
```
10  REM CALCULATE THE CIRCUMFERENCE OF A CIRCLE
15  REM FORMULA FOR CIRCUMFERENCE IS C=2*PI*R
20  LET P = 3.14  ←——— P stands for π.
30  LET R = 2.75  ←——— R stands for radius.
40  PRINT 2*P*R
50  END
```

You may use any letter to stand for a variable. Always be sure that the assigned value of a variable is on the right of the equal sign. The statement LET A + B = C is not valid, but LET C = A + B is.

You can use quotation marks around any group of words, numbers, or symbols that you want printed exactly as they appear.

Example 2   Write the output for the program.
```
10  REM CALCULATE THE AREA OF A RECTANGLE
20  LET L = 12.5
30  LET W = 5.7
40  LET A = L*W
50  PRINT "THE AREA OF THE RECTANGLE IS" A
60  END
```
The computer prints

Solution   The area of the rectangle ←——— *what is in the quotes*
                is 71.25.   *and the value of* A.

If a program contains an error or "bug" the computer cannot complete execution. Some type of error message will be printed to tell you at which line and why execution was halted.

Example 3 Debug the program.
```
10  REM CALCULATE INTEREST
20  LET P = 100
30  LET R = 0.15
40  LET 2 = T
50  PRINT "INTEREST IS $" P*R*T
60  END
```

Solution Line 40 is incorrect. It should read
```
40  LET T = 2
```

Your Turn Write the output for the program.
```
 •  5  REM AREA OF A SQUARE
    10  LET S = 8.4
    20  LET A = S↑2
    30  PRINT "THE AREA IS" A
    40  END
```

Debug the program.
```
 • 10  LET 50 = A
   20  LET B = 65
   30  LET 35 = C
   40  PRINT A+B+C
   50  END
```

Use a comma or a semicolon to separate variables in a PRINT statement when you want more than one calculation printed on the same line.

Example 4 Write the output for the program.
```
 5  REM CIRCUMFERENCE AND AREA OF A CIRCLE
10  LET P = 3.14
20 ·LET R = 7.2
30  LET C = 2*P*R
40  LET A = P*R↑2
50  PRINT C, A
60  END
```

Solution 45.216   162.7776

Class Exercises

Write the output for the program.

1.
```
 5   REM PERIMETER OF A SQUARE
10   LET S = 6.3
20   LET P = 4*S
30   PRINT "THE PERIMETER IS" P
40   END
```

2.
```
 5   REM AREA OF A SQUARE
10   LET S = 11.6
20   LET A = S↑2
30   PRINT "THE AREA IS" A
40   END
```

Exercises

Write the output for the program.

A

1.
```
10   LET X = 5.7+4.8
20   LET Y = X/5
30   LET Z = Y-1.5
40   PRINT X, Y, Z
50   END
```

2.
```
10   LET X = 12.3-9.7
20   LET Y = 4.36/0.02
30   LET Z = X+Y
40   PRINT "THE SUM IS" Z
50   END
```

3.
```
10   LET E = 7.8
20   LET F = 4.5
30   LET X = E-F
40   LET Y = E*F
50   PRINT X, Y, X+Y
60   END
```

4.
```
10   LET X = 51
20   LET Y = 78
30   LET Z = 96
40   LET M = (X+Y+Z)/3
50   PRINT "THE AVERAGE IS" M
60   END
```

B

5.
```
10   LET P = 3.14
20   LET R = 4
30   LET B = P*R↑2
40   LET H = 6
50   LET V = B*H
60   PRINT "THE VOLUME IS" V
70   END
```

6.
```
10   LET N = 125
20   LET P = 12.98
30   LET C = N*P
40   LET T = 0.04*C
50   PRINT "PURCHASE PRICE IS" C
60   PRINT "TAX IS" T
70   PRINT "TOTAL COST IS" T+C
80   END
```

7. Change the program in Exercise 5 so that it computes the volume of a cone. Use the same values for R and H.

8. Change the program in Exercise 6 so that a 25% discount is deducted from the purchase price before the tax is computed.

Debug the program.

9.
```
5   REM AVERAGE SCORE
10  LET A = 96
20  LET B = 85
30  LET C = 73
40  LET AVG = A+B+C/3
50  PRINT "THE AVERAGE SCORE IS" AVG
60  END
```

10.
```
5   REM DISTANCE FORMULA
10  LET R = 85
20  LET T = 17
30  LET R*T = D
40  PRINT "THE DISTANCE IS"
45  PRINT D, "MI"
50  END
```

Write the BASIC program.

C 11. Write a program to compute the area of a circle with a radius of 2.6. Use at least one REM statement and one LET statement and one PRINT statement with quotation marks.

12. Write a program to compute the area of a triangle with a base of 8.2 and a height of 3.5. Use at least one REM statement and one LET statement and one PRINT statement with quotation marks.

13. Write a program to compute the area and perimeter of a rectangle with length 9.4 and width 6.2. Have the values of the area and perimeter printed on one line.

14. Write a program that prints the sum, difference, product, and quotient of two numbers A and B. Use A = 17.2 and B = 12.5. Have the values printed on one line.

Using the Calculator

To find the square root of a number, make a guess and divide the number by your guess. If the quotient is more than your guess, then guess higher. If the quotient is less than your guess, guess lower.

Use this method to find the square root of each of the following perfect squares.

1. 529

2. 1156

3. 2304

4. 11236

5. 5329

6. 4489

14-3 FOR-NEXT

You can use a FOR-NEXT statement when you want the computer to repeat an operation for a certain number of times using the same or different values for a variable.

Example 1 Write the output for the program.

```
10  FOR X = 1 TO 3
20  PRINT "MATH IS FUN"
30  NEXT X
40  END
```

This statement combined with line 30 creates a loop. It tells the computer to loop back and repeat line 20 three times.

Solution

```
MATH IS FUN
MATH IS FUN
MATH IS FUN
```

Example 2 Write the output for the program.

```
10  FOR A = 1 TO 5
20  PRINT "3 PLUS" A "=" 3+A
30  NEXT A
40  END
```

This FOR-NEXT statement uses the changing value of the variable as a counter.

Solution

```
3 PLUS 1 = 4
3 PLUS 2 = 5
3 PLUS 3 = 6
3 PLUS 4 = 7
3 PLUS 5 = 8
```

Your Turn Write the output for the program.

```
• 10  FOR N = 1 TO 4        • 10  FOR N = 6 TO 11
  20  PRINT 5*N               20  PRINT N "TIMES 2 =" N*2
  30  NEXT N                  30  NEXT N
  40  END                     40  END
```

You can also use a FOR-NEXT statement to count by numbers other than 1.

Example 3 Write the output for the program.

```
 5   REM RAISE NUMBERS TO THE THIRD POWER
10   FOR N = 1 TO 10 STEP 3  ← The STEP 3 command tells
20   PRINT N, N↑3                the computer to add 3 to
30   NEXT N                      the previous value of N, to
40   END                        a maximum of 10, each time
                                  a loop is made.
```

Solution

```
1    1
4    64           ← At each step the computer adds 3
7    343            to the variable before raising to
10   1000           the third power.
```

Your Turn Write the output for the program.

●
```
10   FOR N = 10 TO 20 STEP 2
20   PRINT N, N-6
30   NEXT N
40   END
```

●
```
10   FOR X = 10 TO 40 STEP 10
20   PRINT X, X/5
30   NEXT X
40   END
```

Class Exercises

For each program tell how many loops the computer will make.

1.
```
10   FOR N = 1 TO 25
20   PRINT N+3
30   NEXT N
40   END
```

2.
```
10   FOR N = 15 TO 19
20   PRINT N*6
30   NEXT N
40   END
```

3.
```
10   FOR N = 1 TO 20 STEP 5
20   PRINT N/2
30   NEXT N
40   END
```

4.
```
10   FOR N = 50 TO 100 STEP 5
20   PRINT (N/10)*6
30   NEXT N
40   END
```

Exercises

Write the output for the program.

A
1.
```
10  FOR A = 1 TO 6
20  PRINT A+5
30  NEXT A
40  END
```

2.
```
10  FOR M = 5 TO 10
20  PRINT M "SQUARED =" M↑2
30  NEXT M
40  END
```

B
3.
```
10  FOR Y = 7 TO 14 STEP 2
20  PRINT Y
30  NEXT Y
40  END
```

4.
```
10  FOR N = 20 TO 32 STEP 4
20  PRINT N/4
30  NEXT N
40  END
```

5.
```
10  FOR L = 5 TO 10
20  LET W = 15
30  LET A = L*W
40  PRINT A
50  NEXT L
60  END
```

6.
```
10  FOR R = 8 TO 20 STEP 4
20  LET P = 3.14
30  LET A = P*R↑2
40  PRINT A
50  NEXT R
60  END
```

7. Change the program in Exercise 5 so that the value of L varies from 20 to 30.

8. Change the program in Exercise 6 so that the values of R are 5, 10, 15, and 20.

C
9. Write a program to print out the multiples of 6 from six to ninety.

10. Write a program to calculate the cubes of the first ten odd integers.

11. Write a program to calculate the distance traveled at a constant rate of 55 mph for 3, 6, 9, and 12 hours. Use the formula $d = rt$.

14-4 *Problem Solving*

Flowcharts are frequently used by computer programmers to set forth the logical steps involved in completing a program. Flowcharts are also useful in laying out the logical process needed to solve many types of problems.

Example Suppose you want to write a program to have the computer read the lengths and widths of several rectangles and print out the perimeters if they are greater than 50. Draw a flowchart to show the steps needed to write the program.

Solution

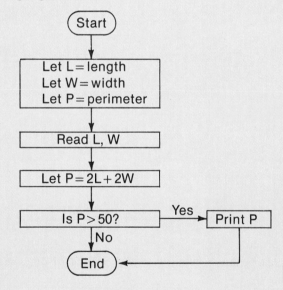

Your Turn • Draw a flowchart for a program that reads the length of the side of a square and prints the area if it is greater than 45.

Chapter 4

Class Exercises

Correct the flowcharts.

1.

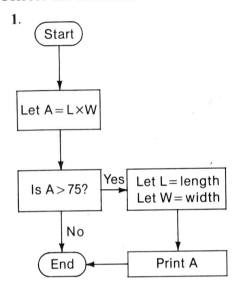

2.

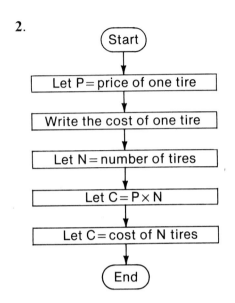

Exercises

Draw a flowchart to show the steps needed to complete the activity or solve the problem.

A 1. Get ready for school.
  2. Read a number *N* and compare it to 100. Then print *N is greater than 100, N equals 100,* or *N is less than 100* depending on the results of the comparison.
  3. Read a temperature *T* (in Fahrenheit degrees). Print *no jacket* if *T* is greater than 60, *light jacket* if *T* is between 40 and 60, and *bundle up* if *T* is less than 40.

B 4. Read the measures of the three angles of a triangle and print *acute* or *obtuse*.
  5. Compute the interest *I* for a loan with principal *P* and rate of interest *R*.
  6. Read three numbers and print the greatest of the three.
  7. Read a test score T and print *excellent* if T is greater than 90, *good* if T is from 80 to 90, *fair* if T is from 70 to 80, and *work harder* if T is less than 70.

Write a program for the flowchart.

C **8.**

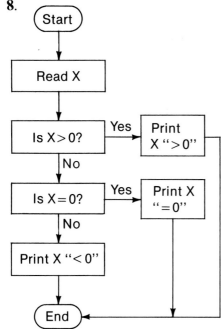

9.

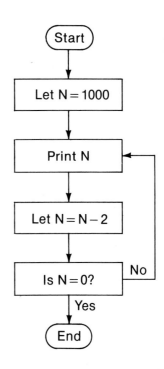

Self Test 1

Write a BASIC program to compute the value of the expression.

1. $48 \div 6$ **2.** $(3 + 7) \div 2$ **3.** $4^2 + 5^3$ 14-1

Write the output for the program.

4.
```
10  LET A = 3.5
20  LET B = 6.8
30  LET C = A*B
40  PRINT A, B, C
50  END
```

5.
```
10  REM CALCULATE PERIMETER
20  LET L = 25
30  LET W = 50
40  LET P = 2*L+2*W
50  PRINT L, W, P
60  END
```
14-2

6. Write a BASIC program to print the even numbers from 20 to 30.   14-3

7. Make a flowchart to show the steps involved in computing a total purchase price for items that qualify for a 20% discount and are subject to a 3% sales tax.   14-4

Self Test answers and Extra Practice are at the back of the book.   **451**

14-5 READ . . . DATA, GOTO

In addition to LET statements you can use a combination of READ and DATA statements to assign values to variables. Whenever a READ statement appears, the computer looks for a DATA statement to determine the value of the variables. It then sequentially assigns one data value to each variable until it runs out of variables or data.

Example 1 Write the output for the program.
```
10  READ X, Y, Z
20  LET A = (X+Y+Z)/3
30  PRINT "YOUR AVERAGE IS" A
40  DATA 126,153,162
50  END
```

The computer assigns 126 to X, 153 to Y, and 162 to Z.

Solution YOUR AVERAGE IS 147

You can use a GOTO statement to form a loop and have the computer repeat the same calculations on several items of data.

Example 2 Write a program using READ . . . DATA, and GOTO statements that calculates and prints the average of three bowling scores for each of 3 bowlers. The bowling scores are:
Bowler 1: 126, 153, 162   Bowler 2: 215, 189, 196
Bowler 3: 143, 165, 170

Solution
```
10  READ X, Y, Z
20  LET A = (X+Y+Z)/3
30  PRINT "THE AVERAGE SCORE IS" A
40  GOTO 10
50  DATA 126,153,162
55  DATA 215,189,196
60  DATA 143,165,170
70  END
```

The computer will continue to loop back to line 10 as long as DATA values are available.

Example 3   Write the output for the program in Example 2.

Solution
```
THE AVERAGE IS 147
THE AVERAGE IS 200
THE AVERAGE IS 159.3333
OUT OF DATA IN LINE 10
```

Your Turn   Write the output for the program.
```
• 10   READ L, W
  15   REM FORMULA FOR PERIMETER IS 2L+2W
  20   LET P = 2*L+2*W
  30   PRINT "PERIMETER IS" P
  40   DATA 15,8
  50   END
```

Class Exercises

Write the output for the program.

1.
```
10   READ L, W
20   LET P = 2*L+2*W
30   PRINT "PERIMETER IS" P
40   DATA 15, 8
50   END
```

2.
```
10   READ P, R
20   LET A = P*R↑2
30   PRINT "AREA IS" A
40   DATA 3.14, 5
50   END
```

Exercises

Rewrite the program so that it uses READ . . . DATA to assign values to variables.

A **1.**
```
10   LET A = 15
20   LET B = 12.7
30   PRINT B-A, B/A, A↑2
40   END
```

2.
```
10   LET L = 10.5
20   LET W = 7.28
30   LET A = L*W
40   PRINT "AREA IS" A
50   END
```

Write the output for the program.

3.
```
10  READ S
20  LET P = 4*S
30  LET A = S↑2
40  PRINT "PERIMETER IS" P
50  PRINT "AREA IS" A
60  GOTO 10
70  DATA 5,13,7.3
80  END
```

4.
```
10  READ L, W
20  LET P = 2*L+2*W
30  PRINT "PERIMETER IS" P
40  GOTO 10
50  DATA 10,7.1,18,12
60  DATA 100,82.5
70  END
```

B 5. Write a program using READ . . . DATA and GOTO that computes the area of four rectangles. Use the following dimensions:

Length	5.2	92.3	17	2.25
Width	3.9	80	11	1.4

6. Write a program using READ . . . DATA AND GOTO that computes the squares of the integers from 1 to 30. Each line of the output should show the integer and its square.

Example The fifth line should be   5   25

C 7. Avon Rental Company rents panel trucks for $45.95 per day plus $.29 per mile. Write a program using READ . . . DATA AND GOTO statements to find the cost for each rental:

Number of days	2	1	1	3	4
Miles	350	120	85	645	2120

14-6 More READ . . . DATA, GOTO

You can use the statements you have learned thus far to write many programs.

Example The following program finds a certain percent of a given number. What is the output?
```
10  READ P, N
20  LET X = P/100*N
30  PRINT P "PERCENT OF" N "IS" X
40  GOTO 10
50  DATA 54,65,125,80,16,40,4.5,152,0.08,1386
60  END
```

Solution
```
54 PERCENT OF 65 IS 35.1
125 PERCENT OF 80 IS 100
16 PERCENT OF 40 IS 6.4
4.5 PERCENT OF 152 IS 6.84
0.08 PERCENT OF 1386 IS 1.1088
OUT OF DATA IN LINE 10
```

Your Turn Write the output for the above program if line 50 is replaced with the following:
```
50  DATA 45,90,12,150,6.8,200
```

Class Exercises

Solve.

1. 58 is what percent of 80?

2. 4.5 is 15% of what number?

3. 3 is what percent of 450?

4. 90 is 225% of what number?

5. 18 is what percent of 24?

6. 375 is 40% of what number?

Exercises

Write the output for the program.

A **1.**
```
10  READ N, D
20  LET P = N/D*100
30  PRINT N "IS" P "PERCENT OF" D
40  GOTO 10
50  DATA 72,48,35,70,75,150,90,120
60  END
```

2.
```
10  READ A, B
20  LET N = A/B*100
30  PRINT A "IS" B "PERCENT OF" N
40  GOTO 10
50  DATA 18,0.15,16,0.25,26.88,0.56
60  END
```

In BASIC, the square root of a number x may be written SQR(x).

Example If $x = 9$, then SQR(x) = 3.

Write the output for the program.

B **3.**
```
10  READ X
20  PRINT X, SQR(X)
30  GOTO 10
40  DATA 1,4,9,16,25,36,49
50  END
```

4.
```
10  READ A, B
20  LET Y = A*B
30  PRINT SQR(Y)
40  GOTO 10
50  DATA 4,9,27,3,5,125,2,8
60  END
```

The following program approximates the length of the hypotenuse, C, for a right triangle with legs of lengths A and B.

5.
```
10  READ A, B
20  LET C = SQR (A↑2+B↑2)
30  PRINT "THE LEG LENGTHS ARE" A, B
40  PRINT "THE LENGTH OF THE HYPOTENUSE IS" C
50  GOTO 10
60  DATA 12,5,8,6,8,15,2,2,15,20
70  END
```

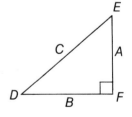

For exercises 6, 7 use the figure and DATA from Exercise 5. You may wish to refer to Chapter 11 to review the trigonometric functions.

C **6.** Write a program to compute tan D. What is the output?

7. Write a program to compute sin D. What is the output?

14-7 Branching: IF . . . THEN

It is possible to tell the computer to jump to some other line in the program under certain conditions. This is called branching. The GOTO statement is one type of branching statement. An example of another kind of branching instruction, IF . . . THEN, is shown below.

Example 1 What is the output for the program?
```
10  READ X, Y, Z
20  LET A = (X+Y+Z)/3
30  PRINT "YOUR AVERAGE TEST SCORE IS" A
40  IF A<60 THEN 70  ◄──────
50  PRINT "YOU ARE PASSING"
60  DATA 80,71,65
70  END
```

When A $<$ 60 is true, the computer will go to line 70; otherwise it will execute the next line in the program.

Solution YOUR AVERAGE TEST SCORE IS 72
YOU ARE PASSING.

Your Turn Write the output for the program in Example 1 if line 60 is changed to the following:
```
60  DATA 53,70,42
```

You can use IF . . . THEN statements to compare data by using any of the following relations.

Symbol	Meaning
$<$	less than
$>$	greater than
$<=$	less than or equal to
$>=$	greater than or equal to
$=$	equal to
$<>$	not equal to

Example 2   Any three numbers *a*, *b*, and *c* that satisfy the equation $a^2 + b^2 = c^2$, are called Pythagorean Triples. Write the output for the program.

```
10  READ A, B, C
20  IF C↑2 = A↑2+B↑2 THEN 40
30  GOTO 10
40  PRINT A, B, C "ARE PYTHAGOREAN TRIPLES"
50  GOTO 10
60  DATA 3,4,5,5,12,14,6,9,10,8,15,17
70  END
```

Solution

```
3, 4, 5 ARE PYTHAGOREAN TRIPLES
8, 15, 17 ARE PYTHAGOREAN TRIPLES
OUT OF DATA IN LINE 10
```

Your Turn   What would be the output above if line 60 had the following data? 60 DATA 6,8,10,9,10,20

Class Exercises

Let *A* = 5 and *B* = 8. Write the line number that will be executed after the IF . . . THEN statement.

1. 20 IF A>6 THEN 40
    30 PRINT A

2. 30 IF A+B>=12 THEN 60
    40 PRINT A+B

Exercises

Write the output for the program.

A 1.
```
10  READ X
20  IF X<=0 THEN 40
30  PRINT X
40  GOTO 10
50  DATA −1.5,0,7.1,−2,4.7,6,−3.2,−4.5,3.9,17.2
60  END
```

2.
```
10   READ A, B
20   IF A>B THEN 60
30   IF A<B THEN 80
40   PRINT A "=", B
50   GOTO 10
60   PRINT A ">", B
70   GOTO 10
80   PRINT A "<", B
90   GOTO 10
100  DATA ⁻5,4,2,2,12,1,9,10,⁻12,15.2,4,18.6,⁻12
110  END
```

Rewrite the program to include an IF . . . THEN statement to accomplish the desired result.

B **3.** Have the computer print "Great Bowling" if the average score is above 220.
```
10   READ X, Y, Z
20   LET A = (X+Y+Z)/3
30   PRINT "AVERAGE IS" A
40   GOTO 10
50   DATA 126,153,162,240,215,232,203,218,239
60   END
```

4. Have the computer print only negative numbers.
```
10   READ N
20   PRINT N
30   GOTO 10
40   DATA −4,9,0,11.3,−9.7,−0.5,8.7
50   END
```

14-8 More Uses of IF . . . THEN

The IF . . . THEN statement can be used in programs to compute weekly payrolls.

Example This program computes the weekly wages for four workers. The rate of pay is $4.95 per hour for the first 40 hours and $7.40 per hour for any hours over 40. What is the output for the program?

```
 5  REM COMPUTE WEEKLY PAY, H = HOURS, P=PAY
10  READ H
20  IF H>40 THEN 50  ←───────────
30  LET P = H*4.95
40  GOTO 60
50  LET P = 40*4.95+(H-40)*7.40
60  PRINT "WEEKLY PAY IS $" P
70  GOTO 10
80  DATA 38, 42, 49, 40
90  END
```

The IF . . . THEN statement separates hours greater than 40 from hours less than or equal to 40.

Solution

```
WEEKLY PAY IS $188.10  ←───────
WEEKLY PAY IS $212.80
WEEKLY PAY IS $264.60  ←───────
WEEKLY PAY IS $198
```

Since H is less than 40, the computer goes to line 30.
Since H is greater than 40, the computer goes to line 50.

Your Turn Write the output for the program in the Example using the following data.
```
80  DATA 35, 39, 48, 50
```

Class Exercises

Use the program that follows to answer the question.

1. What will the computer do at line 40 if A is equal to 10?

2. How many numbers and their squares will the computer print?

3. What would happen if line 40 were changed to read 40 IF A < 100 THEN 20?

```
10  READ A
20  PRINT A, A↑2
30  LET A = A+1
40  IF A<10 THEN 20
50  DATA 1
60  END
```

Exercises

Write the output for the program.

A 1.
```
 10  READ X, Y, Z
 20  LET A = (X+Y+Z)/3
 30  IF A>=90 THEN 70
 40  IF A>=80 THEN 90
 50  IF A>=70 THEN 110
 60  PRINT "TRY AGAIN NEXT YEAR"
 65  GOTO 10
 70  PRINT "BLUE RIBBON"
 80  GOTO 10
 90  PRINT "RED RIBBON"
100  GOTO 10
110  PRINT "YELLOW RIBBON"
120  GOTO 10
130  DATA 71,83,90,85,95,92
135  DATA 64,87,91
140  END
```

Solve.

2. Sample telephone rates from Detroit to Los Angeles are $2.35 for the first three minutes and $.40 for each additional minute. Write a program that will print out the charges for each phone call listed. 2 min, 7 min, 11 min, 3 min, 23 min, 16 min

3. Write a program that READS two numbers at a time from DATA, and prints whether they are equal or unequal (for example, 3 = 3, 4.2 <> 40.2). Use DATA 7.5, 7.5, −1, 1, −16, −16, −18.1, 18.1, 0.62, 6.02.

B **4.** A Freight Company's rules for shipping a package state that width plus height can total no more than 72 in. Write a program that determines if packages with the following dimensions can be carried by air freight. Print "PACKAGE CANNOT BE MAILED" if the package does not meet the requirements. Print nothing if the dimensions are satisfactory. Use READ . . . DATA, GOTO, IF . . . THEN.

Package	Width	Height
1	20 in.	4 in.
2	48 in.	23 in.
3	59 in.	15 in.
4	36 in.	10 in.

Self Test 2

Write the output for the program.

1.

```
10  READ X, Y
20  PRINT X*Y, X/Y
30  GOTO 10
40  DATA 20, 10, 4.8, 2.4, 15, 5
50  END
```
14-5

2.

```
10  READ A, B, C
20  PRINT (A+B)/C
30  GOTO 10
40  DATA 4, 8, 2, 17.5, 15.8, 3
50  END
```
14-6

3. Change the program in Exercise 1 to print the sum of X^2 and Y^2. Write the output for the program. **14-7**

4.

```
10  READ A, B, C
20  LET D = (A+B+C)/3
30  PRINT "YOUR AVERAGE IS" D
40  IF D>=90 THEN 50
45  GOTO 10
50  PRINT "CONGRATULATIONS. YOU'VE MADE HONOR ROLL"
60  GOTO 10
70  DATA 85, 90, 68, 100, 100, 78
80  END
```
14-8

5. Change the program in Exercise 4 to print "SORRY, YOU MISSED HONOR ROLL" if D is less than 90. **14-9**

462 *Self Test answers and Extra Practice are at the back of the book.*

Algebra Practice

Write an equation. Solve.

1. 2803 times 318 is equal to d.

2. The quotient of 5,324,670 and 738 is w.

3. The sum of 12.087, 3.494, and 28.513 is h.

4. The product of 0.018 and 0.62 is a.

5. $7\frac{7}{15}$ minus $2\frac{4}{5}$ is equal to n.

6. $\frac{7}{8}$ divided by $\frac{21}{32}$ is equal to y.

7. The area of a triangle with a base of $3\frac{1}{2}$ in. and a height of 2 in. is A.

8. The volume of a cone with a radius of 5 cm and a height of 9 cm is V.

Use the distributive property to factor each expression.

9. $12a^3b - 15ab^2$ 10. $10p^2q - 5p^2q^2 + 15pq^3$ 11. $2j^3 + 16j^3k - 14j^2k^2$

Solve.

12. $\frac{^-5}{6}s = 15$ 13. $\frac{3}{7}z + 1 = 10$ 14. $^-0.2n = 1.8n - 8$

15. 6.5% of 120 is g.   16. $8\frac{1}{2}$% of 75 is v.   17. 55 is 62.5% of m.

18. $\frac{2}{3}d + {}^-4 = 24$ 19. $20 > \frac{5}{6}k - 4$ 20. $12 - \frac{7}{8}q > \frac{^-5}{6}$

Is the ordered pair a solution of the linear inequality?

21. $(8, 7)$, $x - 3 \leq y$ 22. $(6, 4)$, $y < x - 2$ 23. $(2, 10)$, $y \geq 3x + 5$

Find the lengths of the sides of the triangle.

24. Perimeter 90, sides $3j$, $2j + 6$, $4j - 15$

25. Perimeter 59, sides $6f - 13$, $5f + 2$, $3f$

Write the expression using BASIC symbols.

26. $3a^2 - 5$ 27. $\frac{2r}{s}$ 28. $g^2 + \frac{h}{5}$

Problem Solving for the Consumer

Evaluating Consumer Purchases

When making large purchases it's a good idea to compare the features and prices of several brands. You can start your research by reading consumer magazines, talking to people who already own the item, and visiting stores to have the various options explained and prices quoted.

Michael developed a system of weights to measure the features of several personal computers that met his basic needs for record-keeping, game playing, and word processing.

Here's how his system worked:

1. He chose eight features and weighted each from 1 (not too important) to 5 (very important).

2. He rated each computer (the Essex, Supra, and Whiz) on each feature using the following ratings: 0 = none, 1 = poor, 2 = fair, 3 = good, 4 = very good, 5 = excellent.

3. To get a score for each feature he multiplied the feature weight by the rating.

Complete the table.

	Feature	Weight	Essex		Supra		Whiz	
			Rating	Score	Rating	Score	Rating	Score
1.	Cost	3	2	6	5	?	3	?
2.	Memory	4	5	20	3	?	4	?
3.	Software	5	5	25	3	?	1	?
4.	Display	2	3	?	3	?	3	?
5.	Add-ons	4	5	?	3	?	3	?
6.	Service	3	4	?	3	?	3	?
7.	Keyboard	4	3	?	4	?	4	?
8.	Color	3	3	?	0	?	5	?
9.	Total			?		?		?

Use the information in the table to answer the questions.

10. Which feature was most important?

11. Which feature was least important?

12. The Essex memory score is 20 and the keyboard score is 12, yet both features are weighted 4. Explain the difference in scores.

When visiting the computer store, Michael saw the following ads.

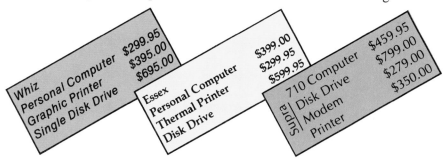

Whiz
Personal Computer $299.95
Graphic Printer $395.00
Single Disk Drive $695.00

Essex
Personal Computer $399.00
Thermal Printer $299.95
Disk Drive $599.95

710 Computer $459.95
Supra Disk Drive $799.00
Modem $279.00
Printer $350.00

Use the information in the ads to solve.

13. What is the total cost for each system as described in the ad?

14. If a sales tax of 7% is applied, what would be the final cost of the Supra system?

15. Suppose you want to buy the Essex System described. You want to put 25% down and pay the rest in twelve monthly installments. If you pay $90.25 per month, how much is the service charge?

16. Suppose you buy a computer system and receive a $100 gift certificate for software. Can you buy a Bouncer game for $24.88, a budgeting program for $37.59, a Cat and Mouse game for $19.95, and an Algebra Program for $23.88? If you owe money, how much? If you get change, how much?

Enrichment

Greatest Integer Function

You can use the integer (INT) command to program a computer to round a decimal to any place.

INT (X) is the greatest integer less than or equal to the number X. The examples below show how this works for various values of X.

INT (2.5) = 2   INT (0.99) = 0   INT (¯2.2) = ¯3   INT (4) = 4

To round a decimal X to the nearest whole number, you first add 0.5 to X and then find INT (X + 0.5). The examples below show how this works.

X	X + 0.5	INT (X + 0.5)
2.6	3.1	3
5.1	5.6	5

When you want to round a decimal to the nearest tenth, you first multiply the number by 10, then add 0.5, and then divide by 10. The command would look like this: INT(X*10+0.5)/10. For the decimal 2.47 the computer would follow these steps:

X	X * 10	X * 10 + 0.5	INT (X * 10 + 0.5)	INT (X * 10 + 0.5)/10
2.47	24.7	25.2	25	2.5

Evaluate the expression.

1. INT (5.2)   **2.** INT (71.78)   **3.** INT (−6.1)   **4.** INT (0.137)

Write the output for the program.

5.
```
10  READ D
20  LET Y = INT(D+0.5)
30  PRINT D" ROUNDED TO THE NEAREST"
40  PRINT "WHOLE NUMBER IS" Y
50  DATA 6.781, 3.925, 4.617, 86.138
60  GOTO 10
70  END
```

6. Change line 20 in Exercise 5 so that numbers are rounded to the nearest tenth, the nearest hundredth. Write the output for each new program.

Chapter Review

True or false? Write *T* or *F*.

1. A PRINT statement is used in a BASIC program to tell a computer to perform operations in a loop.
2. In the statement IF X > = 40 THEN 50 the computer will go to line 50 if X = 40.
3. A computer will execute instructions in numerical order even if a program is written with the steps out of order.
4. REM statements tell the computer to add two numbers.
5. The statement FOR A = 2 TO STEP 2 combined with a NEXT A statement will tell the computer use 2, 4, 6, and 8 for values of A.

What will be the output? Write *a*, *b*, or *c*.

6.
```
10  LET A = 5
20  PRINT A, A↑3
30  END
```
    a. 125
    b. 5 125
    c. 5 15

7.
```
10  LET A = 20
20  LET B = 30
30  LET C = (A+B)/2
40  PRINT C
50  END
```
    a. 25
    b. 20 30 25
    c. 35

Write the output for the program.

8.
```
10  FOR N = 0 TO 25 STEP 5
20  LET M = N*2
30  PRINT M "=" N "TIMES TWO"
40  NEXT N
50  END
```

9.
```
10  READ X, Y, Z
20  LET W = X/Y
30  LET R = W↑Z
40  PRINT R
50  DATA 100,10,3,45,15,2
60  END
```

10.
```
10  READ S
20  LET A = S↑2
30  PRINT "THE AREA OF THE SQUARE IS" A
40  GOTO 10
50  DATA 6,18,32
60  END
```

11. Write a BASIC program to print your name, street address, city, and phone number. Make sure the printout is on four lines.

Chapter Test

Write the expression using BASIC symbols.

1. $3x^2 + 5$ **2.** $(25 + a^2) \div 2$ **3.** $(14 + 6)^2 \div a$ **14-1**

Write the output for the program.

14-2

```
4. 10   LET A = 5
   20   PRINT A "SQUARED =" A↑2
   30   END
5. 20   LET C = 598
   30   LET E = 150
   40   LET P = C−E
   50   PRINT "PROFIT IS" P
   60   END
```

6. Change the program in Exercise 4 to print the squares of the first five multiples of 2.   **14-3**

7. Make a flowchart to show the steps involved in solving a problem to calculate an average of 5 grades.   **14-4**

Write the output for the program.

14-5

```
8. 10   READ B, H
   20   LET A = (B*H)/2
   30   PRINT "THE AREA OF THE TRIANGLE IS" A
   40   GOTO 10
   50   DATA 6,8,12,5,100,60
   60   END
```

9. Change the program in Exercise 8 to calculate the area of a rectangle.   **14-6**

10. Change the program in Exercise 8 so that it includes an IF−THEN statement and prints "THE AREA IS TOO LARGE" if A is greater than 100.   **14-7**

Write the output for the program.

14-8

```
11. 10   READ M, N
    20   IF M>N THEN 40
    30   PRINT "CONGRATULATIONS, NED. YOU WIN"
    35   GOTO 10
    40   PRINT "CONGRATULATIONS MILT. YOU WIN"
    50   GOTO 10
    60   DATA 70,65,108,120,88,87
    70   END
```

Cumulative Review (Chapters 1–14)

Add, subtract, multiply, or divide.

1. $2007 - 14 \cdot 6$

2. $2(4.7 + 0.29)$

3. $10^3 \times 13.7$

4. $2\frac{1}{2} + \frac{5}{8}$

5. $\frac{3}{5} + \frac{7}{9}$

6. $4\frac{1}{9} - \frac{3}{4}$

7. $\frac{5}{6} \times 1\frac{1}{15}$

8. $\frac{2}{9} \times 3\frac{3}{8}$

9. $1\frac{1}{2} \div \frac{4}{5}$

Solve.

10. 30 out of 250 is what percent?

11. 37% of 80 is what number?

12. What is 99% of 350?

13. What percent of 200 is 46?

14. 4% of $350 is what amount?

15. 120% of what amount is $504?

16. What is the simple interest paid on a loan of $2000 for 6 months at a rate of $9\frac{1}{2}\%$?

17. When an item in a store is reduced from $56 to $42, what is the percent of decrease?

Draw a circle graph to show the information.

18. The average American uses 60 gal of water per day in the home. Here is the breakdown of the usage: 40% flushing toilets; 35% washing and bathing; 10% kitchen use; 5% washing clothes; 10% other.

Find the probability.

19. A contest has 5 second prizes and 1 first prize. There are 500 entry blanks in a box. Five are drawn for second prizes, and are not returned to the box again. Then one entry is drawn for the first prize. Rosie has one entry blank in the box. If she doesn't win second prize, what is the probability she'll win first prize?

Write an equation and solve the problem.

20. Bottles of a brand of vitamins are available two ways—with a regular cap and with a child-proof cap. One store has 36 bottles of the vitamins in stock. There are 16 more in child-proof caps than in regular caps. How many bottles of each type are there?

Table of Square Roots of Integers from 1 to 100

Number	Positive Square Root	Number	Positive Square Root	Number	Positive Square Root	Number	Positive Square Root
N	$\sqrt{N}$	N	$\sqrt{N}$	N	$\sqrt{N}$	N	$\sqrt{N}$
1	1	26	5.099	51	7.141	76	8.718
2	1.414	27	5.196	52	7.211	77	8.775
3	1.732	28	5.292	53	7.280	78	8.832
4	2	29	5.385	54	7.348	79	8.888
5	2.236	30	5.477	55	7.416	80	8.944
6	2.449	31	5.568	56	7.483	81	9
7	2.646	32	5.657	57	7.550	82	9.055
8	2.828	33	5.745	58	7.616	83	9.110
9	3	34	5.831	59	7.681	84	9.165
10	3.162	35	5.916	60	7.746	85	9.220
11	3.317	36	6	61	7.810	86	9.274
12	3.464	37	6.083	62	7.874	87	9.327
13	3.606	38	6.164	63	7.937	88	9.381
14	3.742	39	6.245	64	8	89	9.434
15	3.873	40	6.325	65	8.062	90	9.487
16	4	41	6.403	66	8.124	91	9.539
17	4.123	42	6.481	67	8.185	92	9.592
18	4.243	43	6.557	68	8.246	93	9.644
19	4.359	44	6.633	69	8.307	94	9.695
20	4.472	45	6.708	70	8.367	95	9.747
21	4.583	46	6.782	71	8.426	96	9.798
22	4.690	47	6.856	72	8.485	97	9.849
23	4.796	48	6.928	73	8.544	98	9.899
24	4.899	49	7	74	8.602	99	9.950
25	5	50	7.071	75	8.660	100	10

Exact square roots are shown in red. For the others, rational approximations are given correct to three decimal places.

Table of Trigonometric Ratios

Angle	Sine	Cosine	Tangent	Angle	Sine	Cosine	Tangent
1°	.0175	.9998	.0175	46°	.7193	.6947	1.0355
2°	.0349	.9994	.0349	47°	.7314	.6820	1.0724
3°	.0523	.9986	.0524	48°	.7431	.6691	1.1106
4°	.0698	.9976	.0699	49°	.7547	.6561	1.1504
5°	.0872	.9962	.0875	50°	.7660	.6428	1.1918
6°	.1045	.9945	.1051	51°	.7771	.6293	1.2349
7°	.1219	.9925	.1228	52°	.7880	.6157	1.2799
8°	.1392	.9903	.1405	53°	.7986	.6018	1.3270
9°	.1564	.9877	.1584	54°	.8090	.5878	1.3764
10°	.1736	.9848	.1763	55°	.8192	.5736	1.4281
11°	.1908	.9816	.1944	56°	.8290	.5592	1.4826
12°	.2079	.9781	.2126	57°	.8387	.5446	1.5399
13°	.2250	.9744	.2309	58°	.8480	.5299	1.6003
14°	.2419	.9703	.2493	59°	.8572	.5150	1.6643
15°	.2588	.9659	.2679	60°	.8660	.5000	1.7321
16°	.2756	.9613	.2867	61°	.8746	.4848	1.8040
17°	.2924	.9563	.3057	62°	.8829	.4695	1.8807
18°	.3090	.9511	.3249	63°	.8910	.4540	1.9626
19°	.3256	.9455	.3443	64°	.8988	.4384	2.0503
20°	.3420	.9397	.3640	65°	.9063	.4226	2.1445
21°	.3584	.9336	.3839	66°	.9135	.4067	2.2460
22°	.3746	.9272	.4040	67°	.9205	.3907	2.3559
23°	.3907	.9205	.4245	68°	.9272	.3746	2.4751
24°	.4067	.9135	.4452	69°	.9336	.3584	2.6051
25°	.4226	.9063	.4663	70°	.9397	.3420	2.7475
26°	.4384	.8988	.4877	71°	.9455	.3256	2.9042
27°	.4540	.8910	.5095	72°	.9511	.3090	3.0777
28°	.4695	.8829	.5317	73°	.9563	.2924	3.2709
29°	.4848	.8746	.5543	74°	.9613	.2756	3.4874
30°	.5000	.8660	.5774	75°	.9659	.2588	3.7321
31°	.5150	.8572	.6009	76°	.9703	.2419	4.0108
32°	.5299	.8480	.6249	77°	.9744	.2250	4.3315
33°	.5446	.8387	.6494	78°	.9781	.2079	4.7046
34°	.5592	.8290	.6745	79°	.9816	.1908	5.1446
35°	.5736	.8192	.7002	80°	.9848	.1736	5.6713
36°	.5878	.8090	.7265	81°	.9877	.1564	6.3138
37°	.6018	.7986	.7536	82°	.9903	.1392	7.1154
38°	.6157	.7880	.7813	83°	.9925	.1219	8.1443
39°	.6293	.7771	.8098	84°	.9945	.1045	9.5144
40°	.6428	.7660	.8391	85°	.9962	.0872	11.4301
41°	.6561	.7547	.8693	86°	.9976	.0698	14.3007
42°	.6691	.7431	.9004	87°	.9986	.0523	19.0811
43°	.6820	.7314	.9325	88°	.9994	.0349	28.6363
44°	.6947	.7193	.9657	89°	.9998	.0175	57.2900
45°	.7071	.7071	1.0000	90°	1.0000	0.0000	Undefined

Extra Practice – Chapter 1

Evaluate the expression using the given value for the variable.

1. $5x + 7;\ x = 3$ **2.** $2n - 9;\ n = 5$ **3.** $y \div 3;\ y = 27$

4. $14 - 3s;\ s = 4$ **5.** $28 \div y;\ y = 4$ **6.** $m \div 9;\ m = 72$

Solve the equation.

7. $450 + 35 = t$ **8.** $9007 - 138 = c$ **9.** $f = 19 + 189$

10. $k = 370 - 174$ **11.** $28 + 347 + 6 = r$ **12.** $d = 4 + 89 + 275$

Solve the equation.

13. $n = 43 \cdot 9$ **14.** $9^2 = x$ **15.** $p = 38 \cdot 16$

16. $s = 4(6 \cdot 21)$ **17.** $6(5 + 2) = y$ **18.** $n = 18(4 - 1)$

19. $a = 640 \div 4$ **20.** $96 \div 4 = w$ **21.** $121 \div 11 = f$

22. $\dfrac{384}{24} = d$ **23.** $g = \dfrac{156}{13}$ **24.** $\dfrac{1288}{56} = y$

Solve the equation.

25. $5 + 3 \cdot 2 = b$ **26.** $21 \div 7 + 6 = d$ **27.** $(9 - 3)6 = g$

28. $h = 4 + 12 \div 6$ **29.** $0(5 + 18) \div 6 = j$ **30.** $16 \div (5 - 3) - 1 = k$

Combine like terms.

31. $n + 5n$ **32.** $7c - 2 + 4c$ **33.** $12x + 3y - x$

34. $25t - 16t - t$ **35.** $13k + 7 - 12k$ **36.** $28 + 5j + 2$

37. $8a + 4c - 3c$ **38.** $16 + 7r + 8r - 4$ **39.** $9x + 2y + y - 5x$

Write an expression for the phrase.

40. 36 less than a number r **41.** x decreased by 15

42. 45 more than a number x **43.** 42 greater than p

44. w used as a factor 4 times   **45.** the sum of x and y

Complete.

46. Lisa is 8 lb lighter than her dog Dynamo.
    Let w = weight of Dynamo. Then ___?___ = Lisa's weight.

Extra Practice – Chapter 2

Write the decimals in order from greatest to least.

1. 4.7, 4.82, 4.16   **2.** 0.61, 0.35, 0.109   **3.** 6.2, 4.863, 9.7

4. 3.01, 3.018, 3.1   **5.** 6.2, 6.22, 6.02   **6.** 9.56, 8.98, 9.01

Round the number to the place of the underlined digit.

7. $\underline{0}$.8   **8.** $\underline{6}$.3   **9.** 1$\underline{2}$.71   **10.** 35.$\underline{6}$7   **11.** 1.2$\underline{6}$5

12. 16.0$\underline{4}$9   **13.** 0.$\underline{5}$53   **14.** 14.2$\underline{5}$6   **15.** 24$\underline{9}$.9   **16.** 3$\underline{9}$.9

Add, subtract, multiply, or divide.

17. $2.04 + 0.07$ **18.** $36.21 - 0.46$ **19.** $21.09 - 0.672$

20. $5.63 - 0.7$ **21.** 4×0.7 **22.** 0.63×7

23. 5.1×0.6 **24.** 3.2×0.004 **25.** 27.2×10

26. 100×6.24 **27.** 9.6×100 **28.** 4.7×10^2

29. $14.4 \div 12$ **30.** $84 \div 2.1$ **31.** $1008 \div 5.6$

32. $2.313 \div 0.9$ **33.** $49.14 \div 2.1$ **34.** $47.1 \div 10$

35. $3.62 \div 100$ **36.** $5.084 \div 1000$ **37.** $6.4 \div 100$

Complete.

38. $4 \text{ cm} = \underline{\ ?\ } \text{ mm}$ **39.** $12 \text{ mm} = \underline{\ ?\ } \text{ cm}$ **40.** $46 \text{ cm} = \underline{\ ?\ } \text{ m}$

41. $200 \text{ km} = \underline{\ ?\ } \text{ m}$ **42.** $15 \text{ L} = \underline{\ ?\ } \text{ mL}$ **43.** $2.4 \text{ g} = \underline{\ ?\ } \text{ mg}$

Write an expression for the phrase.

44. 6.4 decreased by y **45.** r greater than 3.401

46. c multiplied by 4.6   **47.** 17.4 divided by n

48. p used as a factor 6 times   **49.** 3.07 increased by c

Solve.

50. A traveling salesperson figures her daily driving distance ranges from 60 mi to 130 mi. In the last 7 working days she drove 1085 mi. Did she drive within her estimated range each day?

Extra Practice – Chapter 3

Is the first number divisible by the second number? Write *Yes* or *No*.

1. 480 by 2   **2.** 634 by 4   **3.** 259 by 3   **4.** 260 by 5

5. 1355 by 5   **6.** 432 by 6   **7.** 870 by 3   **8.** 246 by 4

Use exponents to write the prime factorization of the number.

9. 120   **10.** 72   **11.** 625   **12.** 250

13. 68   **14.** 450   **15.** 155   **16.** 270

Write the GCF of the pair of numbers.

17. 8 and 10   **18.** 9 and 6   **19.** 24 and 15

20. 25 and 40   **21.** 12 and 32   **22.** 21 and 35

Write the GCF of the set of expressions.

23. n^2 and n **24.** $2k$ and 10   **25.** $2r^2$ and $2r$

26. $5m$ and $10m^2$ **27.** $6f$ and f^2 **28.** $2c^2$ and $14c$

Use the distributive property to factor the expression.

29. $2a - a$ **30.** $5x + 10y$ **31.** $2h^2 - 6$

32. $12e^2 + 6e$ **33.** $25r + r^2$ **34.** $4ab + 8a$

35. $9b^2 + 3b + 15$ **36.** $ay^2 - a^2y$ **37.** $10n + 16n^2 + 4$

38. $7n - 21 - 14n^2$ **39.** $xy^2 + xy - 3x^2y$ **40.** $5ab^2 - a^2b + 2ab^2x$

Write the LCM of the numbers.

41. 6 and 7   **42.** 4 and 16   **43.** 6 and 9

44. 10 and 4   **45.** 8 and 12   **46.** 18 and 12

Solve.

47. Quality control workers in a factory must inspect every 20th item off the manufacturing line to see if it has any imperfections. The first item to be tested will be the 20th item manufactured, the second item to be tested will be the 40th item manufactured, and so on. How many items must be manufactured in order for the workers to test 15 items?

Extra Practice – Chapter 4

Write the fraction in lowest terms.

1. $\dfrac{16}{20}$ **2.** $\dfrac{4}{8}$ **3.** $\dfrac{18}{24}$ **4.** $\dfrac{20}{35}$ **5.** $\dfrac{5h}{h}$

6. $\dfrac{x^2}{x}$ **7.** $\dfrac{2x}{x}$ **8.** $\dfrac{3a^2}{a}$ **9.** $\dfrac{2a}{2a^2}$ **10.** $\dfrac{15cx}{5c}$

Add or subtract. Write the answer in lowest terms.

11. $\dfrac{3}{8} + \dfrac{3}{8}$ **12.** $\dfrac{2}{3} + 7$ **13.** $\dfrac{11}{12} + \dfrac{8}{12}$ **14.** $\dfrac{4}{9} + \dfrac{2}{3}$

15. $\dfrac{5}{7} + \dfrac{3}{4}$ **16.** $1\dfrac{1}{2} + 2\dfrac{3}{5}$ **17.** $\dfrac{4}{7} + 3\dfrac{5}{8}$ **18.** $\dfrac{6}{7} - \dfrac{1}{3}$

19. $\dfrac{7}{9} - \dfrac{3}{4}$ **20.** $\dfrac{9}{10} - \dfrac{1}{2}$ **21.** $7\dfrac{3}{4} - 1\dfrac{1}{3}$ **22.** $4\dfrac{1}{8} - \dfrac{9}{10}$

23. $\dfrac{4}{a} - \dfrac{3}{a}$ **24.** $\dfrac{6}{c} + \dfrac{4}{c}$ **25.** $\dfrac{2a}{7} + \dfrac{3a}{7}$ **26.** $\dfrac{4n}{10} + \dfrac{n}{10}$

Multiply or divide. Write the answer in lowest terms.

27. $\dfrac{3}{4} \cdot \dfrac{4}{5}$ **28.** $\dfrac{1}{2} \cdot \dfrac{1}{2}$ **29.** $1\dfrac{1}{2} \cdot \dfrac{2}{3}$ **30.** $6 \cdot \dfrac{5}{12}$

31. $\dfrac{5}{9} \div 5$ **32.** $\dfrac{6}{7} \div \dfrac{3}{5}$ **33.** $\dfrac{1}{c} \cdot 8$ **34.** $\dfrac{3}{4} \cdot a$

35. $\dfrac{3n}{7} \cdot \dfrac{21}{9}$ **36.** $\dfrac{2a}{7} \cdot \dfrac{a}{12}$ **37.** $\dfrac{x}{6} \div \dfrac{x}{4}$ **38.** $\dfrac{6}{5r^2} \div \dfrac{3}{r}$

Write the fraction as a terminating or a repeating decimal.

39. $\dfrac{4}{5}$ **40.** $\dfrac{5}{8}$ **41.** $\dfrac{1}{3}$ **42.** $\dfrac{5}{9}$ **43.** $\dfrac{4}{11}$

44. $\dfrac{3}{8}$ **45.** $\dfrac{1}{6}$ **46.** $\dfrac{5}{3}$ **47.** $\dfrac{9}{2}$ **48.** $\dfrac{12}{11}$

Solve.

49. Suppose you and your family are on a cross-country trip and you want to cover 2500 mi in seven days. The first day you intend to drive 300 mi to visit a special spot. How far must you drive each of the other days? Round your answer to the nearest ten miles.

Extra Practice – Chapter 5

Solve the equation.

1. $n + 9 = 14$
2. $4 + s = 12$
3. $52 = x + 19$
4. $k - 16 = 30$
5. $m - 20 = 42$
6. $35 = y - 16$
7. $f - 28 = 7$
8. $61 = y - 52$
9. $n + 16 = 51$
10. $5a = 40$
11. $8c = 48$
12. $15g = 105$
13. $20r = 100$
14. $4t = 76$
15. $\dfrac{n}{4} = 5$
16. $\dfrac{a}{7} = 12$
17. $\dfrac{1}{2}c = 18$
18. $\dfrac{3}{5}n = 6$

Solve the equation.

19. $2a + 5 = 25$
20. $6n + 2 = 20$
21. $3y - 7 = 14$
22. $2(x + 1) = 16$
23. $5(y - 1) = 15$
24. $6(c - 4) = 12$
25. $68 = 4(f + 5)$
26. $45 = 9(g + 2)$
27. $6r = 2r + 12$
28. $7h = 2h + 15$
29. $2a + 1 = a + 17$
30. $5p - 2 = 2p + 7$
31. $14 - s = 7s - 2$
32. $9n + 4 = 5n + 20$
33. $8w - 1 = w + 13$

Complete by writing an algebraic expression.

34. Claire has twice as much money as Cindy.
    Emily has $20 more than Claire.
    Let m = the amount of money Cindy has.
    Then __?__ = the amount of money Claire has,
    and __?__ = the amount of money Emily has.
35. Chip scored four more points than Kevin in the game.
    Kevin scored twice as many points as Mike.
    Let p = the number of points Kevin scored.
    Then __?__ = the number of points Chip scored,
    and __?__ = the number of points Mike scored.

Write an equation. Solve.

36. Franny's brother Jeff is half Franny's age. Jeff is 9 years old. How old is Franny?
37. Pearl has twice as much money this week as she had last week. She has $25. How much money did she have last week?

Extra Practice – Chapter 6

Find the perimeter or circumference. Use 3.14 for π.

1. rectangle
$l = 4.6$ cm
$w = 3.8$ cm

2. square
$s = 12$ in.

3. circle
$r = 4$

The measure of an angle is given below. Tell whether the angle is an acute, obtuse, or a right angle.

4. 160°

5. 90°

6. 35°

7. 85°

8. 110°

Find the area. Use 3.14 for π.

9. rectangle
$b = 5$ ft
$h = 8.4$ ft

10. triangle
$b = 12$ cm
$h = 3.5$ cm

11. circle
$d = 4$ m

Find the volume.

12. rectangular prism
$B = 27$ cm²
$h = 8$ cm

13. triangular prism
$B = 56$ m²
$h = 2.7$ m

14. cube
$l = 12$ mm

15. cylinder
$r = 2$ cm
$h = 15$ cm

16. rectangular pyramid
$B = 150$ in.²
$h = 5.5$ in.

17. cone
$r = 4$ mm
$h = 12$ mm

Find the surface area.

18.
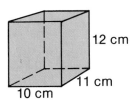
12 cm
11 cm
10 cm

19.
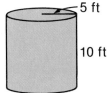
5 ft
10 ft

20.
10 m
4 m
6 m
8 m

A formula for determining distance traveled is $D = rt$. D = distance, r = rate, t = time. Use the formula to solve the problems.

21. A van is traveling down the road at 55 mi per hour. How far will it travel in 4 hours?

22. A truck driver plans to drive 440 mi at a speed of 55 mi per hour. How long will the trip take if the driver takes two breaks, each 15 minutes long?

Extra Practice – Chapter 7

Write the ratio as a fraction in lowest terms.

1. 24 books out of 250 books

2. 70 students absent to 450 total

3. 18 hatchback cars to 50 cars

4. 21 solar heated houses out of 35 houses

Solve the proportion. Round the number to the nearest tenth.

5. $\dfrac{12}{18} = \dfrac{x}{28}$ **6.** $\dfrac{25}{80} = \dfrac{n}{120}$ **7.** $\dfrac{65}{90} = \dfrac{80}{a}$ **8.** $\dfrac{9}{11} = \dfrac{c}{85}$

Write the percent.

9. 7 out of 10

10. 35 out of 50

11. 17 out of 340

12. 288 out of 480

13. What percent of 90 is 45?

14. What percent of 650 is 156?

Solve.

15. 26% of 45 is what number?

16. 35% of 600 is what number?

17. What is 80% of 25?

18. $2\frac{1}{2}$% of \$700 is what amount?

19. What is $\frac{1}{2}$% of 90?

20. What is 200% of \$875?

21. 126 is 20% of what number?

22. 11.55 is 5.5% of what number?

23. 65% of what number is 140?

24. 86% of what number is 172?

Find the percent of increase or decrease.

25. old price: \$350
increase: \$35

26. from: \$700
to: \$455

27. from: 1500
to: 1800

Write an equation. Solve.

28. In the quality control department in a TV factory, workers found that 6 out of 40 TV sets had problems with the on-off switch. How many TV sets out of 250 might be expected to have the same problem? Round your answer to the nearest whole number.

29. A bus pass used to cost \$17 per month. The price went up to \$22. What percent of increase is this? Round the answer to the nearest tenth of a percent.

Extra Practice – Chapter 8

Compare. Write $<$ or $>$.

1. $^-7 \underline{\ ?\ } 0$ 2. $3 \underline{\ ?\ } ^-2$ 3. $^-4 \underline{\ ?\ } ^-7$ 4. $^-8 \underline{\ ?\ } 5$

5. $9 \underline{\ ?\ } ^-2$ 6. $^-7 \underline{\ ?\ } ^-10$ 7. $^-8 \underline{\ ?\ } ^-1$ 8. $12 \underline{\ ?\ } ^-17$

Add or subtract.

9. $4 + ^-6$ 10. $0 + ^-9$ 11. $^-1 + ^-3$ 12. $^-10 + ^-8$

13. $^-12 - 5$ 14. $^-6 - ^-2$ 15. $^-9 - ^-11$ 16. $8 - ^-4$

17. $^-6 - ^-8$ 18. $^-9 + ^-3$ 19. $^-4 - ^-1$ 20. $16 + ^-8$

21. $^-20 + ^-12$ 22. $^-35 - ^-6$ 23. $8 - ^-15$ 24. $^-6 - ^-19$

Multiply or divide.

25. $6(^-4)$ 26. $^-8(^-6)$ 27. $^-5(4)$ 28. $^-6 \cdot ^-3$

29. $7 \cdot ^-8$ 30. $7(^-3)$ 31. $^-9 \cdot ^-8$ 32. $4 \cdot ^-9$

33. $^-21 \div 3$ 34. $35 \div ^-5$ 35. $^-42 \div ^-6$ 36. $^-56 \div ^-8$

37. $100 \div ^-25$ 38. $^-36 \div 9$ 39. $^-81 \div ^-9$ 40. $72 \div ^-8$

Add, subtract, multiply, or divide.

41. $4.6 - ^-2.3$ 42. $8.6 + ^-9$ 43. $^-6.3 \times ^-5$ 44. $^-7.6 \div ^-2$

45. $2.5 + ^-3.7$ 46. $^-8.2 \times ^-3.1$ 47. $^-8.4 \div 0.6$ 48. $^-0.76 - ^-0.235$

49. $1\frac{1}{3} + ^-4\frac{3}{5}$ 50. $2\frac{7}{8} - \frac{^-1}{6}$ 51. $\frac{6}{8} \times ^-1\frac{3}{4}$ 52. $^-1\frac{2}{3} \div ^-4\frac{1}{6}$

53. $8\frac{1}{2} + ^-2\frac{1}{2}$ 54. $^-6\frac{2}{3} - ^-1\frac{1}{3}$ 55. $\frac{^-4}{5} \times ^-1\frac{1}{2}$ 56. $12\frac{1}{4} \div ^-1\frac{3}{4}$

Solve the equation or inequality.

57. $x - ^-2 = 7$ 58. $y + ^-3 = 9$ 59. $^-2a = 12$ 60. $\frac{b}{^-4} = ^-10$

61. $^-6h = 5.4$ 62. $^-8k = ^-6.4$ 63. $c + \frac{^-1}{2} = 7$ 64. $m - \frac{^-2}{3} = 8$

65. $t + 6 > 2$ 66. $x - 5 < 3$ 67. $y - 6 > ^-8$ 68. $^-8y < 2$

69. $^-5n < ^-2$ 70. $\frac{n}{7} < ^-1$ 71. $\frac{a}{^-4} > 10$ 72. $j - 6 \leq 15$

Extra Practice – Chapter 9

Find the next three terms of the sequence.

1. 0, 4, 8, 12, 16

2. 5, 11, 17, 23, 29

3. ⁻5, 0, 5, 10, 15

4. 8, 13, 18, 23, 28

5. ⁻15, ⁻9, ⁻3, 3, 9

6. 8, 3, ⁻2, ⁻7, ⁻12

Graph the ordered pairs on a coordinate plane. Label the point with the given letter.

7. $A(4, 0)$

8. $B(7, {}^-5)$

9. $C({}^-3, 0)$

10. $D({}^-1, {}^-6)$

11. $E(0, 6)$

12. $F(0, {}^-4)$

13. $G({}^-3, {}^-2)$

14. $H({}^-4, 3)$

Graph the following data. Use the graph to find the 6th term.

15.

No. of term	Sequence
1	0
2	7
3	14
4	21

16.

No. of term	Sequence
1	5
2	9
3	13
4	17

17.

No. of term	Sequence
1	⁻9
2	⁻4
3	1
4	6

Find three ordered pairs that are solutions of the given equation. Then graph them and draw the line for the equation.

18. $y = x + 1$

19. $y = x - 2$

20. $y = x + 5$

21. $y = 2x + 2$

22. $y = 3x - 1$

23. $y = {}^-x + 4$

24. $y = {}^-4x + 2$

25. $y = 5x - 5$

26. $y = \frac{1}{2}x + 6$

What are the slope and *y*-intercept of the equation?

27. $y = 2x + 9$

28. $y = 8x + 7$

29. $y = {}^-x - 2$

30. $y = \frac{1}{4}x + 5$

31. $y = \frac{2}{3}x - 6$

32. $y = \frac{3}{2}x + 8$

33. $y = {}^-8x + \frac{5}{3}$

34. $y = \frac{3}{4}x + \frac{3}{5}$

35. $y = \frac{{}^-1}{3}x - \frac{1}{4}$

Is the ordered pair a solution of the linear inequality?

36. $(2, 5); y > x + 2$

37. $({}^-1, 7); y > 2x + 10$

38. $(0, 4); y < x + 7$

39. $({}^-5, {}^-12); y < 3x + 5$

Extra Practice – Chapter 10

1. Find the range, mode, median and mean of the given data: 3.1; 8.4; 5.1; 3.1; 0.8; 1.1; 3.6.

Use the circle graph.

2. There are 2450 students attending the college. How many are in each age group?

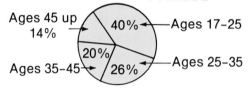

STUDENTS, EAST SIDE COMMUNITY COLLEGE

Ages 45 up 14%
Ages 17–25 40%
Ages 35–45 20%
26% Ages 25–35

Use the bar graph.

3. About what percent of the total sales were red?

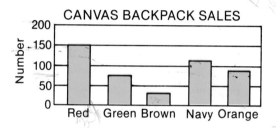

CANVAS BACKPACK SALES

Red Green Brown Navy Orange

Use the line graph.

4. Traffic at __?__ is the heaviest.
5. The decline in traffic after __?__ (time) is sharper than after __?__.

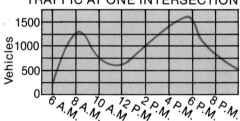

TRAFFIC AT ONE INTERSECTION

Use the histogram.

6. About __?__ watched television 40–59 hours per week.
7. There were approximately __?__ viewers surveyed in all.

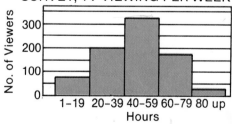

SURVEY, TV VIEWING PER WEEK

1–19 20–39 40–59 60–79 80 up
Hours

Use the scattergram.

8. Using the trend line, how much time would you expect a 17-mile commute to take? a 5-mile commute?

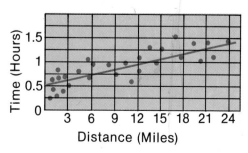

Distance (Miles)

Extra Practice – Chapter 11

For each pair of similar triangles, find *n*.

1.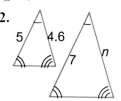

2.

3.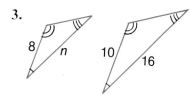

Use the Pythagorean Theorem to find the length of the hypotenuse to the nearest hundredth. Use the table on page 470.

4.

5.

6.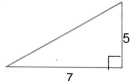

For the triangle, find tan *A*, sin *B*, and cos *A*. Write as a ratio in lowest terms.

7.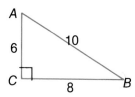

8.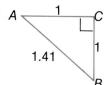

9.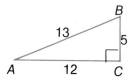

10–12. For Exercises 7–9, use the table on page 471 to find the measures of ∠*A* and ∠*B* to the nearest degree.

Find the measure of ∠*A* and ∠*B* to the nearest degree. Use the table on page 471.

13.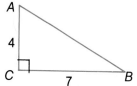

14.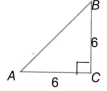

15.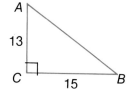

Extra Practice – Chapter 12

The following words are written on individual cards. Words are picked at random. Use this information for Exercises 1–7.

TO	SLOWLY	THE	APPLE	ON
BALLOON	MEOWS	A	MY	RUNNING
WALK	UNDER	KITTEN	CHAIR	WEATHER

1. What is P(CHAIR)?
2. What is P(WALK or APPLE)?
3. What is P(not WEATHER)?
4. What is P(SIDEWALK)?
5. Suppose three words are picked and each time replaced. What is the probability that the words THE KITTEN MEOWS are picked in any order?
6. If three words are picked, what is the probability that THE is picked first, KITTEN next, and MEOWS third? The cards are not replaced as they are picked.
7. The words BARKS, BIG, and DOG are written on a spinner. What is the probability of picking KITTEN from the cards and spinning BARKS on the spinner?

There are 5 yellow cards, 4 blue cards, and 6 green ones in a box. One card is picked at random. Use this information for Exercises 8–9.

CARDS	
yellow	5
blue	4
green	6

8. What are the odds in favor of picking a yellow card?
9. What are the odds against picking a green card?

Chip and his friends each took a random sample of 15 marbles from a bag containing 360 marbles, some red, some white, and some blue, with the following results.

	Red	White	Blue
Chip	4	6	5
Roberto	3	5	7
Peg	5	4	6

10. How many marbles of each color would you predict to be in the bag based on the combined data?

Extra Practice – Chapter 13

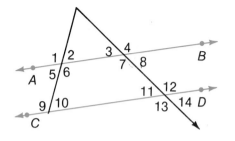

$\overleftrightarrow{AB}$ and $\overleftrightarrow{CD}$ are parallel. Complete.

1. $\angle 1 = 120°$ so $\angle 9 = \underline{?}$

2. $\angle 8 = 75°$ so $\angle 11 = \underline{?}$

3. $\angle 2 = 60°$ so $\angle 10 = \underline{?}$ and $\angle 5 = \underline{?}$

Find the missing measure of the angle in triangle ABC.

4. $\angle A = 45°$, $\angle B = 60°$ 5. $\angle C = 90°$, $\angle A = 35°$ 6. $\angle B = 110°$, $\angle A = 22°$

The pairs of polygons are similar. Find the length of the missing side.

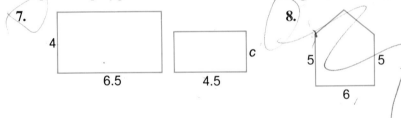

7.

8.

Find the lengths of the sides of the triangle.

9. perimeter 30

10. perimeter 36

11. perimeter 54

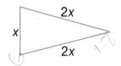

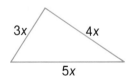

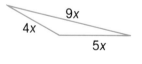

Look at the parallelogram. Complete.

12. $\overline{AB} \parallel \underline{?}$, so $\angle 2 = \angle \underline{?}$.

13. $\overline{AD} \parallel \underline{?}$. If $AD = 8$ cm, then $CB = \underline{?}$ cm.

14. $\triangle ABC \cong \underline{?}$, so $\angle 5 = \angle \underline{?}$.

15. $\angle 1 + \angle 2 = \angle \underline{?} + \angle \underline{?}$

16. $\angle 5 + \angle 3 = \angle \underline{?} + \angle \underline{?}$

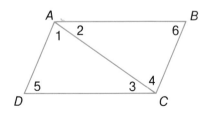

Extra Practice – Chapter 14

Write the output for the program.

1.
```
5   REM AREA OF TRIANGLE
10  LET B = 4
20  LET H = 9
30  PRINT "AREA IS" 1/2*B*H
40  END
```

2.
```
10  FOR A = 10 TO 20
20  PRINT "15% OF" A "IS" .15*A
30  NEXT A
40  END
```

3. Change the program in Exercise 2 so that the computer will print 15% of 10, 15, 20, and 25.

4. Make a flowchart to show the steps to follow when you want to compare two numbers X and Y and write the numbers in order, the greater number second.

Write the output for the program.

5.
```
5   REM BUDGET CHECK
10  READ A, B, C
20  LET N = A+B+C
30  PRINT "YOU'VE SPENT" N
    "SO FAR"
40  DATA 24.98, 6.59, 16.99
50  END
```

6.
```
5   REM FIGURING TOTAL COST
10  READ P
20  LET C = P*.29
30  PRINT "THE TOTAL COST IS" C
40  GOTO 10
50  DATA 5, 6, 7, 8, 9
60  END
```

7.
```
5    REM TRIANGLE CHECK
10   READ A, B, C
20   LET S = A+B+C
30   IF S < > 180 THEN 70
40   PRINT "THE SUM OF THE MEASURES OF THE ANGLES IS 180"
50   PRINT "SO THE POLYGON IS A TRIANGLE"
60   GOTO 10
70   PRINT "NOT A TRIANGLE"
80   GOTO 10
90   DATA 40, 50, 90, 62, 75, 30, 25, 82, 90
100  END
```

8. Change the program in Exercise 7 so that the computer will read data for four angles and determine whether the figure can be a quadrilateral or not.

Glossary

absolute value (p. 274) The distance of a number from 0 on the number line.

acute angle (p. 179) An angle with a measure between 0° and 90°.

additive inverse (p. 241) *See under* opposite of a number.

algebraic factorization (p. 81) An algebraic expression written as the product of its factors.

alternate interior angles (p. 413) *See under* parallel lines.

associative properties (pp. 5, 14)
$(a + b) + c = a + (b + c)$
$(ab)c = a (bc)$

base, in a percent (p. 228) The number represented by 100% to which other quantities are compared.

certain event (p. 376) An event with probability 1, or 100%.

circumference (p. 176) The distance around a circle.

coefficient (p. 24) A number multiplied by a variable in an expression.

collinear points (p. 411) Points that lie on the same line.

commutative properties (pp. 5, 14)
$a + b = b + a$
$ab = ba$

composite number (p. 75) A number with more than two factors.

congruent polygons (p. 419) Polygons that have the same size and the same shape.

consecutive integers (pp. 72, 294) Integers that differ by 1.

coordinates (p. 285) The number paired with a point on the number line; an ordered pair of numbers assigned to a point on the coordinate plane.

corresponding angles (p. 413) *See under* parallel lines.

cosine ratio (p. 358) The cosine of an acute angle of a right triangle is the ratio of the length of the side adjacent to the angle to the length of the hypotenuse.

dependent events (p. 390) Events in which the outcome of one event affects the outcome of the other events.

diagonal (p. 415) A line segment that joins two non-consecutive vertices of a polygon.

diameter (p. 176) A line segment that passes through the center of a circle and whose endpoints are on the circle. Also, the length of such a segment.

distributive properties (p. 14)
$a(b + c) = ab + ac$
$a(b - c) = ab - ac$

equally likely outcomes (p. 375) Two or more outcomes are equally likely if each outcome is as likely to occur as the others.

equiangular triangle (p. 425) A triangle in which the three angles have the same measure.

equilateral triangle (p. 425) A triangle with three equal sides.

equation (p. 4, 135) A statement that two numbers or expressions are equal.

event (p. 375) A set of one or more outcomes in a probability problem.

experimental probability (p. 396) The ratio of the number of favorable outcomes of an experiment to the number of experiments.

exponent (p. 15) A raised number that indicates the number of times a factor is used.

expression (p. 1) An expression represents a number. $x + 2$ and $3d - 4$ are expressions.

factor (p. 75) Any of two or more numbers that are multiplied to obtain a product.

favorable outcome (p. 375) A desired outcome.

flowchart (p. 449) A diagram of the logical steps needed to solve a problem or to write a computer program.

frequency distribution (p. 329) A summary of the number of items in each interval of a tally.

frequency polygon (p. 330) A broken line connecting the midpoint of the top of each bar of a histogram.

graph of a number (p. 241) The point on the number line paired with a number.

greatest common factor (GCF) (p. 79) The greatest number that is a factor of each of two or more numbers or expressions.

histogram (p. 330) A graph used to show a frequency distribution.

hypotenuse (p. 351) The side of a right triangle that is opposite the right angle.

identity number for addition (p. 21) Zero. $0 + a = a$.

identity number for multiplication (p. 21) One. $1 \cdot a = a$.

impossible event (p. 376) An event with probability 0.

independent events (p. 386) Events in which the outcome of one event in no way affects the outcome of the other events.

inequality (p. 267) A statement that two quantities are unequal.

integers (p. 241) The numbers 1, 2, 3, 4, . . ., together with their opposites ⁻1, ⁻2, ⁻3, ⁻4, . . ., and 0.

isosceles triangle (p. 425) A triangle with at least two sides equal.

least common denominator (LCD) (p. 109) The least common multiple of the denominators of two or more fractions.

least common multiple (LCM) (p. 87) The least positive number that is a multiple of two of more given numbers.

leg of a right triangle (p. 351) One of the two sides which form the right angle of a right triangle.

like terms (p. 24) Terms with the same variables.

linear equation (p. 295) An equation whose graph is a straight line.

linear inequality (p. 298) An inequality whose graph is a region above or below a straight line.

mean (p. 309) The sum of a set of numbers divided by the total number of addends.

median (p. 309) In a set of numbers arranged in increasing order, the middle number.

mode (p. 309) The number that occurs most frequently in a set of data. Sometimes a set of data has no mode.

multiple (p. 87) A multiple of a number is any product of that number and a natural number.

natural numbers The numbers 1, 2, 3, 4, and so on.

obtuse angle (p. 179) An angle with a measure greater than 90° and less than 180°.

odds (p. 393) A ratio that compares favorable outcomes to unfavorable outcomes.

opposite of a number (p. 241) A number and its opposite are the same distance from 0 on the number line but on opposite sides. The opposite of 3 is ⁻3. Also called *additive inverse*.

outcome (p. 375) One possible result in a probability problem.

parallel lines (p. 180) Lines in a plane that do not intersect. $\angle 1$ and $\angle 5$ are corresponding angles; $\angle 3$ and $\angle 5$ are alternate interior angles; $\angle 2$ and $\angle 8$ are alternate exterior angles.

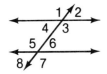

parallelogram (p. 428) A quadrilateral with opposite sides parallel.

percent (p. 216) A notation for a ratio with the denominator 100.

perimeter (p. 173) The distance around a geometric figure.

perpendicular lines (p. 180) Lines that meet to form right angles.

power (p. 15) A product of equal factors. The third power of 7 is 7^3 or $7 \cdot 7 \cdot 7$.

prime number (p. 75) A number greater than 1 whose only factors are itself and 1.

prism (p. 192) A solid figure with two identical bases.

probability (p. 375) The ratio of the number of favorable outcomes to the total number of outcomes.

proportion (p. 212) An equation that shows two equal ratios.

Pythagorean theorem (p. 351) In a right triangle, the square of the length of the hypotenuse (c) equals the sum of the squares of the lengths of the two other sides (a and b); $c^2 = a^2 + b^2$.

radius (p. 177) A segment drawn from the center of a circle to any point on the circle. Also, the length of such a segment.

random sample (p. 399) A sample in which each possible outcome is equally likely to appear.

range (p. 309) The difference between the greatest and least numbers in a set of data.

ratio (p. 209) The ratio of one number to another (not zero) is the quotient of the first number divided by the second.

reciprocals (p. 118) Two numbers whose product is 1.

rhombus (p. 428) A parallelogram with all sides equal in length.

right angle (p. 179) An angle with a measure of 90°.

right prism A prism in which the edges joining the bases are perpendicular to the bases; that is, each edge is perpendicular to all the lines it intersects in the bases.

sample space (p. 379) The set of all possible outcomes in a probability problem.

scalene triangle (p. 425) A triangle with no two sides equal.

scattergram (p. 333) A graph of individual points that shows the relationship between two sets of data. Each point represents an ordered pair with one coordinate from each set of data.

scientific notation (p. 90) The system of writing a number as the product of two factors in which the first factor is a number between 1 and 10 and the second factor is a power of 10. For example, $210{,}000 = 2.1 \times 10^5$.

segment (p. 411) A part of a line consisting of two endpoints and all the points between them.

sequence (p. 85) An ordered list of terms in which each term can be obtained by applying a rule.

similar polygons (p. 345) Polygons in which corresponding angles have equal measures and corresponding sides are in proportion.

sine ratio (p. 358) The sine of an acute angle of a right triangle is the ratio of the length of the side opposite the angle to the length of the hypotenuse.

skew lines (p. 412) Lines that do not intersect and do not lie in the same plane.

slope (p. 294) The steepness of a nonvertical line; the ratio

$$\text{slope} = \frac{\text{difference of } y\text{-coordinates}}{\text{difference of } x\text{-coordinates}}.$$

solution (pp. 135, 292) Replacement values for variables that make the equation true.

square root (p. 352) If $a^2 = n$ then a is a square root of n. The positive square root of n is shown by the symbol $\sqrt{n}$ and the negative square root by $-\sqrt{n}$.

statistics (p. 309) The study of the collection, organization, and interpretation of data.

straight angle (p. 179) An angle with a measure of 180°.

system of equations (p. 304) Two or more equations in the same variables.

term (p. 24) An expression using numerals or variables or both to indicate a product or a quotient.

term of a sequence (p. 85) One of the entries in a sequence.

transversal (p. 413) A line that intersects two other lines.

trapezoid (p. 428) A quadrilateral with exactly one pair of opposite sides parallel.

trend line (p. 333) The line to which the points of a scattergram is closest.

variable (p. 1) A symbol used to represent one or more numbers.

vertex of an angle (p. 180) The common endpoint of the sides of an angle.

vertex of a polygon (p. 415) A point where two adjacent sides of a polygon intersect.

y-intercept (p. 294) The y-coordinate of the point where a line intersects the y-axis.

Index

Credits

Book design by Ligature Publishing Services, Inc.

Cover and title page photo: ''Tiered Orbits'' by Jerome Kirk, at the Phoenix Civic Plaza. Photo by Nick Nicholson/Image Bank.

Technical art by ANCO/Boston

Diagrams on pages *173*, *234*, *235* and *308 - 343* by William DePippo.

Photographs

facing page 1 Carl Frank/Photo Researchers, Inc. **1** Eunice Harris/Photo Researchers, Inc. **8** Australian Information Service. **11** David F. Hughes/The Picture Cube. **13** Gillian Harper. **20** Adelheid Heine-Stillmark/The Image Bank. **29** Milton Feinberg/The Picture Cube. **30** Len DeMunde/The Image Bank. **34** James Ballard. **41** Anthony Howarth/Woodfin Camp. **45** Burton McNeely/The Image Bank. **48** Janeart Ltd./The Image Bank. **60** F. Fontana/The Image Bank. **65** Hank Morgan/Rainbow. **70** Bill Gallery/Stock, Boston. **85** Betty and Dennis Milon. **87** Lou Jones. **95** B. Cole/The Picture Cube. **100** Bill Gallery/Stock, Boston. **108** Owen Franken/Stock, Boston. **115** Lewis Trusty/Animals Animals. **120** Katherine Baldwin. **123** Russell Abraham/Stock, Boston. **129** Barbara Alper/The Picture Group. **134** © Joel Gordon. **140** Edith G. Haun/Stock, Boston. **145** ©Joel Gordon. **161** Bohdan Hrynewych/Stock, Boston. **162** Lou Jones. **167** Lou Jones. **168** G. L. Kooyman/Animals Animals. **172** Timothy Eagan/Woodfin Camp. **185** Focus on Sports. **191** Brett H. Froomer/The Image Bank. **194** Luis Villota/The Stock Market. **202** Katherine Baldwin. **203** Jonathon Rawle/Stock, Boston. **208** Ken Lax/The Stock Shop. **214** courtesy of Megatek Corp. **219** Mike Mazzaschi/Stock, Boston. **220** Stephen Comeau. **225** G. Gladstone/The Image Bank. **230** Michal Heron. **234** Ken Lax/The Stock Shop. **240** Thomas Ives. **243** Joe DiMaggio & JoAnne Kalish/Focus on Sports. **245** Phil Ellin/The Picture Cube. **256** NASA. **259** Michal Heron. **263** Alan G. Nelson/Animals Animals. **273** Marc Romanelli/The Image Bank. **278** Bernard P. Wolff/Photo Researchers, Inc. **284** Stephen Comeau. **297** Michal Heron. **302** Stanley Rowin/The Picture Cube. **308** Focus on Sports. **315** Stephen Comeau. **321** Rick Browne/The Picture Group. **328** Stanley Rowin/The Picture Cube. **338** R. Terry Walker/The Picture Cube. **344** Gary Cralle/The Image Bank. **348** C. T. Seymour/The Picture Cube. **351** Stephen Comeau. **363** Niki Mareschal/The Image Bank. **368** Tom Tracy/The Stock Shop. **374** David Frazier/Photo Researchers, Inc. **375** Michal Heron. **382** Bill Galler/Stock, Boston. **393** Michal Heron. **404** Michal Heron/Monkmeyer Press. **410** Michael James. **411** Michael James. **424** Dave Britzius/The Stock Shop. **438** courtesy of Evans & Sutherland and Rediffusion Simulation. **448** Harald Sund/The Image Bank. **454** Katherine Baldwin. **459** Bill Gallery/Stock, Boston. **461** Cara Moore/The Image Bank. **465** Michal Heron.

Answers to Your Turn Exercises

CHAPTER 1 WHOLE NUMBERS AND VARIABLES
Page 1 •24 •36 **Page 2** •$a + 7$
•$n \div 3$ **Page 4** •$482 + 59 = x$; $x = 541$
•$1535 + 496 = b$; $b = 2031$ **Page 5**
(Example 2) •$35 + 37 = n$; $n = 72$
•$40 + 39 = n$, $n = 79$ •$256 + 460 = n$;
$n = 716$ **(Example 3)** •147 •166 •396
Page 9 •$1500 - 685 = y$; $y = 815$
•$365 - 87 = n$; $n = 278$ **Page 11**
•$75 - 69 = w$; 6; lost 6 kg **Page 15**
•$x = 970$ •$n = 7700$ •$y = 8600$
•$a = 342$ **Page 16** •y^5 •16 •48
Page 18 •$276 \div 3 = 92$ or $276 \div 92 = 3$
•$87 \times 8 = 696$ •$6111 \div 9 = 679$ or
$6111 \div 679 = 9$ **Page 19** •$68 \div 5 = n$;
$n = 13R3$ •$160 \div 12 = g$; $g = 13R4$
Page 24 (Example 3) •$a = 14$ •$b = 18$
•$m = 148$ •$n = 100$ **(Example 4)** •$10a$
•$22n + 2x$ •$6z + y$

CHAPTER 2 DECIMALS
Page 36 •five hundred seventy-four and four
hundred five thousandths •7005.2002
Page 38 •$4.1 < 4.16$ •$30.72 > 30.62$
Page 42 (Example 3) •7.0 •18.786
(Example 4) •0.60 **Page 45** •40
Page 46 •$y = 25.908$ **Page 49**
(Example 2) •38 **(Example 3)**
•$m = 7.89$ •$d = 15.66$ **Page 51** •Yes
•$210 **Page 53 (Example 1)** •$y = 0.0084$
•$x = 0.0186$ **(Example 2)** •$a = 4892.1$
$s = 3700.6$ **Page 54** •15,000 **Page 57**
•7.42 **Page 58 (Example 3)** •8
(Example 4) •$d = 0.2378$ **Page 61**
•7000 mL •0.0095 hL

CHAPTER 3 NUMBER THEORY
Page 72 (Example 2) •2, 3, 4, 6, yes; 5, 10,
no **(Example 3)** •odd •17 •18
Page 76 •$54 = 2 \times 3^3$ •$126 = 2 \times 3^2 \times 7$
Page 77 •$2^3 \times 13$ •$2^2 \times 5 \times 11$
•$2 \times 3^3 \times 5^2$ **Page 79** •GCF of 8 and 12
is 4 •GCF of 15 and 20 is 5 •GCF of 12
and 32 is 4 •GCF of 12 and 18 is 6 •GCF
of 9 and 30 is 3 •GCF of 16 and 24 is 8
Page 81 •$5m$ •3 **Page 82** •$4(2x - 1)$
•$y(5y + 1)$ •$7(a + 2b - c)$ **Page 85**
•12:15 P.M. **Page 87** •24 •180 **Page 88**

•24 •225 •$16x^2$ **Page 90** •Yes •No
•No •120,000 •61,000,000,000
•750,000,000 **Page 91** •5.4×10^4
•2.038×10^9 •3.1×10^6

CHAPTER 4 FRACTIONS
Page 102 (Example 1) •$4\frac{3}{8}$, $2\frac{8}{9}$ **(Example 2)**
•$\frac{44}{7}$, $\frac{49}{6}$ **Page 106** •y •$\frac{1}{2b}$ •$\frac{c}{2}$
Page 109 •$6\frac{5}{12}$ •$\frac{12}{d}$ •$\frac{e}{3}$ **Page 112**
(Example 2) •$\frac{9}{x}$ •$\frac{3}{xy}$ •$\frac{3a}{10}$ **(Example 3)**
•$\frac{b}{6}$ •$\frac{3}{5}$ •$1\frac{9}{10}$ **Page 116** •$\frac{2}{5}$ •$\frac{7a}{20}$
Page 119 •$\frac{1}{7}$ •$\frac{1}{6}$ •$\frac{1}{2m}$ **Page 122**
•$2666.67 **Page 125** •$\frac{1}{4}$ •$0.\overline{6}$ •$0.8\overline{3}$

CHAPTER 5 SOLVING EQUATIONS
Page 136 (Example 2) •Yes •No •Yes
(Example 4) •$h = 5$ •$y = 14$ •$z = 40$
Page 139 •$c = 40$ •$d = 52$ •$e = 9$
Page 141 (Example 2) •$b = 6$ •$c = 9$
•$d = 16\frac{8}{9}$ **(Example 3)** •Divide by 5;
$p = 4$ •Subtract 10; $s = 6$ **Page 143**
•$a = 26$ •$c = 36$ **Page 146** •$w = 35$
•$v = 20$ •$n = 15$ **Page 150 (Example 3)**
•$d = 6$ •$x = 5$ •$e = 39$ **(Example 4)**
•$x = 8$ •$x = 4$ **Page 154** •$d = 3$
•$f = 6$ **Page 155** •$3s$, $3s + 1$ **Page 156**
•$2x + 2$ •$2n$, $2n + 1$ **Page 159** •60 in.
Page 163 •$x + 29 = 57$; 28 people

CHAPTER 6 FORMULAS
Page 173 •84 ft **Page 174** •65
Page 177 •28.26 yd •34.54 in. •66 m
Page 181 •JH •AB **Page 184** •10 in.²
•26 cm² **Page 186** •113.04 cm² **Page 187**
•34.54 sq. units •226.08 sq. units
Page 189 •45 lb **Page 193** •60 cm³
Page 195 •4019.2 in.³ •$1339.7\overline{3}$ in.³
Page 198 •552 in.² **Page 199** •253 in.²

CHAPTER 7 RATIO, PROPORTION, PERCENT
Page 210 •$\frac{3}{1}$ •$\frac{8}{5}$ •$\frac{9}{10}$ **Page 213** •$7\frac{1}{2}$ h
•280 km **Page 217 (Example 1)** •7%

•15%   •14%   **(Example 2)**   •$\frac{37}{100}$; 0.37
•$\frac{9}{100}$; 0.09   •$\frac{83}{100}$; 0.83   **Page 219**   •$7.14
Page 220   •$18.75   •$266.00   **Page 223**
•Diamond Bank   **Page 225**   •1.5   •3.96
Page 226 (Example 2)   •Office supplies, $60;
salaries, $6000 (Foot of page)   •116.1
•150   **Page 228**   •2212   •4288 books
Page 231   •37.5%   •20%

CHAPTER 8 INTEGERS AND RATIONAL NUMBERS

Page 242   •$^-12 < ^-2$ **Page 243**   •$^-9$
Page 246   •$y = ^-4$ **Page 248**   •$^-3$ •14
•$^-7$ •11   **Page 250**   •$y = ^-56$
•$p = ^-30$ **Page 251**   •$p = ^-27$ •$g = ^-56$
Page 253   •$y = ^-6$ **Page 257 (Example 2)**
•$t = ^-8.69$ •$n = \frac{^-5}{8}$ **(Example 4)**
•$h = ^-9\frac{5}{8}$ **Page 265**   •$n = ^-7$ •$k = 2$
•$m = ^-24.8$ •$a = ^-21.7$ **Page 268**
•$t < 4$ •$x > 7$ **Page 269**   •$x \geq ^-1$

CHAPTER 9 GRAPHING IN THE COORDINATE PLANE

Page 279   •32   •300   •2984   •4002
Page 280   •26   •32   •152   •302
Page 282   •$2n + 3$ **Page 286**   •$E(^-5, ^-3)$
•$F(1, 2)$ •$G(^-4, 2)$ •$H(1, ^-5)$ **Page 292**
•Yes   **Page 295 (Example 1)**   •slope, 4; y-
intercept, $^-2$; $y = 4x - 2$ **(Example 2)**
•slope, 6; y-intercept, 9   •slope, 4; y-
intercept, $^-3$ •slope, $^-7$; y-intercept, $^-2$
Page 299   •Not a solution   •Not a solution

CHAPTER 10 STATISTICS

Page 310   •mean = 14 L, range = 5 L,
mode = 12 L, median = 13 L   **Page 312**
•6336   **Page 315**   •$30,000   •$15,000
•$55,000   •$$$$ $$$$ $$ •$$$$ $$$$ $$$$
$$$ •$$$$ $$$$ $$$$ $$$$ $$$$ $
Page 318   •$425   **Page 319**   •345
Page 322   •Television   **Page 326**   •What is
the latest I can leave home and get to the
concert before 7:30 P.M.? 6.50 P.M.
Page 330   •1   **Page 333**   •67 in.   **Page
334**   •About 5

CHAPTER 11 SPECIAL TRIANGLES

Page 346   •6   **Page 348**   •45 ft
Page 352   •2.236   **Page 356**   •0.6634

Page 359   •$\sin P = \frac{3}{5}$, $\cos P = \frac{4}{5}$
•$\sin M = 0.6$, $\cos M = 0.8$ **Page 362**
•2.3096   •6   **Page 364**   •$\angle A = 37°$,
$\angle B = 53°$

CHAPTER 12 PROBABILITY

Page 376   •$\frac{1}{2}$, 0.5, 50%   •$\frac{1}{2}$, 0.5, 50%   •1, 1,
100%   •0, 0, 0%   **Page 380**   •$\frac{5}{6}$
Page 383   •$\frac{1}{24}$ **Page 386**   •$\frac{6}{25}$ •$\frac{4}{25}$
Page 387   •$\frac{1}{216}$ •$\frac{1}{24}$ **Page 390**   •$\frac{2}{7}$ •$\frac{2}{7}$
Page 394   •$\frac{5}{992}$ •$\frac{992}{5}$ **Page 396**   •$0.\overline{6}$
•$0.\overline{13}$ •$0.\overline{2}$ **Page 397**   •0.225   **Page 399**
•$\frac{7}{20}$ •$\frac{1}{5}$ **Page 400**   •2750   •4250   •750

CHAPTER 13 GEOMETRY

Page 412   •$\overleftrightarrow{NB}$ **Page 413 (Example 3)**
•parallel   •skew   **(Example 4)**   •50°; 50°;
50°   **Page 419**   •27   **Page 423**   •Potassium,
0.427 lb; sulfur, 0.305 lb; sodium, 0.183 lb;
chlorine, 0.183 lb; magnesium, 0.061 lb; iron,
0.00488 lb   **Page 425**   •$\triangle ACD$ •$\triangle BAD$
•$\triangle BCD$ **Page 426**   •$XZ = 5$, $ZY = 7$,
$XY = 11$ •Acute

CHAPTER 14 PROGRAMMING IN BASIC

Page 439   •10 PRINT 26.7/15.25, 20 END
•10 PRINT 4↑6, 20 END   **Page 440**   •48,
6, 1.8   **Page 443**   •THE AREA IS 70.56
•10 LET A = 50, 30 LET C = 35
Page 446   •5, 10, 15, 20   •6 TIMES 2 = 12,
7 TIMES 2 = 14, 8 TIMES 2 = 16, 9 TIMES
2 = 18, 10 TIMES 2 = 20, 11 TIMES
2 = 22   **Page 447**   •10, 4; 12, 6; 14, 8; 16,
10; 18, 12; 20, 14   •10, 2; 20, 4; 30, 6; 40,
8   **Page 453**   •PERIMETER IS 46.
Page 455   •45   PERCENT OF 90 IS 40.5,
12 PERCENT OF 150 IS 18, 6.8 PERCENT
OF 200 IS 13.6   **Page 457**   •YOUR
AVERAGE TEST SCORE IS 55.   **Page
458**   •6, 8, 10 ARE PYTHAGOREAN
TRIPLES. OUT OF DATA IN LINE 10
Page 460   •WEEKLY PAY IS $173.25,
WEEKLY PAY IS $193.05, WEEKLY PAY IS
$257.20, WEEKLY PAY IS $272

APPENDIX SIMPLIFYING POLYNOMIALS

Page 516   •$5a^2 + 3a$ •$6x^2 - 3xy$

499

Answers to Self Tests

CHAPTER 1 WHOLE NUMBERS AND VARIABLES

Self Test 1, p. 10   1. 42 n **2.** 33 **3.** 2 **4.** 53
5. 178 **6.** 112 **7.** 3625 **8.** 63 **9.** 62
Self Test 2, p. 26   1. 14.4 mi/gal **2.** 1428
3. 2530 **4.** 405 **5.** 13 **6.** 78 **7.** 27 **8.** 872
9. 475 **10.** 1006 **11.** 45 **12.** 34 **13.** 60

CHAPTER 2 DECIMALS

Self Test 1, p. 44   1. 28.07 **2.** 7013.0062
3. 600.004 **4.** 5.000014 **5.** 7.56, 7.66, 7.67,
8.66 **6.** 82.29, 92.28, 92.29, 92.92 **7.** 0.7,
0.75, 0.756, 0.76 **8.** 8.3, 8.34, 8.341, 8.44
9. 1 **10.** 4 **11.** 36 **12.** 130 **13.** 0.5
14. 8.4 **15.** 41.0 **16.** 112.7
Self Test 2, p. 62   1. 74.3 **2.** 30.65
3. 123.095 **4.** 4.58 **5.** 7.96 **6.** 298.15
7. Yes **8.** 174.64 **9.** 0.0086 **10.** 8650
11. 2.3 **12.** 2.4 **13.** 0.362 **14.** 4.562
15. 841 **16.** 0.724

CHAPTER 3 NUMBER THEORY

Self Test 1, p. 83   1. Yes **2.** No **3.** No
4. Yes **5.** $2^2 \times 3 \times 7$ **6.** $2^3 \times 5 \times 23$
7. 3×7^2 **8.** $2^3 \times 5 \times 47$ **9.** 2 **10.** 5
11. 1 **12.** $3(a - 4)$ **13.** $5b(b + 2)$
14. $2(1 + 3c - 2c^2)$
Self Test 2, p. 92   1. Monday **2.** 360 **3.** $6a$
4. $24b^2$ **5.** 8000 **6.** 170,000 **7.** 940,000,000
8. 2.8×10^6 **9.** 9.6×10^4 **10.** 5.4×10^7

CHAPTER 4 FRACTIONS

Self Test 1, p. 114   1. $\frac{23}{3}$ **2.** $\frac{29}{6}$ **3.** $\frac{49}{9}$ **4.** $\frac{71}{7}$
5. $2\frac{1}{10}$ **6.** 3 **7.** 2 **8.** $3x^2$ **9.** $\frac{c}{2}$ **10.** $\frac{5a}{8b}$
11. $\frac{3x}{4}$ **12.** $\frac{13}{b}$ **13.** $\frac{11}{c}$ **14.** $10\frac{1}{15}$ **15.** $8\frac{17}{28}$
16. $\frac{3n}{r}$ **17.** $\frac{31s}{24}$ **18.** $1\frac{29}{48}$

Self Test 2, p. 126   1. $\frac{7p}{13}$ **2.** $2x$ **3.** $\frac{2d}{5}$
4. $\frac{5n^2}{108}$ **5.** $\frac{18b}{121}$ **6.** $\frac{1}{9}$ **7.** $\frac{2}{3}c$ **8.** $0.0\overline{6}$
9. 0.175 **10.** 0.5625 **11.** $\frac{59}{100}$ **12.** $\frac{1}{100}$ **13.** $5\frac{13}{20}$

CHAPTER 5 SOLVING EQUATIONS

Self Test 1, p. 148   1. Yes **2.** Yes **3.** 58
4. 25 **5.** 18 **6.** 13 **7.** 50 **8.** 231 **9.** 91
10. 123 **11.** 55 **12.** 525 **13.** 39 **14.** 40
Self Test 2, p. 164   1. 15 **2.** 5 **3.** 9 **4.** 11
5. 2 **6.** 3 **7.** $j + 5$ **8.** $j + 2$
9. $2n = 16$; $8 **10.** $(x + 7)3 = 54$; 11

CHAPTER 6 FORMULAS

Self Test 1, p. 182   1. 44 ft **2.** 68 in.
3. 53.8 cm **4.** 65.94 yd **5.** 20.096 m
6. 6.908 mi **7.** right **8.** obtuse **9.** straight
10. acute **11.** acute
Self Test 2, p. 191   1. 16 ft² **2.** 150.96 in.²
3. 120.36 yd² **4.** 19.625 ft² **5.** 379.94 yd²
6. 243.1616 cm² **7.** 625 mi/h
Self-Test 3, p. 200   1. 1200 cm³ **2.** 360 mm³
3. 2260.8 in.² **4.** 615.44 cm³ **5.** 72 m²
6. 18,463.2 mm²

CHAPTER 7 RATIO, PROPORTION, PERCENT

Self Test 1, p. 224   1. Glisten: $\frac{11}{17}$; $\frac{22}{85}$; $\frac{8}{85}$.
Pearly White: $\frac{63}{88}$, $\frac{9}{44}$, $\frac{7}{88}$. Shine'n Smile: $\frac{19}{96}$, $\frac{25}{32}$,
$\frac{1}{48}$ **2.** 4 **3.** 18 **4.** 11 **5.** 21 **6.** 16%
7. 25% **8.** 20% **9.** 25% **10.** $7.65
11. $4.12 **12.** $10.63 **13.** $2.60, $5.20, $7.80,
$10.40, $13, $15.60, $18.20, $20.80, $23.40,
$26. You save $1.56
Self Test 2, p. 232   1. 1.5 **2.** 675 **3.** 3.8
4. 150 **5.** 350 **6.** 75% **7.** 65% **8.** 6.9%

CHAPTER 8 INTEGERS AND RATIONAL NUMBERS

Self Test 1, p. 255   1. $5 > {}^-6$ **2.** $3 > {}^-8$
3. $20 > {}^-12$ **4.** $18 > {}^-4$ **5.** $^-11$ **6.** $^-22$
7. $^-29$ **8.** 3 **9.** $^-6$ **10.** $^-8$ **11.** $^-8$ **12.** $^-18$
13. $^-3$ **14.** $^-56$ **15.** $^-40$ **16.** 66 **17.** $^-8$
18. $^-5$ **19.** 4
Self Test 2, p. 270   1. 0.4 **2.** $^-3.7$ **3.** 5.9
4. $^-6.3$ **5.** $^-1.44$ **6.** 1.3 **7.** 8 ways **8.** 7.7
9. 7.6 **10.** $^-2.1$ **11.** $y > 2$ **12.** $x < 3$
13. $b < {}^-3$

CHAPTER 9 GRAPHING IN THE COORDINATE PLANE

Self Test 1, p. 291 **1.** $^-1$, 0, 1, 2, 3 **2.** 8, 11, 14, 17, 20 **3.** $^-6$, $^-8$, $^-10$, $^-12$, $^-14$ **4.** 4, 3, 2, 1, 0 **5.** $3n - 2$; 28; 148 **6.** $3n - 6$; 24; 144 **7.** $^-4n + 13$; $^-27$; $^-187$ **8.** $5n + 6$; 56; 256 **9.** Move 3 right and up 5. **10.** Move 4 left. **11.** Move 6 right and down 2. **12.** Move 5 left and down 1. **13.** 31 **14.** $^-13$
Self Test 2, p. 300 **1.** The graph passes through (0, 1), ($^-1$, 0). **2.** The graph passes through (0, $^-5$), (2, 1). **3.** The graph passes through (0, 4), (2, 0). **4.** Slope, 4; y-intercept, $^-1$. **5.** Slope, $^-1$; y-intercept, 2 **6.** Slope, 3; y-intercept, 4 **7.** Yes **8.** Yes **9.** No

CHAPTER 10 STATISTICS

Self Test 1, p. 317 **1.** 30.5 mi, 36 mi, 31 mi, 12 mi **2.** 9.525 km, 9.4 km, 9.4 km, 0.9 km **3.** $79 **4.** $144 **5.** 200 **6.** 160
Self Test 2, p. 325 **1.** 10 **2.** 70 **3.** Kim **4.** June to July
Self Test 3, p. 336 **1.** 18–21 **2.** 2 **3.** $30,000/$y$ **4.** 10 years

CHAPTER 11 SPECIAL TRIANGLES

Self Test 1, p. 354 **1.** 4 **2.** 5.6 **3.** 9 m **4.** 11.2 m **5.** 7.8 in. **6.** 7.2 cm **7.** 9.9 in.
Self Test 2, p. 366 **1.** $\frac{3}{5}$ **2.** $\frac{3}{4}$ **3.** 1.6639 **4.** 0.4472 **5.** 0.8945 **6.** 0.4472 **7.** 0.8945 **8.** 4.2 **9.** 5.1 **10.** 14.0 **11.** 53°; 37° **12.** 53°: 37°

CHAPTER 12 PROBABILITY

Self Test 1, p. 389 **1.** $\frac{1}{5}$ **2.** $\frac{2}{5}$ **3.** $\frac{4}{5}$ **4.** $\frac{2}{5}$ **7.** 60 **8.** 24 **9.** $\frac{1}{10}$ **10.** $\frac{1}{5}$ **11.** $\frac{1}{5}$
Self Test 2, p. 402 **1.** $\frac{5}{92}$ **2.** $\frac{3}{46}$ **3.** $\frac{2}{23}$ **4.** $\frac{1}{3}$ **5.** $\frac{7}{5}$ **6.** $\frac{5}{1}$ **7.** 0.3 **8.** 0.2 **9.** 3600 **10.** 2400

CHAPTER 13 GEOMETRY

Self Test 1, p. 424 **1.** $\angle 4$, $\angle 6$; $\angle 3$, $\angle 5$ **2.** 146° **3.** 34° **4.** 540°; 43° **5.** 720°; 119° **6.** $2\frac{1}{2}$ **7.** 3 **8.** 5056 gal
Self Test 2, p. 430 **1.** 15, 5, 15 **2.** 17, 17, 17 **3.** 21, 30, 37 **4.** *APL* **5.** 60° **6.** 15

CHAPTER 14 PROGRAMMING IN BASIC

Self Test 1, p. 451 **1.** 10 PRINT 48/6, 20 END **2.** 10 PRINT (3+7)/2, 20 END **3.** 10 PRINT 4↑2+5↑3, 20 END **4.** 3.5, 6.8, 23.8 **5.** 25, 50, 150 **6.** 10 FOR N=20 TO 30 STEP 2, 20 PRINT N, 30 NEXT N, 40 END **7.** Answers will vary
Self Test 2, p. 462 **1.** 200, 2; 11.52, 2; 75, 3 **2.** 6, 11.1 **3.** 20 PRINT X↑2+Y↑2; 500, 28.8, 250 **4.** YOUR AVERAGE IS 81; YOUR AVERAGE IS 92.6667; CONGRATULATIONS. YOU'VE MADE HONOR ROLL **5.** 40 IF D<90 THEN 50, 50 PRINT "SORRY, YOU MISSED HONOR ROLL"

Answers to Selected Exercises

CHAPTER 1. WHOLE NUMBERS AND VARIABLES

Page 3 Exercises **1.** 18 **3.** 11 **5.** 50 **7.** 8
9. 6 **11.** 11 **13.** 16 **15.** 15 **17.** 57
19. $x + 15$ **21.** $y - 65$ **23.** $r \div 26$ **25.** $24n$
27. $s \div 8$ **29.** $n + 52$ **31.** $y + 8$
33. $125 - t$ **35.** $\$19.35 \div n$

Page 3 Problems **1.** $20b$

Pages 6-7 Exercises **1.** 55 **3.** 189 **5.** 456
7. 589 **9.** 1397 **11.** 1562 **13.** 1662
15. 4785 **17.** $598 + 63 = x$; 661
19. $886 + 749 = v$; 1635
21. $158 + 6839 = z$; 6997 **23.** 89 **25.** 256
27. 1881 **29.** $15 + 26 + 56 + 46 = d$;
$d = 143$ **31.** $56 + 26 + 15 + 15 + 10 = d$;
$d = 122$ **33.** 274

Pages 9-10 Exercises **1.** 1689 **3.** 1002
5. $3.41 **7.** 22,148 **9.** 92 **11.** 1230
13. 389 **15.** 2715 **17.** 5542
19. $4080 - 67 = x$ **21.** $x = 6000 - 87$
23. $a = 75 + 46$ **25.** 18 **27.** 374 **29.** 259
31. 97 **33.** $897 - 255 = n$; 642
35. $426 - 399 = n$; 27 **37.** No; for example,
$(6 - 4) - 2 \neq 6 - (4 - 2)$

Pages 12-13 Exercises
1. $480 + 521 + 511 + 609 = t$; 2121
students **3.** $94 - 72 = t$; 22 seconds
5. $(350 + 158) - 198 = m$; $310
7. $(475 - 298) + 965 = b$; $1142

Pages 16-17 Exercises **1.** 988 **3.** 26,169
5. 3,047,404 **7.** 1728 **9.** 9 **11.** 7161
13. 24,320 **15.** 54 **17.** 238 **19.** 406
21. 343 **23.** 360 **25.** 86 **27.** $5086b$
29. $r + 3000$ **31.** $a + 7215$ **33.** $9n$

Page 17 Problems **1.** $132 **3.** 3 tickets; 2nd
balcony matinee; $5

Pages 19-20 Exercises **1.** 12 **3.** 128
5. 1559R1 **7.** 92R34 **9.** 16 **11.** 9
13. 364R10 **15.** 11 **17.** 15 **19.** 11 **21.** 15
23. 58 **25.** $468 \div x$ **27.** $r + 76$
29. $w \div 339$ **31.** $21 \div n$ **33.** No, for

example $6 \div 3 \neq 3 \div 6$

Page 20 Problems **1.** $25.50

Page 22 Exercises **1.** 1 **3.** 23 **5.** 76 **7.** 0
9. 9865 **11.** 675 **13.** 96 **15.** 9 **17.** 50
19. 1000 **21.** 0 **23.** 0 **25.** 1 **27.** 1 **29.** 1
31. 1 **33.** 37 **35.** 76 **37.** Yes, 0; $n - 0 = n$

Page 22 Using the Calculator **1.** 732,323
3. 6761 **5.** 22,761

Page 25 Exercises **1. a.** 22 **b.** 26 **3. a.** 26
b. 42 **5. a.** 39 **b.** 114 **7. a.** 31 **b.** 72
9. a. 210 **b.** 1728 **11.** $4; b$ **13.** $87; c$
15. $963; z$ **17.** 4, 7; p, q **19.** $4n$ **21.** $16x$
23. $47r - 17x$ **25.** $30s - 5t$ **27.** $11e - 27b$
29. 60 **31.** 20 **33.** 1 **35.** 72 **37.** 2
39. 438 **41.** $19a + 6b$ **43.** $6b - 4y + 75$

Page 27 Algebra Practice **1.** 14 **3.** 9 **5.** 60
7. $406 + 97 = s$; 503 **9.** $89 + 19 = l$; 108
11. $623 + 94 = f$; 717 **13.** $42h$
15. $6273 \div n$ **17.** $c + 50$ **19.** 42 **21.** 683
23. 1 **25.** 4 **27.** 6 **29.** 25 **31.** 16 **33.** 0
35. 0 **37.** $2n$ **39.** $5s + t$ **41.** $8x + y$
43. $10a + 8$ **45.** $3x + y$

Pages 28-29 Problem Solving on the Job
1. $228 **3.** 42 **5.** $8.75 **7.** $285
9. 7500 bricks **11.** $14,280 **13.** $4140
15. $2745

Page 30 Enrichment **1.** 1, 1, 2, 3, 5, 8, 13,
21, 34, 55, 89, 144, 233, 377, 610, 987, 1597,
2584, 4181, 6765 **3.** 54; 143; 1596

Page 31 Chapter Review **1.** c **3.** a **5.** d
7. c **9.** a **11.** a **13.** b **15.** T **17.** T
19. F **21.** F **23.** T **25.** T **27.** T **29.** T

Page 33 Cumulative Review (Ch. 1) **1.** 486
3. 7267 **5.** 2232 **7.** 35 **9.** 80 **11.** 133
13. 15 **15.** 1 **17.** 9 **19.** 16 **21.** 43 **23.** 40
25. 2 **27.** 3 **29.** $4x + 4$ **31.** $7x$ **33.** y
35. $6p + 6$ **37.** $7b$ **39.** $14c + 21$ **41.** 0
43. $7g + 9h$ **45.** $x - 16$ **47.** $54 + 16 + w$
49. $12c$

CHAPTER 2. DECIMALS

Pages 36-37 Exercises 1. 24.45 **3.** 24.22
5. 24.61 **7.** 86.79; 86.790 **9.** 86.77; 86.770
11. c **13.** b **15.** b **17.** 0.7 **19.** 0.2401
21. 17.006 **23.** 4001.0023 **25.** 0.009
27. 4.0006 **29.** 640.050201 **31.** 0.1 **33.** 0.1
35. 0.1 **37.** 0.01

Pages 39-40 Exercises 1. 3.7, 5.2, 6.8
3. 3.2, 3.3, 4.2 **5.** 52.2, 52.3, 53.3 **7.** 87.77,
87.78, 87.87 **9.** 9.013, 9.103, 9.301
11. 413.12, 413.23, 413.32, 431.32 **13.** 2.1349,
21.349, 213.49, 2134.9 **15.** 0.8, 0.08, 0.008
17. 39.954, 29.95, 29.9 **19.** 71.455, 71.45, 71.4
21. 7.2913, 7.291, 7.29, 7.28 **23.** 190.9,
190.099, 190.09, 19.009 **25.** 610.8, 61.0813,
61.081, 61.008 **27.** > **29.** > **31.** >
33. > **35.** > **37.** <

Page 40 Problems 1. $a > 4.1$ **3.** $s = 5.21$
5. $7.45 < c$ **7.** $h > 7.04$ **9.** $7.6 < f$

Page 40 Challenge 4 dimes, 1 quarter, 1 half
dollar

Pages 43-44 Exercises 1. 5 cm **3.** 10 cm
5. 12 cm **7.** 0 **9.** 47 **11.** 89 **13.** 493
15. 470 **17.** 0.3 **19.** 38.1 **21.** 69.0
23. 863.0 **25.** 766.3 **27.** 0.57 **29.** 31.21
31. 38.49 **33.** 840.40 **35.** 590.01 **37.** 0.131
39. 18.579 **41.** 86.005 **43.** 877.056
45. 440.321 **47.** 99.5, 100.4 **49.** 108.5, 109.4
51. 501.5, 502.4

Pages 46-47 Exercises 1. 47.06 **3.** 84.05
5. 57.44 **7.** 701.72 **9.** 807.71 **11.** 40
13. 770 **15.** 80 **17.** 99 **19.** 7.89
21. 32.343 **23.** 117.257 **25.** 36.38
27. 31.278 **29.** 17.742 **31.** $12b$ **33.** $33.84a$
35. $25.797g$ **37.** $36.2406m$ **39.** $193.51t$
41. $22.31g + 3.77h$ **43.** $24.91s + 57t$
45. $7.94j + 3.9k$

Page 47 Problems 1. $x + 3.7$ **3.** $b + 0.92$
5. $5.61 + a$ **7.** $b + 2.3$ **9.** $4.85 + 31.7 = b$;
36.55 **11.** $71.6 + 2.13 = g$; 73.73
13. $n = 3.074 + 12.62$; 15.694

Pages 49-50 Exercises 1. 0.242 **3.** 4.17
5. 244.87 **7.** 18.75 **9.** 475.7514 **11.** 2
13. 537 **15.** 246 **17.** 6 **19.** 1.91 **21.** 0.661
23. 29.92 **25.** 0.287 **27.** 4.738 **29.** 220.69
31. $2.2a$ **33.** $7.68x$ **35.** $51.74b$ **37.** $3.6v$

39. $168.108z$ **41.** 13.93 **43.** 23.93 **45.** 73.15
47. 826.531 **49.** $388.596 - 68.079 = 320.517$
51. $17.0036 - 8.2462 = 8.7574$

Page 50 Problems 1. $8.3 - t$ **3.** $x - 18.4$
5. $b + 7.41$ **7.** $21.073 - 4.68 = r$; 16.393
9. $40.61 + 321.7 = a$; 362.31

Page 52 Exercises 1. 2.66 **3.** 0.91, 0.9
5. 0.91, 0.1 **7.** 4:15 P.M.; 5:00 P.M.

Page 55 Exercises 1. 20.8 **3.** 84.36 **5.** 6.84
7. 0.007 **9.** 0.3609 **11.** 150 **13.** 32
15. 240 **17.** 320 **19.** 42.91 **21.** 22.32
23. 2299.18 **25.** 591.2 **27.** 36.82 **29.** 0.006
31. 4 **33.** 36.8 **35.** 746.3 **37.** 63 **39.** 963.4
41. 640 **43.** 571 **45.** 9812.4 **47.** 29,340
49. 2.108 **51.** 79.19 **53.** 22.03

Page 56 Problems 1. $3.7b$ **3.** $4.82d$
5. $3.712 + m$ **7.** $2h$ **9.** \$3.02 **11.** \$3.80
13. \$.65 **15.** \$2.69 **17.** \$12.05

Pages 59-60 Exercises 1. 4.5 **3.** 0.2
5. 182 **7.** 0.68 **9.** 0.04 **11.** 12.4 **13.** 4
15. 42 **17.** 30 **19.** 60 **21.** 2 **23.** 2
25. 45.236 **27.** 0.623 **29.** 0.055 **31.** 0.06486
33. 5.8861 **35.** 0.8216 **37.** 0.004312
39. 0.000921 **41.** 1.5308 **43.** 1.46 **45.** 2.18
47. 3.5 **49.** 40 **51.** 25 **53.** 32 **55.** 29

Page 60 Problems 1. $7.3a$ **3.** $81.1b$ **5.** $3.8s$
7. $47.2 - d$ **9.** $5.18 \div b$ **11.** $(2.13)(1.5) = z$;
3.195 **13.** $80.08 - 14.7 = v$; 65.38

Page 62 Exercises 1. 1000 **3.** 10 **5.** 0.01
7. 600 **9.** 2.4 **11.** 86 **13.** 85 **15.** 0.0743
17. 0.304 **19.** 8.342 **21.** 45 **23.** 724.3

Page 63 Algebra Practice 1. 12 **3.** 3 **5.** 1
7. $16.18r$ **9.** $13c + 6.2f$ **11.** $13.177b$
13. 7.436 **15.** $a^2 + 4a$ **17.** h^4 **19.** $g - 2.6$
21. $7.5 \div v$ **23.** $196 + 505 = n$; 701
25. $36.409 - 19.7 = h$; 16.709
27. $18 \div 3.6 = b$; 5
29. $11.097 + 230.28 + 1.973 = s$; 243.35
31. $21 + 804 = y$; 825 **33.** $37 + 27.4 = q$;
64.4 **35.** $j = 6^2$; 36 **37.** $86 - 2.71 = h$;
83.29 **39.** $n = (2.01)(0.009)$; 0.01809

**Pages 64-65 Problem Solving for the
Consumer 1.** \$9125.60 **3.** \$6042.60
5. \$1396.90 **7.** \$575.82; \$164.22 **9.** \$7218.66,
\$415.80, \$8257, \$7634.46, \$622.54
11. \$304.23 **13.** \$466.93, \$504.28 **15.** 10

Page 66 Enrichment 1. 100101 110100 100100 010010 010101 101001 **3.** 010010 011001 100101 010001 101001 111000 **5.** 100100 110100 100011 110011 011001 100111 100011 111000 **7.** 011000 100110 100100 010101 110110 100110 101001 100010 **9.** John von Neumann

Page 67 Chapter Review 1. C **3.** A **5.** F **7.** B **9.** No **11.** Yes **13.** Yes **15.** b **17.** a

19. c **21.** 27.15 **23.** 0.42 **25.** 219.75 **27.** 731.235 **29.** 0.423 **31.** 76.2 **33.** 0.0058

Page 69 Cumulative Review (Chapters 1–2)
1. 87,019 **3.** 0 **5.** 35 **7.** 74 **9.** 480.7
11. 84.213 **13.** 4 **15.** 81 **17.** 5290 **19.** 322
21. 30,000 **23.** 5300 **25.** 4.0 **27.** a **29.** b
31. a^2 **33.** $5a^3$ **35.** $5xyz$ **37.** $4t^2w$ **39.** 1
41. $7j + j^2$ **43.** $83 - 47.1 = g$; 35.9
45. $d = 55(2)$; 110 mi

CHAPTER 3. NUMBER THEORY

Pages 73–74 Exercises 1. Yes **3.** Yes
5. Yes **7.** Yes **9.** No **11.** No **13.** Yes
15. No **17.** Divisible by 2, 3, 4, 5, 6, 8, and 10 **19.** Divisible by 2, 3, 4, 5, 6, 9, and 10 **21.** Divisible by 2, 3, 6, 9 **23.** Divisible by 2, 3, 6 **25.** Divisible by 2, 3, 4, 6, 9
27. Divisible by 2, 3, 6 **29.** Divisible by all
31. Divisible by 2, 4, 8 **33.** Divisible by 2, 4, 8 **35.** Divisible by 2, 3, 4, 6, 9, **37.** T
39. T **41.** T **43.** 2 **45.** 4, 6 **47.** $2n + 1$
49. $n + 2$ **51.** Yes **53.** No **55.** Yes

Page 74 Problems 1. Yes **3.** 8 teams of 9, 9 teams of 8, 2 teams of 36, 3 teams of 24, 4 teams of 18, 6 teams of 12

Page 74 Using the Calculator 1. 6
3. 3,628,800 **5.** 29,030,400

Page 78 Exercises 1. $90 = 3 \times 3 \times 2 \times 5$
3. $2^3 \times 3$ **5.** $3^2 \times 7$ **7.** $2 \times 3 \times 5$
9. $2 \times 3 \times 11$ **11.** 2^7 **13.** $2 \times 5 \times 13$
15. $2^3 \times 3^2 \times 5^2 \times 7$ **17.** $2^4 \times 11$
19. $2 \times 3^3 \times 7$ **21.** $2^2 \times 3 \times 5 \times 11$
23. $2^6 \times 5^2$ **25.** $2^4 \times 7^2$ **27.** $5^2 \times 7^2$
29. 15, 6

Page 78 Problems 1. 6; 3, 5; 5, 7; 11, 13; 17, 19; 29, 31; 41, 43

Page 80 Exercises 1. 1, 2, 3, 6 **3.** 1, 3, 5, 15 **5.** 1, 2, 3, 4, 6, 8, 12, 24 **7.** 3 **9.** 6
11. 14 **13.** 4 **15.** 6 **17.** 4 **19.** 3 **21.** 18
23. 6 **25.** 16 **27.** 9 **29.** 12 **31.** 36
33. Yes **35.** No

Page 83 Exercises 1. h **3.** 2 **5.** n **7.** q
9. $3(s^2 - 1)$ **11.** $5(v^2 - 2)$ **13.** $4(3x^3 - 1)$

15. $a(a + 2)$ **17.** $c(c - 2)$ **19.** $e(1 - 2e)$
21. $3g(2g - 1)$ **23.** $5j(3 + 4j)$
25. $5m(6 + 5m)$ **27.** $pq(6 + pq)$
29. $4uv(3u - 2v)$ **31.** $a(y^2 + y + 1)$
33. $2(6 + 4a - 3a^2)$ **35.** $4c(3c - 2 - d)$

Page 86 Exercises 1. 10 **3.** 4 **5.** 204

Page 89 Exercises 1. 8, 16, 24, 32
3. 12, 16 **5.** 30, 60, 90; LCM = 30
7. 14, 28, 42, 56, 70, 84, 98; LCM = 14
9. 30, 60, 90; LCM = 30 **11.** 60;
LCM = 60 **13.** 4 **15.** 9 **17.** 36 **19.** 20
21. 30 **23.** 30 **25.** 90 **27.** 240 **29.** $2a^2$
31. $8c^2$ **33.** $30e^3$ **35.** hj **37.** 100 **39.** 360
41. 300 **43.** $75d^3$ **45.** $24x^2y^2$

Pages 91–92 Exercises 1. a. 1000 **b.** 3000
c. 3100 **3. a.** 100,000 **b.** 200,000
c. 210,000 **5. a.** 1.0×10^4 **b.** 5.0×10^4
c. 5.1×10^4 **7. a.** 1.0×10^6 **b.** 7.0×10^6
c. 7.2×10^6 **9.** 9300 **11.** 17,000 **13.** 180
15. 380,000 **17.** 3,900,000,000,000
19. 7,100,000,000 **21.** 300,000,000
23. 1,230,000 **25.** 7.6×10^4

Page 93 Algebra Practice
1. $2398 + 9746 = r$; 12,144 **3.** $780(67) = m$;
52,260 **5.** $2784 \div 87 = f$; 32
7. $1986.21 + 8.005 + 497.6 = d$; 2491.815
9. $(0.41)(0.25) = y$; 0.1025 **11.** $11.1a + 5.2b$
13. $6.9p - 4.8q$ **15.** 5.356 **17.** $2r^2$ **19.** $4m$
21. $2y$ **23.** $2s(s^2 - 3st + 5)$ **25.** $5b(b^2 - 3)$
27. $fg(fg + f - g)$ **29.** $4x(2xy - x + y^2)$
31. $3w^2$ **33.** $15q^2$ **35.** $3k^2n$ **37.** $60d^2$
39. $20r^2s$ **41.** $48a^2b^2$

Pages 94–95 Problem Solving on the Job
1. 10 days **3.** 9 days **5.** $9\frac{1}{2}$ days **7.** 9 days

9. 9 days **11.** $24.45 **13.** $30.30 **15.** $443
17. $447.50

Page 96 Enrichment 1. 50 **3.** 400 **5.** 199

Page 97 Chapter Review 1. c **3.** c **5.** c
7. d **9.** d **11.** e **13.** $2(3a + 1)$
15. $2(c - 2c^2 + 3)$ **17.** F **19.** T
21. 9:35 P.M.

Page 99 Cumulative Review (Chapters 1–3)
1. 670 **3.** 8 **5.** 100,000 **7.** 125 **9.** 143
11. 10 **13.** $8a + 4$ **15.** $p + 14y - 2$
17. $5a - b$ **19.** $3(x + 1)$ **21.** $2(4 - c)$
23. $5(y^2 + 2)$ **25.** $3(2s^2 - s + 5)$
27. $5(2f^2 + f - 5)$ **29.** $35 - n$ **31.** $12 + 5y$
33. $3d$

CHAPTER 4. FRACTIONS

Pages 103–104 Exercises 1. $\frac{6}{1}$ **3.** $\frac{8}{1}$ **5.** $\frac{10}{1}$
7. $\frac{12}{1}$ **9.** $\frac{14}{1}$ **11.** $\frac{16}{1}$ **13.** $\frac{4}{1}$ **15.** $\frac{7}{2}$ **17.** $\frac{9}{4}$
19. $\frac{8}{11}$ **21.** $\frac{5}{13}$ **23.** $\frac{3}{2}$ **25.** $\frac{5}{2}$ **27.** $\frac{5}{4}$ **29.** $\frac{7}{4}$
31. $\frac{7}{2}$ **33.** $1\frac{1}{2}$ **35.** $1\frac{1}{5}$ **37.** 2 **39.** 7 **41.** 8
43. $\frac{21}{4}$ **45.** $\frac{49}{6}$ **47.** $\frac{44}{7}$ **49.** $\frac{53}{9}$ **51.** $\frac{129}{10}$
53. $1\frac{1}{4}$ **55.** $2\frac{2}{5}$ **57.** $3\frac{2}{7}$ **59.** 2 **61.** $1\frac{11}{20}$

Page 104 Problems 1. $\frac{n}{21}$ **3.** $\frac{m}{30}$ **5.** $\frac{28}{81}$
7. $\frac{65}{81}$ **9.** $\frac{81}{81}$

Page 107 Exercises 1. 6 **3.** 49 **5.** 42 **7.** 3
9. $\frac{2}{3}$ **11.** $\frac{1}{2}$ **13.** $\frac{1}{4}$ **15.** $\frac{3}{2}$ **17.** 1 **19.** b
21. $\frac{1}{2c}$ **23.** $\frac{1}{2}$ **25.** y **27.** $\frac{a}{3}$ **29.** $\frac{2}{3}$ **31.** $\frac{2c}{3}$
33. $\frac{3}{4t}$ **35.** $\frac{3mb}{2}$ **37.** $x + 1$ **39.** 8 **41.** 3
43. $\frac{1}{3}$ **45.** $\frac{3r}{5r + 2}$ **47.** $\frac{3e + 1}{2}$

Page 110 Exercises 1. $9\frac{43}{72}$ **3.** $6\frac{17}{24}$ **5.** $10\frac{73}{88}$
7. $13\frac{3}{7}$ **9.** $10\frac{25}{33}$ **11.** Yes **13.** No
15. $\frac{8w + 2c}{cw}$ **17.** $\frac{3c + 2b}{bc}$ **19.** $\frac{3b}{8}$
21. $\frac{m + 2f}{10}$

Page 110 Problems 1. $4\frac{1}{2} + \frac{3}{5} = n; 5\frac{1}{10}$
3. $\frac{3}{4} + \frac{1}{2} = p; 1\frac{1}{4}$ lb **5.** $\frac{5}{8} + \frac{5}{6} + \frac{2}{3} + \frac{1}{4} = m;$
$2\frac{3}{8}$ mi

Pages 113 Exercises 1. $\frac{1}{3}$ **3.** $\frac{1}{10}$ **5.** $\frac{2}{5}$ **7.** $\frac{2}{3}$
9. $\frac{9}{20}$ **11.** $\frac{5}{14}$ **13.** $\frac{5}{y}$ **15.** $\frac{10}{s}$ **17.** $\frac{a}{n}$ **19.** 0
21. $\frac{2b}{9}$ **23.** $\frac{m}{24}$ **25.** $\frac{3}{4}$ **27.** $\frac{9}{20}$ **29.** $\frac{4}{9}$
31. $1\frac{7}{12}$ **33.** $2\frac{1}{36}$ **35.** $4\frac{8}{15}$ **37.** $\frac{7a + 4}{8}$

39. $\frac{7c + 5}{12}$ **41.** $2a - 2b$

Page 114 Problems 1. $4\frac{2}{3} - 1\frac{1}{3} = n; 3\frac{1}{3}$
3. $\frac{5}{8} - \frac{1}{3} = n; \frac{7}{24}$ **5.** $\frac{7}{8} - \frac{7}{10} = n; \frac{7}{40}$ mi

Page 117 Exercises 1. $2\frac{1}{2}$ **3.** $3\frac{1}{2}$ **5.** 4 **7.** 1
9. $\frac{1}{3}$ **11.** $\frac{1}{4}$ **13.** $\frac{8}{b}$ **15.** $\frac{7f}{13}$ **17.** $\frac{1}{3}$ **19.** 8
21. $\frac{22m}{21}$ **23.** $\frac{2b}{3}$ **25.** $\frac{b}{3}$ **27.** $\frac{g}{6}$ **29.** $\frac{q}{8}$

Page 117 Problems 1. $2 \cdot \frac{3}{4} = n; 1\frac{1}{2}$
3. $25 \cdot \frac{4}{5} = n; 20$ **5.** $25 \cdot \frac{2}{3} = n; 16\frac{2}{3}$ lb

Pages 119–120 Exercises 1. $\frac{1}{6}$ **3.** $\frac{1}{20}$ **5.** $3\frac{1}{3}$
7. $5\frac{1}{4}$ **9.** 1 **11.** $1\frac{1}{2}$ **13.** $\frac{x}{2}$ **15.** $\frac{17c}{64}$ **17.** $\frac{5e}{3}$
19. $\frac{17g}{13}$ **21.** $\frac{11}{35r}$ **23.** $\frac{11}{12e}$ **25.** $\frac{4}{5}b$ **27.** $\frac{3}{8}m$
29. $\frac{1}{3}m$ **31.** $\frac{4}{7}r$

Page 120 Problems 1. $\frac{24}{x}$ **3.** $4\frac{5}{8}$ ft **5.** 4
7. $8\frac{1}{3}$ yd

Pages 122–123 Exercises 1. $1\frac{9}{16}c$ **3.** $60; $
$80 **5.** yes **7.** $368 **9.** 3 mi

Page 125 Exercises 1. 2.5 **3.** 0.24 **5.** 0.625
7. $0.\overline{5}$ **9.** 1.25 **11.** 0.875 **13.** 0.2 **15.** $0.4\overline{6}$
17. $\frac{24}{25}$ **19.** $\frac{11}{50}$ **21.** $\frac{37}{100}$ **23.** $2\frac{7}{10}$ **25.** $1\frac{1}{8}$
27. $1\frac{5}{8}$ **29.** $7\frac{3}{125}$ **31.** 3.2 **33.** $1.\overline{72}$ **35.** $3.41\overline{6}$
37. 4.35 **39.** 1.54 **41.** $0.\overline{6}$ **43.** $0.\overline{3}$ **45.** 0.2

Page 127 Algebra Practice 1. 1 **3.** 13 **5.** 1
7. $3b(b^2 + 2c + 5)$ **9.** $jk(3k + jk + 2j)$
11. $12a^2b$ **13.** 2 **15.** $\frac{6}{b}$ **17.** $\frac{d}{e}$
19. $4073 - 2981 = a; 1092$

21. $9632 \div 112 = u$; 86
23. $86.28 + 100.5 + 1.097 = t$; 187.877
25. $n = 2\frac{1}{3} + 1\frac{1}{4}$; $3\frac{7}{12}$ **27.** $\frac{3}{4} \cdot 18 = n$; $13\frac{1}{2}$
29. $2\frac{2}{5} \cdot 1\frac{7}{8} = n$; $4\frac{1}{2}$

Pages 128–129 Problem Solving for the Consumer 1. $98\frac{1}{8} **3.** $64\frac{3}{8} **5.** $1\frac{1}{8}
7. $1055 **9.** 11 shares **11** $250 **13.** $980

Page 130 Enrichment 1. $\frac{1}{3} **3.** $\frac{11}{9} **5.** $\frac{233}{99}
7. $\frac{62}{99} **9.** $\frac{711}{333}

Pages 131 Chapter Review 1. f **3.** e **5.** c

7. d **9.** b **11.** c **13.** $\frac{c}{2}$ **15.** $\frac{11r}{24}$ **17.** 16

19. $\frac{m}{r}$ **21.** $2\frac{1}{3}$ **23.** $1\frac{5}{14}$ **25.** $9\frac{7}{20}$

Page 133 Cumulative Review 1. 1.74 **3.** 18
5. 21 **7.** 1 **9.** $\frac{1}{5}$ **11.** $8\frac{1}{4}$ **13.** $1\frac{2}{5}$ **15.** $\frac{5}{9}$
17. $\frac{1}{3}$ **19.** $2y + 11x$ **21.** 12 **23.** $6n$
25. $4x^2 + 4x$ **27.** $9x - 18$ **29.** $\frac{d}{8}$ **31.** $24x$

33. $\frac{1}{4a}$ **35.** $17t$ **37.** $18; $22

CHAPTER 5. SOLVING EQUATIONS

Page 137 Exercises 1. yes **3.** no **5.** no
7. no **9.** *no* **11.** 16 **13.** 12 **15.** 20 **17.** 10
19. 82 **21.** 160 **23.** 3 **25.** 2 **27.** 5 **29.** 23
31. 96 **33.** 13 **35.** 30 **37.** 7 **39.** 85
41. 2.2 **43.** 1.6 **45.** $\frac{1}{2}$ **47.** 3
49. $e + 36 = 60$ **51.** $d + 7 = 8$

Page 139 Exercises 1. 9 **3.** 5 **5.** 4 **7.** 6
9. 2 **11.** 1 **13.** 4 **15.** 11 **17.** 3 **19.** 39
21. 16 **23.** 18 **25.** 197 **27.** 843 **29.** 19
31. 13 **33.** 5.5 **35.** 0.1 **37.** $5\frac{1}{4}$ **39.** 3
41. 3.24

Page 142 Exercises 1. 3 **3.** 6 **5.** 9 **7.** 4
9. 7 **11.** 6.5 **13.** 30 **15.** 5 **17.** 61 **19.** 30
21. 7 **23.** 32 **25.** 2 **27.** 20 **29.** 1.2
31. 200 **33.** 9 **35.** 20 **37.** 16 **39.** 12
41. 40 **43.** 13 **45.** 8 **47.** 0 **49.** $247\frac{1}{2}$
51. 9 **53.** $1\frac{1}{3}$

Page 144 Exercises 1. 6 **3.** 5 **5.** 14 **7.** 18
9. 13 **11.** 17 **13.** 20 **15.** 20 **17.** 46
19. 211 **21.** 60 **23.** 100 **25.** 126 **27.** 141
29. 1 **31.** $5\frac{1}{4}$ **33.** 5 **35.** 5 **37.** 16 **39.** 74
41. 94 **43.** 73 **45.** 11 **47.** 0.7

Page 144 Problems 1. $5n$ **3.** $x + 5$ **5.** 55

Page 147 Exercises 1. 2 **3.** 4 **5.** 2 **7.** 4
9. 8 **11.** 24 **13.** 10 **15.** 28 **17.** 24
19. 300 **21.** 16 **23.** 120 **25.** 36 **27.** 76
29. 17.5 **31.** 7.4 **33.** 3 **35.** 5.1 **37.** 60
39. 38 **41.** 62 **43.** 180 **45.** 159 **47.** 12
49. 3 **51.** 3

Pages 151–152 Exercises 1. $a = 6$
3. $x = 6$ **5.** $d = 20$ **7.** $f = 5$ **9.** 36 **11.** 4
13. 18 **15.** 5 **17.** 5 **19.** 3 **21.** 16 **23.** 2
25. 9 **27.** 1 **29.** 4 **31.** 24 **33.** 32 **35.** 0
37. 5 **39.** 9 **41.** 5 **43.** 7 **45.** 7 **47.** 5
49. $1\frac{2}{3}$ **51.** $3\frac{5}{6}$ **53.** Yes **55.** No **57.** Yes

Page 152 Challenge Plan B

Page 154 Exercises 1. $e = 4$ **3.** $g = 2$
5. 11 **7.** 4 **9.** 9 **11.** $1\frac{4}{5}$ **13.** 4 **15.** 1
17. 16 **19.** 2 **21.** 6 **23.** 7 **25.** 25 **27.** 2
29. 2 **31.** 4 **33.** 1 **35.** 1 **37.** 1 **39.** 11
41. 5 **43.** 6

Pages 157–158 Exercises 1. d **3.** k **5.** c
7. g **9.** b **11. a.** $n + 1$; **b.** $2x + 1$;
c. $3p + 1$; **d.** $2r + 1$ **13. a.** $50 - 10$;
b. $m - 10$; **c.** $2x - 10$ **15. a.** $s + 3$;
b. $s + 2$ **17. a.** $\frac{1}{2}a$; **b.** $\frac{1}{2}a + 3$ **19. a.** $2c$;
b. $2c + 5$ **21.** $0.1d$ **23.** $\frac{h}{10}$

Pages 160–161 Exercises 1. a; 16 years
3. d; 192 in. **5.** $s + 13 = 35$; 22
7. $t - 3 = 62$; 65 in. **9.** $x + x + 5 = 27$;
16 games **11.** $3s = 18$; 6 tickets
13. $x + x + 38 = 952$; 457 boys, 495 girls

Pages 163–164 Exercises 1. $x - 14 = 29$;
43 people **3.** $2n = 54$; 27 computers
5. $6x = 486$; 81 **7.** $2t + t = 687$; 229,458
9. $c + c + 20 = 78$; 49 customers;
29 customers **11.** $\frac{1}{2}x + x = 1272$; 424 pets;
106 pets

506

Page 165 Algebra Practice **1.** $308 \cdot 280 = k$; 86,240 **3.** $56.07 - 9.806 = w$; 46.264 **5.** $52.635 \div 8.7 = f$; 6.05 **7.** $7\frac{1}{2} - 3\frac{5}{6} = k$; $3\frac{2}{3}$ **9.** $3\frac{1}{2} + 1\frac{1}{8} + 4\frac{3}{4} = h$; $9\frac{3}{8}$ **11.** $uv(8v + 2uv + u^2)$ **13.** $qr(7 + qr + r)$ **15.** $12gh^2$ **17.** 7 **19.** 23.7 **21.** 1.8 **23.** 6.7 **25.** $\frac{5}{6}$ **27.** 24 **29.** 7 **31.** 11

Pages 166-167 Problem Solving on the Job **1.** $37.43 **2.** $37.80 **5.** $43.37 **7.** $241.36 **9.** $3952 **11.** $697.24

Page 169 Chapter Review **1.** a **3.** 19 **5.** 4 **7.** 26 **9.** 94 **11.** 150 **13.** 42 **15.** 12 **17.** 16 **19.** 3 **21.** c **23.** e **25.** d **27.** f **29.** $6n + 9 = 345$; 56

Page 171 Cumulative Review **1.** 388.2 **3.** 157 **5.** 7965.79 **7.** 81 **9.** 16 **11.** $\frac{1}{4}$ **13.** 68 **15.** 0 **17.** 1670 **19.** $\frac{1}{2}(a + b + c)$ **21.** $5y(y - 2)$ **23.** $5(n^2 - 5)$ **25.** $n - 47$ **27.** $(9.50)(8) + (9.50)(1.5)(4) = x$; $133 **29.** $56 - 24 = x$; 32

CHAPTER 6. FORMULAS

Page 175 Exercises **1.** 44 **3.** 28.4 **5.** 16.8 mi **7.** 64.5 **9.** $28\,s$ **11.** $78.9\,m$ **13.** 16

Page 178 Exercises **1.** 31.4 **3.** 15.7 **5.** $238.64\,m$ **7.** $94.2\,m$ **9.** 19.5936 km **11.** 132 **13.** $2\frac{2}{3}$ **15.** 3

Page 178 Problems **1.** Neither.

Pages 181-182 Exercises **1.** Yes **3.** No **5.** No **7.** acute **9.** straight **11.** obtuse **13.** obtuse **15.** obtuse **17.** *ST, TV* **19.** 180° **21.** 270° **23.** 180° **25.** No. They do not form right angles.

Page 185 Exercises **1.** 17.5 ft² **3.** 110.25 m² **5.** $43\frac{1}{20}$ in. **7.** 72 cm² **9.** 288 square units

Page 185 Problems **1.** 9 yd **3.** 20 yd **5.** 9 ft

Page 188 Exercises **1.** 200.96 in.² **3.** 5538.96 yd² **5.** 38.465 in.² **7.** 254.34 yd² **9.** 314 mi² **11.** circle; 481.5 in.² **13.** circle; 15.94 ft² **15.** 18.84 sq. units **17.** 100.48 sq. units **19.** 1316.16 yd²

Pages 190-191 Exercises **1.** 132 **3.** 104.5 **5.** 52.25 **7.** $6.\overline{6}$ ft **9.** 595 mi/h

Pages 193-194 Exercises **1.** 154 in.³ **3.** 330 yd³ **5.** 200 ft³ **7.** 6331.625 yd³ **9.** 29.75 m³ **11.** 7644 in.³ **13.** 1521 ft³

Page 194 Problems **1.** 16 yd² **3.** 6 m **5.** 11 **7.** $47,282\frac{2}{3}$ ft³

Pages 196-197 Exercises **1.** 2769.48 in.³ **3.** 2207.8125 m³ **5.** 4521.6 m³ **7.** 1959.36 cm³ **9.** 221.37 ft³ **11.** 602.88 cm³

Page 200 Exercises **1.** 384 sq. units **3.** 533.8 sq. units **5.** 465 ft² **7.** 468 cm² **9.** 1628 cm²

Page 201 Algebra Practice **1.** $r = 355 \times 982$; 348,610 **3.** $0.0861 \div 0.07 = x$; 1.23 **5.** $(\frac{1}{8})(4\frac{4}{5}) = q$; $\frac{3}{5}$ **7.** $2b(3c + b + 2bc)$ **9.** $\frac{3}{8}$ **11.** $\frac{7}{8}$ **13.** 6 **15.** $P = (2)(4.8) + (2)(3.6)$; 16.8 m **17.** $C = (3.14)(28)$; 87.92 yd **19.** $A = (3.14)(5^2)$; 78.5 m² **21.** $V = (10)(6) \div 3$; 20 m³ **23.** $S = 2(5 \times 3) + 2(5 \times 10) + 2(3 \times 10)$; 190 cm²

Pages 202-203 Problem Solving for the Consumer **1.** 130 ft² **3.** 2 qt; 4 qt **5.** $15.50 **7.** $47.21; Yes **9.** $19.81

Page 204 Enrichment **1.** 9, 13, 17, 21, 25, 29, 33 **3.** 7, 12, 19, 28, 39, 52, 67 **5.** $n \to n^2 + 2n$; 24, 35, 48, 63, 80

Page 205 Chapter Review **1.** d **3.** h **5.** l **7.** j **9.** o **11.** n **13.** c **15.** m **17.** c **19.** a

Page 207 Cumulative Review **1.** 2608 **3.** 62 **5.** 213 **7.** 44.4 **9.** 31 **11.** 67 **13.** 46 **15.** 8920 **17.** 56 **19.** 165 **21.** 4000 **23.** 16 **25.** 16 **27.** 486 **29.** 6.6 **31.** $n \cdot 10 = 4.6$; 0.46 **33.** $(\frac{2}{3})(63) = n$; 42 **35.** $\frac{1}{4}n = 35$; **37.** 64K

CHAPTER 7. RATIO, PROPORTION, PERCENT

Pages 210–211 Exercises 1. $\frac{7}{2}$ **3.** $\frac{4}{1}$ **5.** $\frac{9}{1}$
7. $\frac{3}{1}$ **9.** 2 : 3 **11.** 19 : 20 **13.** 47 : 92
15. $x : y$ **17.** 23 : 40 **19.** $\frac{7}{15}$; $\frac{2}{5}$; $\frac{2}{15}$

Page 211 Problems 1. a. 48 : 120, 72 : 48
b. $\frac{2}{5}$, 2 : 5; $\frac{3}{2}$, 3 : 2 **3.** $\frac{2325}{4481}$; $\frac{4481}{1192}$; $\frac{2325}{1192}$

Page 214 Exercises 1. 3 **3.** 1 **5.** 27 **7.** 18
9. 45 **11.** 70 **13.** 720 **15.** 54.125 **17.** 123
19. The perimeter is doubled.

Page 215 Problems 1. 800 mi **3.** $133.33
5. 12 runs **7.** 48 games

Page 215 Using the Calculator 1. 120 cm
3. 153.6 cm **5.** 140 cm **7.** 179.2 cm

Page 218 Exercises 1. 7% **3.** 99% **5.** 12%
7. 3% **9.** 15% **11.** 90% **13.** $\frac{1}{20}$; 0.05 **15.** $\frac{9}{20}$;
0.45 **17.** $\frac{2}{25}$; 0.08 **19.** $\frac{91}{100}$; 0.91 **21.** $\frac{1}{2}$; 0.50
23. 20% **25.** 90% **27.** 38% **29.** 12%
31. 33% **33.** 14% **35.** 24% **37.** 50%
39. 75% **41.** 43% **43.** 45%

Page 221 Exercises 1. 288 **3.** 18 **5.** 140
7. 1064 **9.** 138 **11.** 12 **13.** $31.98
15. $510.00 **17.** $111.18 **19.** $38.95
21. $25.66 **23.** $135.75

Page 221 Problems 1. $1200; $1000; $720;
$520; $400; $320; $240; $120; The total
amount is more than $4000 because the sum
of the percents is greater than 100%.

Page 227 Exercises 1. 8% **3.** 750%
5. 512.5% **7.** 0.3% **9.** 362% **11.** 1.25%
13. 0.8% **15.** 0.5% **17.** 2 **19.** 1241.25
21. 45.10

Page 227 Problems 1. 4786 mixers **3.** 8090

Page 229 Exercises 1. 1500 **3.** 79,300
5. 1200 **7.** 1200 **9.** 10,000 **11.** 52 **13.** 263

Page 229 Problems 1. $2300 **3.** 5000
transmissions

Pages 231–232 Exercises 1. 26% **3.** 18%
5. 10% **7.** 15% **9.** 20% **11.** 15% **13.** 13%
15. 16% **17.** 133.3% **19.** 40.7% **21.** 28.2%
23. 17.4% **25.** 27.6% **27.** 11.8%

Page 232 Problems 1. $11.\overline{1}\%$ **3.** 20%

Page 233 Algebra Practice 1. $487 \cdot 503 = s$;
244,961 **3.** $20 - 5.683 = d$; 14.317
5. $\frac{2}{15} + \frac{29}{30} = w$; $1\frac{1}{10}$ **7.** $2(2\frac{5}{8}) + 2(1\frac{7}{8}) = p$;
$7\frac{3}{4}$ in. **9.** $(4\frac{1}{2})(2\frac{1}{4})(\frac{1}{2}) = a$; $5\frac{1}{16}$ in.^{2}
11. $(2)(2^2)(3.14) + (4)(4)(3.14) = s$; 75.36 ft^2
13. $6rs^2$ **15.** $\frac{3}{14}$ **17.** 3.5 **19.** 576 **21.** 4
23. 1.6 **25.** 200 **27.** 72 **29.** 2.5

Pages 234–235 Problem Solving on the Job
1. $13,500 **3.** $200; $5200 **5.** $4044
7. $400 **9.** $13,229 **11.** $55.00, $170.00

Page 236 Enrichment 1. 3, 15, 75, 375
3. direct; inverse

Page 237 Chapter Review 1. T **3.** F **5.** T
7. a **9.** e **11.** c **13.** b

Page 239 Cumulative Review 1. $3\frac{2}{3}$ **3.** $7\frac{11}{24}$
5. 0 **7.** 20 **9.** 5.3 **11.** $5\frac{5}{8}$ **13.** 0 **15.** 8340
17. 33 ft^2 **19.** 28.08 m^2 **21.** 28.26 m^2 **23.** 4
seconds **25.** $58.00

CHAPTER 8. INTEGERS AND RATIONAL NUMBERS

Page 242 Exercises 1. $^-8$ **3.** 23 **5.** 17
7. $^-3$ **9.** 4 **15.** $12 > 0$ **17.** $67 > ^-57$
19. 7 units to left **21.** 9 units to left **23.** 9
units to left **25.** True **27.** False

Page 244 Exercises 1. $^-7$ **3.** $^-12$ **5.** $^-12$
7. $^-14$ **9.** $^-13$ **11.** $^-14$ **13.** $^-20$ **15.** $^-34$

17. $^-20$ **19.** $^-35$ **21.** $^-44$ **23.** $^-51$ **25.** $^-81$
27. $^-73$ **29.** ^-7x **31.** ^-20t **33.** ^-25g
35. ^-20y **37.** $^-9n + ^-12p$ **39.** $^-20a + ^-18x$

Page 247 Exercises 1. 6 **3.** $^-2$ **5.** 0 **7.** $^-3$
9. 1 **11.** $^-5$ **13.** 0 **15.** 2 **17.** $^-4$ **19.** 2
21. 11 **23.** $^-5$ **25.** 15 **27.** 27 **29.** $^-8$
31. 4 **33.** a **35.** ^-y **37.** $4x$ **39.** $6m$
41. ^-p **43.** $30n$ **45.** $x + 7y + 6$
47. $^-2t + 6p + 10$ **49.** $3g + ^-3n + ^-8$

Page 249 Exercises 1. $12 + {}^-3 = 9$
3. $4 + 10 = 14$ **5.** ${}^-9 + {}^-6 = {}^-15$
7. $17 + 8 = 25$ **9.** ${}^-8 + 19 = 11$
11. $16 + 23 = 39$ **13.** 10 **15.** ${}^-10$ **17.** ${}^-27$
19. ${}^-18$ **21.** 0 **23.** 42 **25.** ${}^-12$ **27.** 22
29. 0 **31.** 47 **33.** ${}^-47$ **35.** ${}^-9d$ **37.** $18x$
39. ${}^-11h$ **41.** $2p$ **43.** $10x + 7y$
45. $19g + {}^-3n$

Page 252 Exercises 1. ${}^-88$ **3.** ${}^-48$ **5.** 42
7. 30 **9.** ${}^-32$ **11.** 15 **13.** ${}^-120x$ **15.** ${}^-28h$
17. 88 **19.** ${}^-120$ **21.** ${}^-48$ **23.** ${}^-12$ **25.** ${}^-30$
27. 35 **29.** ${}^-28$ **31.** 8 **33.** ${}^-18$ **35.** 30
37. ${}^-40$ **39.** 12 **41.** ${}^-120$ **43.** 120
45. ${}^-16x + 28$ **47.** $15d + 12$
49. ${}^-24k + {}^-32$ **51.** ${}^-66t + 28$

Pages 254–255 Exercises 1. ${}^-7$ **3.** 11
5. ${}^-7$ **7.** ${}^-7$ **9.** ${}^-3$ **11.** ${}^-11$ **13.** $2x$
15. ${}^-2d$ **17.** ${}^-2m$ **19.** ${}^-13k$ **21.** ${}^-2x$
23. ${}^-20y$ **25.** $9h$ **27.** ${}^-8n$ **29.** ${}^-8$ **31.** ${}^-12$
33. ${}^-10$ **35.** ${}^-8$ **37.** ${}^-35$ **39.** 3 **41.** 11
43. ${}^-32$ **45.** 20 **47.** ${}^-39$ **49.** ${}^-24$ **51.** 0

Page 255 Problems 1. ${}^-7°C$ **3.** All; $1641
profit

Page 258 Exercises 1. Negative **3.** Positive
5. Negative **7.** Positive **9.** 0.6 **11.** 8.3
13. 12.3 **15.** 1.51 **17.** 0 **19.** 6.42 **21.** $1\frac{1}{3}$
23. ${}^-1\frac{1}{4}$ **25.** ${}^-14$ **27.** $1\frac{1}{2}$ **29.** ${}^-9\frac{5}{8}$ **31.** ${}^-26\frac{1}{9}$
33. ${}^-11.2$

Page 258 Using the Calculator Final
balance = $61.48. The club has lost money.

Pages 260–261 Exercises 1. ${}^-9.2$ **3.** 17.36
5. ${}^-\frac{1}{4}$ **7.** ${}^-\frac{2}{5}$ **9.** ${}^-1.9$ **11.** ${}^-4.6$ **13.** ${}^-\frac{4}{5}$
15. $1\frac{1}{6}$ **17.** 50.22 **19.** 42.78 **21.** 15.61
23. ${}^-\frac{1}{2}$ **25.** ${}^-\frac{1}{8}$ **27.** ${}^-\frac{3}{16}$ **29.** ${}^-4$ **31.** ${}^-4$
33. $1\frac{1}{4}$ **35.** $3a$ **37.** ${}^-15n$

Page 264 Problems 1. 10 handshakes **3.** 6
ways **5.** 16 ways **7.** $\frac{1}{2}$ ft; $46\frac{1}{2}$ ft **9.** $5\frac{5}{9}$ ft,
$16\frac{2}{3}$ ft, $27\frac{7}{9}$ ft

Page 266 Exercises 1. ${}^-1$ **3.** ${}^-12$ **5.** ${}^-12$
7. 7 **9.** ${}^-12$ **11.** 15 **13.** ${}^-\frac{3}{10}$ **15.** 7 **17.** 7.9

19. ${}^-1.1$ **21.** 12.6 **23.** 10 **25.** ${}^-4.8$ **27.** 6.6
29. ${}^-2.5$ **31.** 0.9 **33.** $\frac{1}{2}$ **35.** ${}^-\frac{5}{6}$ **37.** ${}^-1\frac{1}{2}$
39. $1\frac{1}{2}$ **41.** ${}^-1$ **43.** ${}^-3$

Pages 269–270 Exercises 1. Add 7
3. Divide by 6 **5.** No **7.** No **9.** Yes
11. Yes **13.** $a < {}^-3$ **15.** $p > 6$ **17.** $b < {}^-4$
19. $n < 27$ **21.** $n > {}^-10$ **23.** $y > \frac{3}{5}$
25. $a \geq {}^-20$ **27.** $x \leq 5$ **29.** $d < {}^-30$
31. $h \leq 4$ **33.** $x > {}^-2$ **35.** $x \geq 3$
37. $x > 4$ **39.** $m < \frac{13}{3}$

Page 271 Algebra Practice 1. $761(1002) = r$;
$762{,}522$ **3.** $364.8 + 1.065 + 98.37 = d$;
464.235 **5.** $17\frac{8}{9} - 5\frac{3}{8} = m$; $12\frac{37}{72}$
7. $2\frac{1}{2} + 3\frac{5}{8} + 5\frac{3}{4} = P$; $11\frac{7}{8}$
9. $(3.14)(2^2)4 = V$; $50.24\ m^3$
11. $5q(r^2 + rq + 2q)$ **13.** $\frac{4}{9}$ **15.** 1.2
17. 8.75 **19.** ${}^-9$ **21.** ${}^-11$ **23.** 12 **25.** 27
27. ${}^-63$ **29.** ${}^-14$ **31.** ${}^-13.31$ **33.** $5\frac{2}{3}$ **35.** $1\frac{5}{9}$
37. $x > 4$ **39.** $m < 20$

**Pages 272–273 Problem Solving for the
Consumer 1.** $198 **3.** $354, one way
5. $481.85 **7.** $1201

Page 274 Enrichment 1. ${}^-7$ **3.** ${}^-12$ **5.** ${}^-\frac{7}{10}$
7. $|{}^-8|$ **9.** $|{}^-15|$ **11.** $|{}^-50|$ **13.** $|{}^-24|$ **15.** $|{}^-4|$,
$|0|, |{}^-3|, |{}^-4|, |5|$ **17.** ${}^-|3.8|, |{}^-4.3|, |5.6|, |{}^-6.2|$,
$|7.1|$ **19.** 3 **21.** 15 **23.** 39 **25.** 17
27. $a = \pm b$ **29.** $|a| > |b|$, $a \geq 0$ and $b \leq 0$,
or $|a| > |b|$, $a \leq 0$ and $b \geq 0$

Page 275 Chapter Review 1. $<$ **3.** $>$ **5.** c
7. b **9.** No **11.** Yes **13.** No **15.** No
17. No **19.** Yes **21.** ${}^-10$ **23.** 10 **25.** ${}^-35$
27. 10.7 **29.** ${}^-10$ **31.** 1.2 **33.** $y < 3$
35. $g > {}^-14$ **37.** $a < {}^-54$ **39.** $x > {}^-20$
41. $y > {}^-3$

**Page 277 Cumulative Review (Chapters 1–8)
1.** 5.735 **3.** 13.167 **5.** 120 **7.** $\frac{8}{15}$ **9.** 160
11. $27\frac{1}{2}$ **13.** $1\frac{3}{5}$ **15.** $1\frac{1}{4}$ **17.** 50.5 **19.** 38
21. 24 **23.** 8.1 **25.** 250 **27.** 48
29. $120\ in.^3$ **31.** $50.24\ cm^3$ **33.** 4.9 points, 9.8
points

CHAPTER 9. GRAPHING IN THE COORDINATE PLANE

Page 281 Exercises 1. 24; 132, 156, 180, 204, 228 **3.** 127; 638, 765, 892, 1019, 1146 **5.** $^-$24; 444, 420, 396, 372, 348 **7.** 9, 10, 11, 12, 13 **9.** 0, 5, 10, 15, 20 **11.** 10, 8, 6, 4, 2 **13.** $^-$7, $^-$20, $^-$33, $^-$46, $^-$59 **15.** 258, 758 **17.** 1215, 825

Page 283 Exercises 1. $4n$; 40; 60 **3.** $3n + 4$; 34; 49 **5.** $7n + 3$; 73; 108 **7.** $^-$1; 2, 1, 0, $^-$1, $^-$2 **9.** 5; 0, 5, 10, 15, 20 **11.** 3; 6, 9, 12, 15, 18 **13.** $-n + 11$; 1; $^-$39 **15.** $4a - 9$; 31; 191 **17.** $5n - 2$; 48; 248 **19.** $\frac{5}{2}a + \frac{11}{2}$; $30\frac{1}{2}$; $130\frac{1}{2}$ **21.** $^-\frac{8}{5}n + \frac{88}{5}$; $1\frac{3}{5}$; $^-62\frac{2}{5}$

Page 284 Problems 1. a. $1.75; $3; $13 **b.** 78 weeks **3.** 6 **5.** 10 **7.** 22 **9.** 7 tables

Page 287 Exercises 13. $(^-6, 3)$ **15.** $(5, ^-3)$ **17.** $(0, ^-3)$ **19.** $(^-4, 0)$ **21.** square **23.** triangle **25.** square **27.** parallelogram

Page 290 Exercises 1. 11 **3.** 15 **5.** 14 **7.** 23 **9.** $^-$1 **11.** 16 **13.** $15\frac{1}{2}$ **15.** $^-5\frac{3}{4}$

Page 291 Problems 1. $70\,h$

Page 293 Exercises 1. $(^-2, ^-5)$ **3.** $(3, 5)$ **5.** $^-$18; $(^-5, ^-18)$ **7.** 2; $(0, 2)$

Pages 296–297 Exercises 1. 1; 3; $y = x + 3$ **3.** $^-$4; 20; $y = ^-4x + 20$ **5.** $\frac{3}{5}$; $^-$4; upward

7. 1.5; 7.2; upward **9.** 7.3; $^-$12.6; upward **11.** same; parallel **13.** product is $^-$1; perpendicular **15.** horizontal **17.** vertical

Page 297 Problems 1. $\frac{3}{16}$

Pages 299–300 Exercises 1. No **3.** Yes **5.** No **7.** Yes **9.** Yes **11.** Yes **13.** Yes

Page 301 Algebra Practice
1. $49{,}452 \div 78 = c$; 634 **3.** $103 + 99.44 = v$; 202.44 **5.** $4\frac{2}{3} + 4\frac{7}{16} = t$; $9\frac{5}{48}$ **7.** $4(6\frac{7}{8}) = P$; $27\frac{1}{2}$ in. **9.** $(20)(6) = V$; 120 ft^3 **11.** $st(3t + st + 2s)$ **13.** $^-$2 **15.** $^-$4 **17.** $^-$2 **19.** 6, 7, 8, 9, 10 **21.** 1, $^-$3, $^-$7, $^-$11, $^-$15 **23.** $^-\frac{16}{7}n + 15$; $^-12\frac{3}{7}$; $^-30\frac{5}{7}$ **25** Answers vary **27.** 5; 3 **29.** $^-$3; 6 **31.** Yes

Pages 302–303 Problem Solving on the Job 1. d **3.** d **5.** c **7.** d **9.** d

Page 304 Enrichment 1. Yes **3.** Yes **5.** $(10, ^-3)$ **7.** $(1, 1)$ **9.** $(^-1, ^-6)$

Page 305 Chapter Review 1. d **3.** c **5.** a **7.** $5n - 4$; 71; 96 **9.** $3n - 8$; 37; 52 **15.** Yes **17.** No **19.** No **25.** 4; $^-$5 **27.** No **29.** No **31.** No

Page 307 Cumulative Review (Chapters 1–9) 1. $3\frac{15}{16}$ **3.** $\frac{1}{6}$ **5.** 1.96 **7.** 21 **9.** 190 **11.** $^-$10 **13.** $^-$27 **15.** 2 **17.** 24 **19.** $^-$8 **21.** $^-$12 **23.** 11 **25.** 9 **27.** 36 **29.** 10 **31.** 4 **33.** 3 **35.** 1 **37.** 864 cm^2

CHAPTER 10. STATISTICS

Page 311 Exercises 1. 19.25, 18 **3.** $\frac{5}{12}$, $\frac{1}{8}$ **5.** 3.9, 1.3 **7.** 71°, 72°, 72°, 5° **9.** $57.45, $49.95, $49.95, $100

Page 311 Problems 1. 62,612 **3.** $4

Page 313 Exercises 1. $4000 **3.** $1000 **5.** $4350 **7.** $100 **9.** WROQ, 45%; WCLS, 5%; WCAW, 20%; WANY, 30%

Pages 316–317 Exercises 1. $400 **3.** $100, $50 **5.** $2\frac{1}{2}$; $250 **7.** $2\frac{1}{4}$, $225 **9.** $150 **11.** 3 **13.** $5\frac{1}{2}$

Pages 319–320 Exercises 1. 90 **3.** 615 **5.** 300 **7.** 60 **9.** 25% **11.** Political action **13.** $15,000 **15.** $13,000

Pages 323–324 Exercises 1. Decrease **3.** 1960 to 1980 **5.** Magazine **7.** 35% **9.** 15% **11.** West **13.** After **15.** 41 million

Pages 327–328 Exercises (Answers may vary) **1.** How long will it be before rental charges equal the cost of buying the TV? 12 months **3.** What is the greatest amount of milk and detergent you can buy for $4? $\frac{3}{4}$ gal milk, 32 oz detergent **5.** When would you start dinner so that it would be ready at 6:30? 4:20 **7.** How long will it be before the cost of buying batteries equals the cost of buying a charger and rechargeable batteries? 13 months

Pages 331–332 Exercises 1. 76–79, 3; 80–83, 1; 84–87, 2; 88–91, 1; 92–95, 2; 96–99, 3 **3.** $1\frac{3}{4}$ times greater **5.** 15 **7.** Between 8–9 and 10–11 **9.** 2 **11.** 7 **13.** 3

Page 335 Exercises 1. *L, R, P* **3.** *R* **5.** Negative **7.** 150 lb **9.** Negative **11.** Under 3:40

Page 337 Algebra Practice 1. $385(124) = y$; 47,740 **3.** $70.54 + 8.346 + 21.43 = n$; 100.316 **5.** $5\frac{1}{2} - 3\frac{4}{5} = s$; $1\frac{7}{10}$ **7.** $C = 2(3.14)(6.2)$; 38.936 cm **9.** $2r^3s$ **11.** $12j^3k^2$ **13.** 2 **15.** 128 **7.** 75

19. $n < {}^-7$ **21.** 6, 7, 8, 9, 10 **23.** 9, 12, 15, 18, 21

Pages 338–339 Problem Solving for the Consumer 1. $65 **3.** $183.50 **5.** E **7.** A, B, E, F **9.** 4 **11.** C

Page 340 Enrichment 1. 2 **3.** 14 **5.** 12 **7.** 82 **9.** 10 at or below; 61 **11.** 30 at or below; 91 **13.** 24 at or below; 84

Page 341 Chapter Review 1. B, 30 jeans **3.** A, $30 **5.** D, 136 **7.** C; 600

Page 343 Cumulative Review (Chapters 1–10) 1. 25 **3.** 56 **5.** 0.27263 **7.** $1\frac{7}{20}$ **9.** $5\frac{1}{6}$ **11.** $18\frac{13}{36}$ **13.** 3 **15.** 0 **17.** 5 **19.** 4 **21.** 2 **23.** 64.35 **25.** 46.2 **27.** 81.25% **29.** $^-12$ **31.** $^-210$ **33.** $^-22$ **35.** $^-5$ **37.** $^-5$; 0 **39.** $\frac{1}{2}$; 8 **41.** $^-16$; 4 **43.** 63

CHAPTER 11. SPECIAL TRIANGLES

Page 347 Exercises 1. 4.5 **3.** 3 **5.** 14 **7.** 6 **9.** 12

Page 349 Exercises 1. 7.5 ft **3.** 69 ft **5.** 21.8 ft

Page 350 Problems 1. 200 m **3.** 10 ft 4 in.

Page 353 Exercises 1. 6 **3.** 9 **5.** 1 **7.** 2.449 **9.** 3.606 **11.** Yes **13.** No **15.** Yes **17.** 3.606 km **19.** 10 cm **21.** 8.602 m **23.** 6.325 **25.** 7.483 **27.** 8.062

Page 354 Problems 1. 5.4 ft **3.** 6.7 mi **5.** 7 ft

Pages 356–357 Exercises 1. $\frac{1}{2}$; 2 **3.** $\frac{3}{2}$; $\frac{2}{3}$ **5.** 4; $\frac{1}{4}$ **7.** 0.8 **9.** 0.9 **11.** 1.9 **13.** 1.8 **15.** 4.4 **17.** 1.3 **19.** 0.9

Page 357 Using the Calculator 1. 3 **3.** 15 **5.** 49

Pages 359–360 Exercises 1. $\frac{5}{13}$; $\frac{12}{13}$; $\frac{12}{13}$; $\frac{5}{13}$ **3.** $\frac{3}{5}$; $\frac{4}{5}$; $\frac{4}{5}$; $\frac{3}{5}$ **5.** $\frac{70}{99}$; $\frac{70}{99}$; $\frac{70}{99}$; $\frac{70}{99}$ **7.** 40; $\frac{40}{41}$ **9.** 8; $\frac{8}{17}$ **11.** 8.602; 0.8138 **13.** $\frac{4}{5}$ **15.** 0.7071

17. 0.7071 **19.** $\frac{4}{5}$ **21.** $\frac{3}{4}$ **23.** $\frac{3}{5}$ **25.** $\frac{3}{5}$ **27.** $\frac{3}{5}$; $\frac{3}{5}$; sin $M = \cos N$; Yes

Page 360 Challenge 1. $\dfrac{\sqrt{6}}{6}$ **3.** $\dfrac{2\sqrt{5}}{5}$ **5.** $\dfrac{7\sqrt{3}}{3}$

Page 362 Exercises 1. 6 **3.** 4.6 **5.** 13.0 **7.** 5.7 **9.** 18.1

Page 363 Problems 1. 477 m **3.** 55 mi

Page 365 Exercises 1. 34° **3.** 64° **5.** 61° **7.** 63°; 27° **9.** 48°; 42° **11.** 34°; 56° **13.** 66°; 24° **15.** 30°; 60° **17.** 37° **19.** 53° **21.** 26°; 64°; 8

Page 366 Problems 1. 23°

Page 367 Algebra Practice 1. $293,675 \div 425 = m$; 691 **3.** $700 - 469.89 = C$; 230.11 **5.** $12\frac{3}{5} + 5\frac{9}{10} = g$; $18\frac{1}{2}$ **7.** $(3.14)(4.2)^2 = A$; 55.3896 in.² **9.** $2g(g - 1 + 3g)$ **11.** $3cd(4c + cd - d^2)$ **13.** $^-6$ **15.** 9 **17.** 312 **19.** $V > {}^-3$ **21.** $5n - 12$; 63; 138 **23.** $5.3 + 2.8 = b$; 8.1 cm

511

Pages 368-369 Problem Solving on the Job
1. 43.75%; 51.43%; 48.84%; 46.55%; 43.75%
3. 24% **5.** 51.06% **7.** 5d **9.** 14d
11. $15,500; $20,133.33 **13.** 30 h; 67.5 h;
60 h; 22.5 h

Page 370 Enrichment **1.** $3\sqrt{2}$ **3.** 5; 5
5. 2; 4

Page 371 Chapter Review **1.** b **3.** 7.2 m

5. a **7.** a **9.** 16 **11.** 37°; 53° **13.** 41°;
49° **15.** $\frac{4}{5}$ **17.** $\frac{3}{5}$

Page 373 Cumulative Review **1.** $\frac{2}{9}$ **3.** $\frac{1}{4}$
5. $\frac{9}{16}$ **7.** $\frac{14}{27}$ **9.** $\frac{2}{5}$ **11.** $\frac{7}{9}$ **13.** $\frac{7}{8}$ **15.** $\frac{1}{3}$
17. 60% **19.** 55.04 **21.** 35 **23.** $150
25. ⁻2 **27.** ⁻7 **29.** ⁻11 **31.** 8 **33.** ⁻48
35. ⁻5 **37-41.** Check students' graphs.
43. $24.75; $54.75

CHAPTER 12. PROBABILITY

Pages 377-378 Exercises **1.** $\frac{1}{2}$ **3.** $\frac{100}{500} = \frac{1}{5}$
5. (Ex. 1) 50% **7.** (Ex. 3) 20% **9.** $\frac{1}{4}$ **11.** $\frac{1}{2}$
13. $\frac{1}{4}$ **15.** $\frac{1}{2}$ **17.** $\frac{3}{4}$ **19.** $\frac{3}{4}$ **21.** 30% **23.** 0%
25. 80% **27.** $\frac{1}{4}$ **29.** $\frac{2}{3}$

Page 378 Problems **1.** 50% **3.** 48%

Pages 380-381 Exercises **1.** $\frac{1}{8}$ **3.** $\frac{1}{4}$ **5.** $\frac{1}{8}$
7. $\frac{1}{6}$ **9.** 0 **11.** $\frac{1}{3}$ **13.** $\frac{1}{12}$ **15.** 0 **17.** $\frac{1}{6}$ **19.** $\frac{1}{4}$
21. $\frac{1}{6}$ **23.** $\frac{1}{2}$ **25.** $\frac{2}{3}$

Page 381 Problems **1.** $\frac{1}{8}$

Pages 384-385 Exercises **1.** 10
3. $2 \times 2 \times 5 = 20$ **5.** 125 **7.** 90 **9.** $\frac{1}{15}$
11. $\frac{2}{15}$ **13.** $\frac{1}{18}$ **15.** $\frac{1}{90}$ **17.** $\frac{1}{30}$ **19.** $\frac{1}{90}$ **21.** 16
23. $\frac{1}{4}$

Page 385 Problems **1.** 60 **3.** 150 **5.** 240

Pages 387-388 **1.** $\frac{1}{36}$ **3.** $\frac{1}{12}$ **5.** $\frac{5}{36}$ **7.** $\frac{1}{32}$
9. $\frac{1}{15}$ **11.** $\frac{21}{32}$ **13.** $\frac{8}{15}$ **15.** $\frac{1}{720}$ **17.** $\frac{7}{18}$
19. $\frac{1}{288,000}$ **21.** Freshman

Page 388 Problems **1.** 0.64

Page 389 Using the Calculator **1.** 0.512
3. 0.250 **5.** 0.062

Pages 391-392 Exercises **1.** $\frac{2}{3}$ **3.** 0.02
5. $\frac{4}{11}$ **7.** $\frac{5}{22}$ **9.** $\frac{1}{66}$ **11.** $\frac{2}{11}$ **13.** $\frac{3}{28}$ **15.** $\frac{15}{56}$
17. $\frac{1}{56}$ **19.** $\frac{1}{56}$ **21.** $\frac{1}{132}$ **23.** $\frac{1}{12}$ **25.** $\frac{5}{22}$

Page 392 Problems **1.** $\dfrac{9 \cdot 8 \cdot 7 \cdot 6 \cdot 5}{10^5}$ **3.** $\dfrac{1}{10^5}$
5. 0

Pages 394-395 Exercises **1.** $\frac{1}{1}$ **3.** $\frac{1}{5}$ **5.** $\frac{1}{2}$
7. $\frac{2}{1}$ **9.** $\frac{2}{1}$ **11.** $\frac{1}{1}$ **13.** $\frac{2}{1}$ **15.** $\frac{3}{2}$ **17.** $\frac{5}{4}$ **19.** $\dfrac{n}{1}$
21. $\frac{3}{7}$ **23.** $\frac{4}{1}$ **25.** $\dfrac{n}{100 - n}$ **27.** A **29.** $\frac{4}{1}$
31. $\frac{1}{2}$ **33.** $\dfrac{1}{n - 1}$ **35.** 16; $\frac{3}{8}$

Page 398 Exercises **1.** 0.105 **3.** 0.215
5. 0.075 **7.** 0.1$\overline{6}$ **9.** 0.2 **11.** 0.25 **13.** 0.3
15. 0.58$\overline{3}$ **17.** 0.5 **19.** $\frac{23}{82}$

Page 398 Problems **1.** 0.400 **3.** Season.
Greater number of times at bat gives a better
approximation.

Page 401 Exercises **1.** 0.3 **3.** 0.6 **5.** 0.42
7. 360 **9.** 336 **11.** 1333 **13.** 162 **15.** 1160

Page 403 Algebra Practice **1.** $(304)(465) = b$;
141,360 **3.** $5.84 + 9.473 + 6.958 = j$;
22.271 **5.** $20\frac{3}{8} - 12\frac{5}{6} = h$; $7\frac{13}{24}$
7. $2(6.2) + 2(5.9) = P$; 24.2 m
9. $6x^2(8x - 7)$ **11.** $2r(r^2 - 3rs + 2s^3)$
13. ⁻2 **15.** 3.84 **17.** 78 **19.** $k < 6$
21. (⁻1, ⁻7), (0, ⁻6), (1, ⁻5). (Answers will
vary.) **23.** (⁻1, 13), (0, 8); (1, 3). (Answers will
vary.) **25.** $g - 18.79 = 58.42$; $77.21

**Pages 404-405 Problem Solving for the
Consumer** **1.** $\frac{4}{1}$ **3.** 800 **5.** Sum of the
results from Polls 1-4. **7.** Telephone book

Page 406 Enrichment **1.** It is cold. **3.** 13 is
a prime number. **5.** 48 is a multiple of 4
7. You will receive $1000. You will be on TV.
You will be a celebrity.

Page 407 Chapter Review 1. $\frac{3}{5}$ **3.** 0 **5.** $\frac{9}{25}$
7. $\frac{4}{25}$ **9.** $\frac{12}{49}$ **11.** $\frac{87}{245}$ **15.** $\frac{3}{5}$ **17.** 720 **19.** $\frac{3}{10}$

Page 409 Cumulative Review 1. $4\frac{5}{12}$ **3.** $1\frac{23}{45}$
5. 83 **7.** 3196 **9.** 17 **11.** 7 **13.** 112°

CHAPTER 13. GEOMETRY

Page 414 Exercises 1. F **3.** T **5.** T **7.** F
9. T **11.** Alternate interior
13. Corresponding **15.** $\angle 7$, $\angle 13$; $\angle 6$, $\angle 16$;
$\angle 3$, $\angle 9$; $\angle 2$, $\angle 12$ **17.** 125° **19.** Yes
21. No **23.** No

Page 417 Exercises 1. F **3.** F **5.** T
7. No **9.** Yes. No. **11.** 360°; 70° **13.** 80°,
140°, 40°, 100° **15.** 108°, 60°, 90°, 132°,
150°

Pages 420–421 Exercises 1. Yes.
Corresponding sides are in proportion.
3. Yes. Corresponding sides are in
proportion. **5.** No **7.** Yes **9.** No **11.** No
13. 5; 1.75 **15.** $\angle DAB$, $\angle BEF$; $\angle ABC$,
$\angle EBC$; $\angle BCD$, $\angle BCF$; $\angle CDA$, $\angle CFE$;
$\overline{AB}$, $\overline{BE}$; $\overline{BC}$, $\overline{BC}$; $\overline{CD}$, $\overline{CF}$; $\overline{AD}$, $\overline{EF}$.
17. $\angle EMO$, $\angle ETR$; $\angle MOU$, $\angle TRA$;
$\angle OUS$, $\angle RAP$; $\angle USE$, $\angle APE$; $\angle SEM$,
$\angle PET$; $\overline{MO}$, $\overline{TR}$; $\overline{OU}$, $\overline{RA}$; $\overline{US}$, $\overline{AP}$; $\overline{SE}$,
$\overline{PE}$; $\overline{EM}$, $\overline{ET}$ **19.** Yes; SSS

Pages 423–424 Exercises 1. 15.25 in.
3. Carbon monoxide, 21.652.4 g; nitrogen,
1687.2 g; hydrocarbons, 1124.8 g; solid
particles, 98.8 g; sulfur, 76 g

15. 50° **17.** 99° **19.** 32.4 cm²
21. 17.64 m² **23.** 28.26 ft³
25. $420 + m = 725$; 305 mi
27. $x + x + 160 + x - 70 = 1860$; Team A,
$750; Team B, $590; Team C, $520

Page 427 1. Acute **3.** Equiangular
5. Isosceles **7.** Equilateral **9.** 9, 21, 14
11. 20, 20, 26 **13.** 70°

Page 430 Exercises 1. NAU **3.** 50 **5.** 40°
7. 85° **9.** 98 **11.** 2

Page 431 Algebra Practice
1. $1{,}336{,}064 \div 136 = p$; 9824
3. $152 - 107.98 = z$; 44.02 **5.** $6\frac{1}{4} + 1\frac{2}{3} = m$;
$7\frac{11}{12}$ **7.** $3.14(10.4) = c$; $32.656\ m$ **9.** $4xy^2$
11. $20a^2b^2$ **13.** 10 **15.** 21.6 **17.** 2.5
19. $p < 2\frac{1}{4}$ **21.** 2; 7 **23.** $^-5$; 12 **25.** 28, 28,
26 **27.** $(x + 129) + x = 815$; 343 trucks, 472
cars

Pages 432–433 Problem Solving on the Job
1. $222 **3.** $104, $58.40, $143.20, $305.60
5. $93.20, $8.40, $55.20, $156.80 **7.** $55.50
9. $28.20

Page 434 Enrichment 1. 120 **3.** 240 **5.** 336

Page 435 Chapter Review 1. F **3.** T **5.** T
7. T **9.** c **11.** FE **13.** EH **15.** $\overline{MI}$, $\overline{ST}$, or
$\overline{IS}$ **17.** T **19.** $\angle KNM$ **21.** $\angle NKM$

Page 437 Cumulative Review (Chapters 1–13)
1. c **3.** b **5.** a **7.** 3.75 **9.** No **11.** No
13. $\frac{1}{2}$; 50% **15.** 3 small, 6 medium, 7 large

CHAPTER 14. PROGRAMMING IN BASIC

Page 441 Exercises 1. $2*x - 7$ **3.** $1/2*b*h$
5. $a/2 + b/3$ **7.** $x*y/z$ **9.** 6; 33.5; 8.75; 31
11. 10 PRINT $1*2*3$, 20 END **13.** 10
PRINT $512 - 367$, 20 END **15.** 10 PRINT
$3.14*2*7.2$, 20 END

Page 441 Challenge Output $= 27$; 10 PRINT
$(8+4)\uparrow2/2*3 - 5$; 10 PRINT
$8+4\uparrow2/2*(3-5)$; 10 PRINT $8+4\uparrow2/(2*3-5)$

Pages 444–445 Exercises 1. 10.5; 2.1; 0.6
3. 3.3; 35.1; 38.4 **5.** THE VOLUME IS
301.44 **7.** Change line 50 to 50 LET
$V = B*H/3$ **9.** 40 LET $M = (A+B+C)/3$
50 PRINT "THE AVERAGE SCORE IS" M

Page 445 Using the Calculator 1. 23 **3.** 48
5. 73

Page 448 Exercises 1. 6, 7, 8, 9, 10, 11
3. 7, 9, 11, 13 **5.** 75, 90, 105, 120, 135, 150
7. 10 FOR $L = 20$ TO 30

Pages 453–454 Exercises 1. Delete lines 10 and 20; 5 READ A B; 35 DATA 15, 12.7 **3.** PERIMETER IS 20; AREA IS 25; PERIMETER IS 52; AREA IS 169; PERIMETER IS 29.2; AREA IS 53.29; OUT OF DATA IN LINE 10 **5.** 10 READ L, W; 20 LET A=L*W; 30 PRINT "AREA IS" A; 40 GO TO 10; 50 DATA 5.2, 3.9, 92.3, 80, 60 DATA 17, 11, 2.25, 1.4 70 END **7.** 10 READ D, M; 20 LET T=45.95*D+0.29*M; 30 PRINT "TOTAL COST IS" T; 40 GOTO 10; 50 DATA 2, 350, 1, 120, 1, 85; 60 DATA 3, 645, 4, 2120; 70 END

Page 456 Exercises 1. 72 IS 150 PERCENT OF 48; 35 IS 50 PERCENT OF 70; 75 IS 50 PERCENT OF 150; 90 IS 75 PERCENT OF 120; OUT OF DATA IN LINE 10 **3.** 1, 1; 4, 2; 9, 3; 16, 4; 25, 5; 36, 6; 49, 7, OUT OF DATA IN LINE 10 **5.** THE LEG LENGTHS ARE 12, 5; THE LENGTH OF THE HYPOTENUSE IS 13; THE LEG LENGTHS ARE 8, 6; THE LENGTH OF THE HYPOTENUSE IS 10; THE LEG LENGTHS ARE 8, 15; THE LENGTH OF THE HYPOTENUSE IS 17; THE LEG LENGTHS ARE 2, 2; THE LENGTH OF THE HYPOTENUSE IS 2.82843; THE LEG LENGTHS ARE 15, 20; THE LENGTH OF THE HYPOTENUSE IS 25 **7.** 10 READ A, B; 20 LET C = SQR(A↑2 + B↑2); 30 LET S = A/C; 40 PRINT "SIN D IS" S; 50 GOTO 10; 60 DATA 12, 5, 8, 6, 8, 15, 2, 2, 15, 20; 70 END

Pages 458–459 Exercises 1. 7.1, 4.7, 6, 3.9, 17.2 **3.** 35 IF $A<=220$ THEN 40; 36 PRINT "GREAT BOWLING"

EXTRA PRACTICE

Page 472 Extra Practice—Chapter 1 1. 22 **3.** 9 **5.** 7 **7.** 485 **9.** 208 **11.** 381 **13.** 387 **15.** 608 **17.** 42 **19.** 160 **21.** 11 **23.** 12 **25.** 11 **27.** 36 **29.** 0 **31.** $6n$ **33.** $11x + 3y$ **35.** $k + 7$ **37.** $8a + c$ **39.** $4x + 3y$ **41.** $x - 15$ **43.** $p + 42$ **45.** $x + y$

Page 473 Extra Practice—Chapter 2 1. 4.82, 4.7, 4.16 **3.** 9.7, 6.2, 4.863 **5.** 6.22, 6.2, 6.02 **7.** 1 **9.** 13 **11.** 1.3 **13.** 0.6 **15.** 250 **17.** 2.11 **19.** 20.418 **21.** 2.8 **23.** 3.06

Pages 461–462 Exercises 1. RED RIBBON; BLUE RIBBON; RED RIBBON **3.** 10 READ X, Y 20 IF X=Y THEN 40 30 IF $X<>Y$ THEN 60 40 PRINT X; "="; Y 50 GO TO 10 60 PRINT X; "$<>$"; Y 70 GO TO 10 80 DATA 7.5, 7.5, $^-1$, 1, $^-16$, $^-16$, $^-18.1$, 18.1, 0.62, 6.02 90 END

Page 463 Algebra Practice 1. $2803(318) = d$; 891,354 **3.** $12.087 + 3.494 + 28.513 = h$; 44.094 **5.** $7\frac{7}{15} - 2\frac{4}{5} = n$; $4\frac{2}{3}$ **7.** A $= \frac{1}{2}(3\frac{1}{2})(2)$; $3\frac{1}{2}$ in.^{2} **9.** $3ab(4a^2 - 5b)$ **11.** $2j^2(j + 8jk - 7k^2)$ **13.** 21 **15.** 7.8 **17.** 88 **19.** $28\frac{4}{5} > k$ **21.** Yes **23.** No **25.** 17, 27, 15 **27.** $2*r/s$

Page 464 Problem Solving for the Consumer 1. 15, 9 **3.** 15, 5 **5.** 20, 12, 12 **7.** 12, 16, 16 **9.** 110, 85, 88 **11.** Display **13.** Whiz, $1389.95; Essex, $1298.90; Supra, $1887.95 **15.** $108.83

Page 466 Enrichment 1. 5 **3.** $^-7$ **5.** 6.781 ROUNDED TO THE NEAREST WHOLE NUMBER IS 7; 3.925 ROUNDED TO THE NEAREST WHOLE NUMBER IS 4; 4.617 ROUNDED TO THE NEAREST WHOLE NUMBER IS 5; 86.138 ROUNDED TO THE NEAREST WHOLE NUMBER IS 86

Page 467 Chapter Review 1. F **3.** T **5.** F **7.** a **9.** 1000, 9

Page 469 Cumulative Review (Chapters 1–14) 1. 1923 **3.** 13,700 **5.** $1\frac{17}{45}$ **7.** $\frac{8}{9}$ **9.** $1\frac{7}{8}$ **11.** 29.6 **13.** 23% **15.** $420 **17.** 25% **19.** $\frac{1}{495}$

25. 272 **27.** 960 **29.** 1.2 **31.** 180 **33.** 23.4 **35.** 0.0362 **37.** 0.064 **39.** 1.2 **41.** 200,000 **43.** 2400 **45.** $3.401 + r$ **47.** $17.4 \div n$ **49.** $3.07 + c$

Page 474 Extra Practice—Chapter 3 1. Yes **3.** No **5.** Yes **7.** Yes **9.** $2^3 \cdot 3 \cdot 5$ **11.** 5^4 **13.** $2^2 \cdot 17$ **15.** $5 \cdot 31$ **17.** 2 **19.** 3 **21.** 4 **23.** n **25.** $2r$ **27.** f **29.** $a(2 - 1)$ **31.** $2(h^2 - 3)$ **33.** $r(25 + r)$ **35.** $3(3b^2 + b + 5)$ **37.** $2(5n + 8n^2 + 2)$ **39.** $xy(y + 1 - 3x)$ **41.** 42 **43.** 18 **45.** 24 **47.** 300 items

Page 475 Extra Practice—Chapter 4 **1.** $\frac{4}{5}$
3. $\frac{3}{4}$ **5.** 5 **7.** 2 **9.** $\frac{1}{a}$ **11.** $\frac{3}{4}$ **13.** $1\frac{7}{12}$

15. $1\frac{13}{28}$ **17.** $4\frac{11}{56}$ **19.** $\frac{1}{36}$ **21.** $6\frac{5}{12}$ **23.** $\frac{1}{a}$

25. $\frac{5a}{7}$ **27.** $\frac{3}{5}$ **29.** 1 **31.** $\frac{1}{9}$ **33.** $\frac{8}{c}$ **35.** n

37. $\frac{2}{3}$ **39.** 0.8 **41.** $0.\overline{3}$ **43.** $0.\overline{36}$ **45.** $0.1\overline{6}$
47. 4.5 **49.** 370 mi

Page 476 Extra Practice—Chapter 5 **1.** 5
3. 33 **5.** 62 **7.** 35 **9.** 35 **11.** 6 **13.** 5
15. 20 **17.** 36 **19.** 10 **21.** 7 **23.** 4 **25.** 12
27. 3 **29.** 16 **31.** 2 **33.** 2 **35.** $p + 4$; $\frac{1}{2}p$
37. $2n = 25$; \$12.50

Page 477 Extra Practice—Chapter 6
1. 16.8 cm **3.** 25.12 **5.** Right **7.** Acute
9. 42 ft^2 **11.** 12.56 m^2 **13.** 151.2 m^3
15. 188.4 cm^3 **17.** 200.96 mm^3 **19.** 471 ft^2
21. 220 mi

Page 478 Extra Practice—Chapter 7 **1.** $\frac{12}{125}$
3. $\frac{9}{25}$ **5.** 18.7 **7.** 110.8 **9.** 70% **11.** 5%
13. 50% **15.** 11.7 **17.** 20 **19.** 0.45 **21.** 630
23. $215\frac{5}{13}$ **25.** 10% **27.** 20% **29.** 29.4%

Page 479 Extra Practice—Chapter 8 **1.** $<$
3. $>$ **5.** $>$ **7.** $<$ **9.** $^-2$ **11.** $^-4$ **13.** $^-17$
15. 2 **17.** 2 **19.** $^-3$ **21.** $^-32$ **23.** 23
25. $^-24$ **27.** $^-20$ **29.** $^-56$ **31.** 72 **33.** $^-7$
35. 7 **37.** $^-4$ **39.** 9 **41.** 6.9 **43.** 31.5
45. $^-1.2$ **47.** $^-14$ **49.** $^-3\frac{4}{15}$ **51.** $^-1\frac{5}{16}$ **53.** 6

55. $1\frac{1}{5}$ **57.** 5 **59.** $^-6$ **61.** $^-0.9$ **63.** $7\frac{1}{2}$
65. $t > ^-4$ **67.** $y > ^-2$ **69.** $n > \frac{2}{5}$
71. $a < ^-40$

Page 480 Extra Practice—Chapter 9 **1.** 20,
24, 28 **3.** 20, 25, 30 **5.** 15, 21, 27 **15.** 35
17. 16 **27.** 2; $^-9$ **29.** $^-1$; $^-2$ **31.** $\frac{2}{3}$; $^-6$
33. $^-8$; $\frac{5}{3}$ **35.** $\frac{^-1}{3}$; $\frac{^-1}{4}$ **37.** No **39.** Yes

Page 481 Extra Practice—Chapter 10 **1.** 7.6;
3.1; 3.1; 3.6 **3.** 33% **5.** 6 P.M.; 8 A.M.
7. 800

Page 482 Extra Practice—Chapter 11 **1.** 8
3. 12.8 **5.** 9.22 **7.** $\frac{4}{3}$, $\frac{3}{5}$, $\frac{3}{5}$ **9.** $\frac{5}{12}$, $\frac{12}{13}$, $\frac{12}{13}$
11. 45°, 45° **13.** 60°, 30° **15.** 49°, 41°

Page 483 Extra Practice—Chapter 12 **1.** $\frac{1}{15}$
3. $\frac{14}{15}$ **5.** $\frac{2}{1125}$ **7.** $\frac{1}{45}$ **9.** $\frac{3}{2}$

Page 484 Extra Practice—Chapter 13
1. 120° **3.** 60°; 60° **5.** 55° **7.** $2\frac{10}{13}$ **9.** 6,
12, 12 **11.** 12, 27, 15 **13.** $\overline{BC}$; 8 **15.** 3; 4

Page 485 Extra Practice—Chapter 14
1. AREA IS 18 **3.** 10 FOR A=10 TO 25
STEP 5 **5.** YOU'VE SPENT 48.56 SO FAR
7. THE SUM OF THE MEASURES OF
THE ANGLES IS 180 SO THE POLYGON
IS A TRIANGLE; NOT A TRIANGLE;
NOT A TRIANGLE

APPENDIX

Page 516 Exercises **1.** $6a$ **3.** $8y$
5. $3c + 4d - 5c^3$ **7.** $m + 4n$
9. $5c^4 + 2c - c^2$ **11.** $x + y + 2z$
13. $3a^2b - 3ab^2 + b^2$
15. $r^4 - 4r^2 + 2r + r^3$ **17.** $9w + 6 - 4wz$
19. $^-x^2y^2 + x^2y - 2x^2$

Simplifying Polynomials

The expressions below are called **monomials.**

$$6x \quad z^2 \quad 10 \quad {}^-4m^3n$$

The number 6 in the monomial $6x$ is called the **coefficient.** In the monomial z^2, the coefficient is understood to be 1.

An expression whose terms are monomials is called a **polynomial.** For example, $4x^2 + 3xy + y^2$ is a polynomial whose terms are $4x^2$, $3xy$, and y^2. Polynomials with two and three terms have special names.

Binomials (two terms): $4y + 9$ $2x^2 - 5y$
Trinomials (three terms): $x^2 + 7x + 12$ $3a^2 - 2ab^2 + b^2$

You have learned that two or more terms are called **like terms** if their variable parts are the same. Like terms can be added or subtracted by adding or subtracting their numerical coefficients. Unlike terms cannot be added or subtracted.

$$2x + 6x = 8x \qquad x^2 + x$$

like terms   unlike terms

We say that a polynomial is **simplified** when none of its terms are like terms. To simplify a polynomial, rearrange its terms so that like terms are next to each other. Then combine like terms.

Example 1   Simplify $4y^2 + 7y - 2y^2$.

Solution
$$\begin{aligned}
4y^2 + 7y - 2y^2 &= 4y^2 + 7y + {}^-2y^2 \\
&= 4y^2 + {}^-2y^2 + 7y \quad \longleftarrow \textit{Rearrange the terms.} \\
&= 2y^2 + 7y \quad \longleftarrow \textit{Combine like terms.}
\end{aligned}$$

In practice, we usually do not write the difference as a sum as in Example 1. Instead, we rearrange the terms and subtract coefficients as shown in the example on the next page.

Example 2 Simplify $2x^2 + 9x - 8x^2$.

Solution $\begin{aligned} 2x^2 + 9x - 8x^2 &= 2x^2 - 8x^2 + 9x \longleftarrow \text{\textit{Rearrange the terms.}} \\ &= -6x^2 + 9x \longleftarrow \text{\textit{Combine like terms.}} \end{aligned}$

Example 3 Simplify $4a^2 + 5ab - 4a^2 - 7ab$.

Rearrange the terms.

Solution $\begin{aligned} 4a^2 + 5ab - 4a^2 - 7ab &= 4a^2 - 4a^2 + 5ab - 7ab \longleftarrow \\ &= 0a^2 - 2ab \longleftarrow \text{\textit{Combine like terms.}} \\ &= {}^-2ab \end{aligned}$

Your Turn Simplify.
- $a^2 + 3a + 4a^2$
- $2xy^2 - 3xy + 4xy^2$

Exercises

Simplify.

A **1.** $2a + 5a - a$

2. $^-7c + 5c + 4c$

3. $5y + 6y - 3y$

4. $4x - 3x^2 + 7x$

5. $3c + 4d - 5c^3$

6. $^-m^3 - 6m + 3m^3$

7. $^-2m + 4n + 7m - 4m$

8. $3r^3 + 4r^2 + 2r^3 + r$

9. $4c^4 + 2c + c^4 - c^2$

10. $m^3 + 3m - 4m^2 + m^3$

11. $x + 3y + 2z - 2y$

12. $3a^3 - 7 - 2a^2 - 6a^2$

B **13.** $2a^2b - 3ab^2 + a^2b + b^2$

14. $9x^2y + 5y + 5x^2y - 10x^2y$

15. $r^4 - 3r^2 + 2r - r^2 + r^3$

16. $a^3 + 3a^2b - 6ab^2 + 3a^3 - 2a^2b$

17. $3w + 6 + 2w - 5wz + 4w + wz$

18. $xy - 5y + 7 - 6xy + y + 9$

19. $^-2x^2y^2 + x^2y + 3x^2 + x^2y^2 - 5x^2$

20. $^-7a^3 + a + 5a^2 - 3a + a^3 + a^2$

Algebra Readiness Test

Evaluate the expression using the given value for the variable.

Review work: pages 1–20

1. $n + 6$; $n = 11$ **2.** $x - 9$; $x = 20$

3. $4y$; $y = 9$ **4.** $3 + n^2$; $n = 5$

5. $m \div 6$; $m = 54$ **6.** $10 - 2c$; $c = 4$

7. $12r^2$; $r = 4$ **8.** $z^2 - 14$; $z = 9$

Combine like terms

pages 23–25, pages 45–50

9. $a + 4a$ **10.** $3x + 7 + 4x$

11. $c + 25x - c$ **12.** $3n + 4n + n$

13. $5h - 2.4h + 6$ **14.** $6.3y + 4x + y - x$

15. $0.3p + 21p - 6 + n$ **16.** $1.1s + 5t - 0.1s + t$

Write an expression for the given phrase.

pages 1–3

17. 7 added to x

18. n decreased by 9

19. c more than 17

20. the product of m and 35

21. n used as a factor 4 times

22. 54 divided by r

23. 26.4 subtracted from g

Write an equation for the given sentence. Solve the equation.

pages 4–10, pages 18–20, pages 45–50

24. The sum of 8247 and 16 is x.

25. The product of 14 and 35 is c.

26. 86 increased by 135 is y.

27. 4.7 decreased by 3.9 is equal to t.

28. n is equal to 4.92 greater than 6.5.

518

Use the distributive property to factor the expression.

pages 81–83

29. $c^2 + 4c$

30. $4m + 3m^2$

31. $5t + 7t^2$

32. $k - 5k^2$

33. $4mn + 4n$

34. $3y^2 - 6y + 9yz$

Write the fraction in lowest terms.

pages 105–107

35. $\dfrac{5}{10}$

36. $\dfrac{6}{18}$

37. $\dfrac{8}{24}$

38. $\dfrac{x}{x}$

39. $\dfrac{2x}{4}$

40. $\dfrac{10a}{5a}$

41. $\dfrac{21cd}{7c}$

42. $\dfrac{9m^2}{3m}$

43. $\dfrac{8ab}{4ab^2}$

Add, subtract, multiply, or divide.

pages 108–120

44. $\dfrac{2}{8} + \dfrac{1}{8}$

45. $\dfrac{a}{9} + \dfrac{3a}{9}$

46. $\dfrac{8}{6} - \dfrac{3}{6}$

47. $\dfrac{10}{x} - \dfrac{3}{x}$

48. $\dfrac{m}{2} - \dfrac{m}{4}$

49. $\dfrac{1}{3} \cdot \dfrac{7}{8}$

50. $\dfrac{1}{2} \cdot n$

51. $\dfrac{3}{c} \cdot \dfrac{x}{7}$

52. $\dfrac{6}{n} \div 11$

Solve the equation.

pages 138–148

53. $c + 7 = 12$

54. $5 + x = 9$

55. $21 = p + 16$

56. $9 + d = 25$

57. $6x = 24$

58. $40e = 1200$

59. $0.2f = 8$

60. $12g = 48$

61. $d - 9 = 12$

62. $m - 12 = 52$

63. $\dfrac{a}{10} = 5$

64. $\dfrac{e}{9} = 7$

Solve the multi-step equation.

pages 149–154

65. $2a + 5 = 21$

66. $4c - 2 = 14$

67. $5f - 9 = 21$

68. $6m - 1 = 35$

69. $6 + 8d = 62$

70. $9j - 7 = 65$

71. $2x = x + 8$

72. $5r = 7 - 2r$

73. $3y = y + 18$

Write an expression.

pages 155–158

74. Emily is 4 in. taller than Carl.
Let h = Carl's height in inches.
Emily's height is __?__ in.

75. Fred has twice as much money as Al.
Let m = amount of money Al has.
Amount of money Fred has = __?__ .

76. Jean's term paper has 7 fewer pages than Don's.
Let p = number of pages in Don's term paper.
Number of pages in Jean's term paper = __?__ .

Write an equation. Solve.

pages 159–164

77. The length of a picture frame is 14 in.
The length is 2 times the width.
What is the width?

78. Mark is 2 in. taller than Bob.
Mark is 70 in. tall.
How tall is Bob?

79. In Houston, Texas, the average yearly rainfall
is 15.8 in. greater than in Dallas. Houston's
average rainfall is 48.2 in. What is Dallas'?

80. One number is three times another.
The sum of the numbers is 36.
What are the two numbers?

Add, subtract, multiply, or divide.

pages 243–257

81. $^-8 + {}^-6$ | **82.** $6 + {}^-11$ | **83.** $^-4 + 9$

84. $^-6 + 2$ | **85.** $^-12 - 3$ | **86.** $^-18 - {}^-4$

87. $^-20 - {}^-9$ | **88.** $^-5(^-8)$ | **89.** $^-9 \cdot 6$

90. $^-3 \cdot 2d$ | **91.** $^-21 \div {}^-3$ | **92.** $^-40 \div 10$

**Find three ordered pairs that are solutions of the
given equation, then graph them and draw the line
for the equation.**

pages 292–297

93. $y = x - 4$ | **94.** $y = x + 8$

95. $y = 2x + 1$ | **96.** $y = 3x - 4$